Debating
EVOLUTION
before
DARWINISM

An Exploration of Science and Religion in America, 1844–1859

Ryan C. MacPherson, Ph.D.
author of *Rediscovering the American Republic*

MANKATO, MINNESOTA
INTO YOUR HANDS LLC
2015

Into Your Hands LLC

Mankato, Minnesota

www.intoyourhandsllc.com

Debating Evolution before Darwinism: An Exploration of Science and Religion in America, 1844–1859, by Ryan C. MacPherson, Ph.D.

Adapted, with revisions, from Ryan C. MacPherson, "The *Vestiges of Creation* and America's Pre-Darwinian Evolution Debates: Interpreting Theology and the Natural Sciences in Three Academic Communities," Ph.D. diss., University of Notre Dame, 2003. Copyright ©2003 by Ryan C. MacPherson. All rights reserved.

Cover art: Background image © Shutterstock.com. Used by permission. Foreground images from *Vestiges of the Natural History of Creation* (1844). Public Domain.

ISBN-10: 098575432X

ISBN-13: 978-0-9857543-2-7

Library of Congress Subject Headings

Chambers, Robert, 1802–1871—*Vestiges of the Natural History of Creation*

Religion and Science—United States—History—19th Century

Natural History—Religious Aspects—Christianity

Evolution (Biology)—Religious Aspects—Christianity

First printing, July 2015.

CONTENTS

INTRODUCTION
Theology and the Natural Sciences in the Decades Preceding Darwinism

CHAPTER ONE
America: The *Vestiges of Creation* Crosses the Atlantic

CHAPTER TWO
Princeton: *Vestiges* Meets the Scientific Sovereignty of God

CHAPTER THREE
Harvard: *Vestiges* Meets the Unitarian Character of Science

CHAPTER FOUR
Yale: *Vestiges* Meets the Moral Government of God

CONCLUSION
Vestiges of Creation Leaves a Legacy

METHODOLOGICAL POSTSCRIPT
Socially Situated Systematic Theologies

APPENDICES

BIBLIOGRAPHIES

INTRODUCTION

Theology and the Natural Sciences in the Decades Preceding Darwinism

"the Vestiges you have read? the Vestiges, the Vestiges?"

Ralph Waldo Emerson, writing to his cousin, 1845[1]

Recalling the Era of Vestiges

"We knew nothing of Evolution beyond what we gleaned from the *Vestiges of Creation*," recalled Julia Wedgwood in 1900, as she reflected upon British intellectual life of the mid 1800s.[2] Those who read one of the American reprints of her article would have readily understood what she meant, if only they, too, could think back to the late 1840s and 1850s. During that decade and a half—a fifteen-year period preceding the 1859 publication of her uncle Charles Darwin's *Origin of Species*—hundreds of articles and scores of books came forth from American presses concerning an earlier and highly controversial "development hypothesis," or what Wedgwood later learned to call "evolution." The attention

---

1. Ralph Waldo Emerson to Elizabeth Hoar, 17 June 1845, in *The Letters of Ralph Waldo Emerson*, ed. Ralph L. Rusk (vols. 1–6) and Eleanor M. Tilton (vols. 7–10), 10 vols. (New York: Columbia University Press, 1939–1999), 3:289–90, at 90.

2. Julia Wedgwood, "John Ruskin," rpt. from *Contemporary Review*, in *Living Age* 225, no. 2909 (7 Apr. 1900): 1–8, at 2; also rpt. as Julia Wedgwood, *John Ruskin* (New York: Tucker, 1900).

centered around an anonymously published book, first appearing in London in 1844 and soon coming to America in reprint after reprint: *Vestiges of the Natural History of Creation.* College presidents mentioned it in public addresses, college and seminary professors prepared their students to critique it, and ministers debunked it in sermons. Meanwhile, in letters and diaries others privately engaged in the nation's first widespread contemplation of evolution.

From its opening page to the last, *Vestiges* took readers like Wedgwood on two parallel journeys. First, the audience passed through virtually every part of the known universe, from the distant stars revealed by recently enhanced telescopes to the microscopic life forms known as animalcules; meanwhile, readers also surveyed the latest literature of nearly every science known to Victorian readers, including astronomy, comparative anatomy, embryology, chemistry, geology, mineralogy, natural history, natural philosophy, natural theology, philology, phrenology, and zoology. The author posited a comprehensive "law of development" that would unify the cosmological, geological, and biological phenomena peppering the pages of Britain's and America's major quarterlies. The result was an historical narrative of nature's development, spanning innumerable millennia and tracing human civilization back to a primordial "fire mist," or glowing ball of gas in a distant corner of the universe. From this nebula, clusters of matter congealed into stars and planets, forming the sun and the earth. As regular and as predetermined as Newton's law of gravitation, the law of development gave birth not only to the earth but also to its earliest life forms. These simple sea plants, zoophytes, and fish eventually evolved according to the same law of development into diverse species by means of prolonged gestation. The record of these transformations could be found in fossil strata of plants, fishes, reptiles, birds, and mammals of increasing complexity. The entire development had occurred with precision as planned by the Deity, the Creative Intelligence who fashioned the first nebula, endowed it with a law of development, and then stood back as the universe brought its diverse forms to light—all in accordance with that inner law of development.[3]

3. See Robert Chambers, *Vestiges of the Natural History of Creation and Other Evolutionary Writings, Including Facsimile Reproductions of the First Editions of "Vestiges" and Its Sequel "Explanations,"* ed. with an introduction by James A. Secord (Chicago: Chicago University Press, 1994). Appendix B lists American reprints. For convenience, the Secord edition will be cited throughout this book, unless otherwise noted.

As Wedgwood reminded her audience in 1900, *Vestiges* had popularized evolutionary thinking at a time when "we knew Charles Darwin as the writer of an interesting book of travels," but not yet as the author of *On the Origin of Species*.[4] Indeed, the juxtaposition of "Darwin" with "evolution" was rare prior to the November 1859 publication of *Origin of Species*. For one thing, "evolution" in the 1840s and 1850s more often referred to the development of an embryo than to the transmutation of a species. It was, in fact, *Vestiges* that help to expand the common meaning of "evolution" from embryology into phylogeny, or species development.[5] Prior to *Vestiges*, when one dropped the name "Darwin" into a conversation concerning species transmutation, it generally brought to mind the romantic botanical poetry of Charles's grandfather Erasmus Darwin or else the controversial geological theories of Henry Darwin Rogers, a sympathetic reader of *Vestiges*, who in 1846 was denied a post at Harvard College for his "speculative" views on nature's history.[6] Charles Darwin was known in Boston as a traveler who had noted that boiling a potato at a high altitude will not heat it sufficiently for cooking.[7] Back in London, the young naturalist was keeping his own ideas on species transmutation private, in part for fear of being lumped with the author of *Vestiges*.[8]

It was, in fact, not the 1859 appearance of Darwin's *Origin of Species*, but rather the publication of *Vestiges* in London in November 1844 and its republication in New York early in 1845, that marked the beginning of an era of

4. Wedgwood, 2.

5. Peter J. Bowler, "The Changing Meaning of 'Evolution,'" *Journal of the History of Ideas* 36, no. 1 (1975): 95–114.

6. Erasmus Darwin, *Zoonomia; or, The Laws of Organic Life*, facsimile rpt., with a new preface by Throm Verhave and Paul R. Bindler, 2 vols. (1794–1796; New York: AMS Press, 1974), 1:500–10. A failure to distinguish between Erasmus and Charles Darwin led John Stewart to conclude mistakenly that Albert Dod's reference to "Darwin" in his 1845 review of *Vestiges* (which Stewart mistakenly dates to 1846) was to Charles rather than Erasmus. See [Albert Baldwin Dod], "Vestiges of the Natural History of Creation," *Biblical Repertory and Princeton Review* 17, no. 4 (1845): 505–57, at 531; John W. Stewart, *Mediating the Center: Charles Hodge on American Science, Language, Literature, and Politics*, Studies in Reformed Theology and History series, vol. 3, no. 1 (Princeton, NJ: Princeton Theological Seminary, Winter 1995), 31. For a comparison of the two Darwins' views, see James Harrison, "Erasmus Darwin's Views of Evolution," *Journal of the History of Ideas* 32 (1971): 247–64. Dod's review will be discussed further in chapter 2, and the Rogers affair will be treated at length in chapter 3.

7. "Potato Jelly," *Boston Daily Evening Transcript*, 2 Dec. 1845, n.p.

8. Adrian Desmond and James Moore, *Darwin* (New York: Norton, 1992), esp. chaps. 21 and 23; Secord, 429–33.

widespread evolutionary discussions in Britain and the United States—a subsequently forgotten era to which Wedgwood was among the last living links.[9] Wedgwood struggled to convey to her turn-of-the-century audience what Anglophone intellectual life had been like prior to Herbert Spencer's popularized combination of Darwin's evolutionism with August Comte's positivism. "The name of Comte was so unfamiliar," she wrote, "that I remember a young man fresh from college, not at all stupid, informing his cousins that it was the French way of writing and pronouncing Kant."[10]

As distant as the 1840s seemed by 1900, Wedgwood's memory that her childhood neighbors had first become familiar with evolution through *Vestiges*, rather than through Darwin or Spencer, does not stand alone. University of California geology professor Joseph LeConte, whose non-sectarian heart warmed to evolution gradually, culminating in ardent apologetics for the "gospel" of theistic evolution during the 1870s and 1880s, recalled in 1903 that *Vestiges*, which he had read as a young man in South Carolina during the spring of 1845, "was my first introduction to the doctrine of evolution" and "formed an epoch in my intellectual history."[11] *Vestiges* was no less significant—and, indeed, probably more memorable—for those who opposed evolution in the post-Darwinian era. William George Williams, arguing against "modern rationalism," "modern skepticism," "modern atheist materialism," and—to continue borrowing his own words—"other isms exposed in these brief lectures" at Ohio Wesleyan University in 1873, recalled that *Vestiges* had once been read as widely as the several more recent proponents of "infidelity": Charles Darwin, T. H. Huxley, Herbert Spencer, and John Tyndall.[12] During the final three decades of the century, writers of all shades of opinion—including spiritualists, Christian apologeticists, and agnostic positivists—could agree at least on this: *Vestiges* had brought before a wide

9. For a late-nineteenth-century confession that *Vestiges* (not Darwin's *Origin*) "converted me to materialism," that is, a grand history of the universe predicated on materialism, see W. Holman Hunt, "The Pre-Raphaelite Brotherhood: A Fight for Art," reprinted from *Contemporary Review* in *Living Age* 170, no. 2195 (17 July 1886): 131–39, at 134.

10. Wedgwood, 2.

11. Joseph LeConte, *The Autobiography of Joseph LeConte*, ed. William Armes (New York: D. Appleton and Company, 1903), 105, 286.

12. William George Williams, *The Ingham Lectures. A Course of Lectures on the Evidences of Natural and Revealed Religion*, delivered before the Ohio Wesleyan University, Delaware, OH (Cleveland: Ingham, Clarke and Company; New York: Nelson & Phillips, 1873), 49, 109.

audience some bold ideas adapted from the nebular hypotheses of William Herschel and Pierre Simon Laplace, the spontaneous generation experiments of Andrew Crosse and Henry Weekes, and the organic development theories of Jean Baptiste de Lamarck, Étienne Geoffroy St. Hilaire, Lorenz Oken, and other Europeans.[13]

Even the publication of Darwin's work in 1859—though later regarded as a sharp turning point in history—did not immediately upstage the attention Americans had been directing toward *Vestiges*. Reviews of *On the Origin of Species* became another occasion for writers to articulate their evaluations of *Vestiges* and whatever larger project to which they attributed both *Vestiges* and *Origin*.[14] Those Americans not *au courant* with scientific developments also continued to discuss *Vestiges* on its own terms, apart from any relation to Darwin's theory. Two articles in the *Ladies' Repository*—one of the nation's leading monthlies—addressed the evolutionary theory of *Vestiges* in 1862 and 1863 without mentioning *Origin of Species*.[15] At an 1866 celebration of the sixteenth anniversary of California's admission to statehood, the Hon. John Whipple Dwinelle criticized the author of *Vestiges* for denying that God actively intervenes in natural and human history, but he said nothing of Darwin.[16] As late as 1868, a reviewer could ignore *On the Origin of Species* and suggest that John

13. St. George Jackson Mivart, *On the Genesis of Species* (New York: D. Appleton, 1871), 15; Martyn Paine, *Physiology of the Soul and Instinct, as Distinguished from Materialism* (New York: Harper & Brothers, 1872), 108; Robert Patterson, *Fables of Infidelity and Facts of Faith* (Cincinnati: Western Tract Society, 1875), 45; James P. C. Southall, *The Recent Origin of Man, as Illustrated by Geology and the Modern Science of Pre-Historic Archæology* (Philadelphia: J. B. Lippincott & Co., 1875), 61; Joseph Haven, *A History of Philosophy* (New York: Sheldon & Company, 1876), 13–14; John Tyndall, "Inaugural Address of Professor John Tyndall," rpt. from *Nature* in *Living Age* 122, no. 1581 (26 Sept. 1874): 802–24, at 813–14; R. B. Welch, "The Modern Theory of Forces," *Princeton Review* 4, no. 13 (Jan. 1875): 28–38, at 33; E. L. Youmans, "Spencer's Evolution Philosophy," *North American Review* 129, no. 275 (Oct. 1879): 389–404, at 392; Joseph LeConte, "Evolution in Relation to Materialism," *Princeton Review* 1 (Jan.-June 1881): 149–74, at 152.

14. See, for example, Review of *On the Origin of Species*, by Charles Darwin, *Dial* 1, no. 3 (Mar. 1860): 196–97; "Darwin on the Origin of Species," Review of *On the Origin of Species*, by Charles Darwin, *North American Review* 90, no. 187 (Apr. 1850): 474–507; "Darwin on the Origin of Species," review of ten works, rpt. from the *Edinburgh Review*, in *Living Age* 66, no. 840 (7 July 1860): 3–26; [Francis Bowen], Review of *On the Origin of Species*, by Charles Darwin, *North American Review* 90, no. 187 (Apr. 1860): 474–506; Francis Bowen, *Remarks on the Latest Form of the Development Theory*, rpt. from *Memoirs of the American Academy*, n.s., vol. 5 (Cambridge: Welch, Bigelow, and Co., 1860).

15. Isaac Smucker, "Who Were the Aboriginals of North America?," *Ladies' Repository* 22, no. 4 (Apr. 1862): 239–43, at 239; Alexander Winchell, "Voices from Nature," *Ladies' Repository* 23, no. 10 (Oct. 1863): 625–28, at 626.

Robert Seeley's recently (and, like *Vestiges*, anonymously) published *Ecce Homo* was the most controversial work since *Vestiges of the Natural History of Creation*.[17] Even during the 1870s, advertisers still promoted reprints of pre-Darwinian refutations of *Vestiges*.[18] Eventually, *Vestiges* did fade into near oblivion. In 1897, when a writer for *Harper's* looked a back at a century's progress in biological science, *Vestiges* was portrayed as but a tiny "oasis" in the large desert that inquiring minds had traveled *en route* to the paradise opened by Darwin's *Origin of Species*.[19] But that was 1897, and Julia Wedgwood was writing in 1900; half a century earlier, American science was caught up in the era of *Vestiges*.

If the people who lived through the nineteenth century's debates concerning evolution were so fixated on *Vestiges*, then why is *Vestiges* so unknown to those who today think back to that time? Scholarship on popular evolutionary thinking in America has previously focused on later developments, such as discussions of social Darwinism during the Gilded Age and debates between modernism and fundamentalism during the Progressive Era.[20] Some historical surveys of science and religion in the immediately pre-Darwinian period have neglected even to mention *Vestiges*.[21] When historians have

16. John Whipple Dwinelle, *Address on the Acquisition of California by the United States* (San Francisco: Sterett and Cubery, 1866), 30.

17. W. E. Gladstone, "Ecce Homo," review of *Ecce Homo: A Survey of the Life and Work of Jesus Christ, Living Age* 96, no. 1235 (1 Feb. 1868): 259–66, at 259. Like many articles in *Living Age*, this appeared originally in Britain.

18. An advertisement for James Barr Walker's *God Revealed in Nature and in Christ; Including a Refutation of the Developmental Theory Contained in the "Vestiges of the Natural History of Creation"* appeared in Z. N. Morrell, *Flowers and Fruits from the Wilderness* (Boston: Gould and Lincoln, 1872), backmatter; moreover, the following periodicals at times discussed *Vestiges* without mentioning Darwin: *Ladies Repository* 22, no. 2 (Feb. 1862): 113; 22, no. 4 (April 1862): 239; 23, no. 10 (Oct. 1863): 626; *Princeton Review* 31, no. 1 (Jan. 1860): 38; new series, 4, no. 13 (Jan. 1875): 33; new series, 1 (Jan.–June 1879): 387.

19. Henry Smith Williams, "The Century's Progress in Biology," *Harper's* 95, no. 570 (Nov. 1897): 930–42, at 934.

20. For example, Richard Hofstadter, *Social Darwinism in American Thought*, with a new introduction by Eric Foner (Boston: Beacon Press, 1992); George M. Marsden, *Fundamentalism and American Culture: The Shaping of Twentieth Century Evangelicalism, 1870–1925* (New York: Oxford University Press, 1980), chaps. 17–21; Edward J. Larson, *Summer for the Gods: The Scopes Trial and America's Continuing Debate over Science and Religion* (Cambridge: Harvard University Press, 1998).

21. Paul K. Conkin, *The Uneasy Center: Reformed Christianity in Antebellum America* (Chapel Hill, NC: University of North Carolina Press, 1995), discusses Darwinism and other controversial issues in

mentioned *Vestiges*, they generally have regarded it as a precursor to the "real" story, which concerns the impact of Darwin's theory on nineteenth-century American natural science and theology.[22] But before 1859 (and even for some years following), the center of controversy was not Charles Darwin but rather the anonymous person whom Americans called "the author of *Vestiges*."[23] One must, therefore, question the accuracy and adequacy of the common historical portrait of American intellectual life for the middle of the nineteenth century.

Historian of science James Secord has uncovered a vast range of reactions to *Vestiges* in Victorian Britain, to the extent that what may be conveniently labeled "the history of nineteenth-century evolutionism" now seems to have had a much looser relation, than previously supposed, to Darwin's *Origin of Species*, a book that did not surpass *Vestiges* in total British sales until 1889. In an unprecedented analysis of readership experiences, Secord has unveiled the

science, such as Tayler Lewis's *Six Days of Creation*, but makes no mention of *Vestiges*. Walter H. Conser, Jr., *God and the Natural World: Religion and Science in Antebellum America* (Columbia, SC: University of South Carolina Press, 1993), discusses "Issues and Controversies" concerning the age and history of the earth, and the origin of the human races (at 18–32), but he does not even so much as allude to *Vestiges*; this is particularly odd considering that Conser does cite an 1846 article entitled "Vestiges of Creation and Its Reviewers" (150n19) in an endnote supporting his claim (at 16) that Americans generally thought theology and the natural sciences were in harmony—but even there Cosner neither mentions *Vestiges* nor the debates that it occasioned concerning organic evolution. More understandable is the omission of *Vestiges* from introductory survey texts, such as Grant Wacker, *Religion in Nineteenth Century America* (New York: Oxford University Press, 2000).

22. For example, Sally Gregory Kohlstedt, *The Formation of the American Scientific Community: The American Association for the Advancement of Science 1848–60* (Urbana: University of Illinois Press, 1976), 114; Theodore Dwight Bozeman, *Protestants in an Age of Science: The Baconian Ideal and Antebellum American Religious Thought* (Chapel Hill: University of North Carolina Press, 1977), 95, 106, 107; Herbert Hovenkamp, *Science and Religion in America, 1800–1860* (Philadelphia: University of Pennsylvania Press, 1978), 190–210; Robert V. Bruce, *The Launching of Modern American Science 1846–1876*, the Impact of the Civil War series, ed. by Harold M. Hyman (New York: Alfred A. Knopf, 1987), 123–24, 126–27; Edward J. Pfeifer, "United States," in *The Comparative Reception of Darwinism,* ed. Thomas F. Glick (Austin: University of Texas, 1972), 168–206, at 169–72; David N. Livingstone, *Darwin's Forgotten Defenders: The Encounter between Evangelical Theology and Evolutionary Thought* (Grand Rapids: William B. Eerdmans, 1987), 9–13, 53–54; Jon H. Roberts, *Darwinism and the Divine in America: Protestant Intellectuals and Organic Evolution, 1859–1900*, paperback ed., with a new foreword (Notre Dame, IN: University of Notre Dame Press, 2001), esp. 14–33.

23. For some earlier investigations of American reactions to *Vestiges*, interpreted on its own terms, see Milton Milhauser, *Just Before Darwin: Robert Chambers and Vestiges* (Middletown, CT: Wesleyan University Press, 1959), *passim*; and, Ronald L. Numbers, *Creation by Natural Law: Laplace's Nebular Hypothesis in American Thought* (Seattle: University of Washington Press, 1977), 28–35. Milhauser, focusing on British reactions, included only occasional references to American responses; Numbers, focusing on the nebular hypothesis, did not analyze American reactions to *Vestiges*' hypotheses of spontaneous generation and organic evolution.

sundry ways in which *Vestiges* captured the imaginations of wide variety of British readers. He has discovered that no two people read *Vestiges* in quite the same way. To the pioneer of modern nursing, Florence Nightingale, it was a like a guidebook to the Hunterian Collection of anatomical specimens at the College of Surgeons. Art commentator Anna Jameson took "spiritual comfort" from the "progressive principle" in *Vestiges*, which she interpreted as the backbone of constitutional government. Poet Alfred Lord Tennyson framed his poem "In Memoriam" on an evolutionary natural history, as he pondered the implications of "nature red in tooth and claw," a phrase later to be applied to Darwinism but first inspired by *Vestiges*. Cambridge geology lecturer Adam Sedgwick objected to the work's materialist tendencies and warned of the ramifications this could have on the moral character of his students, the future of the gentry. British Museum anatomist Richard Owen hesitated to make such a strong objection because he feared the anonymous author might be someone in a position to influence the direction of his career. These and other examples accumulate in Secord's study as a mass of evidence revealing that it was *Vestiges* more than any other work that brought discussions of the "development hypothesis" into lecture halls, churches, parlors, and salons, as well as to the meetings of the British Association for the Advancement of Science, thus prompting people from a broad spectrum of social classes and religious affiliations to contemplate evolution before hardly any of them heard the topic linked to Darwin's name.[24]

Tracing this era of *Vestiges* from the science journalism of Robert Chambers of Edinburgh (the book's secret author) during the 1830s through the successive editions of *Vestiges* from 1844 onward, Secord brought his readers into face-to-face contact with the several controversies that this work occasioned. Five-hundred pages of thick description—gleaned from newspapers, diaries, letters, and marginal notations in personal copies of *Vestiges*—leave one wondering where, in Secord's analysis, Darwin would ever fit in. "The *Origin*," answered Secord at last, "was important in resolving a crisis, not in creating one."

24. James A. Secord, *Victorian Sensation: The Extraordinary Publication, Reception, and Secret Authorship of "Vestiges of the Natural History of Creation"* (Chicago: University of Chicago Press, 2000). For a detailed analysis of Secord's work and its bearing upon historical understandings of the Darwinian revolution, see Ryan Cameron MacPherson, "When Evolution Became Conversation: *Vestiges of Creation*, Its Readers, and Its Respondents in Victorian Britain," essay review of *Victorian Sensation*, by James A. Secord, and *"Vestiges" and the Debate before Darwin*, ed. by John M. Lynch, *Journal of the History of Biology* 34, no. 3 (2001): 565–79.

The Darwinian resolution, in Secord's view, was that science became both evolutionary (like *Vestiges* and *Origin*) and professional (a vocation for the young salaried scientists led by Darwin's chief defender T. H. Huxley). Science was no longer to be the domain of popular journalists (like Chambers) or of the minister-scientists who had been critical of both *Vestiges* and *Origin*. As the first generation of science professionals championed what Huxley dubbed "Darwinism," the era of *Vestiges* faded to a close, relegating the first grand evolutionary work for the English-speaking world to the lower status of "popular science," or—even worse—popular *pseudo*science.

Secord does not stand alone in expanding scholarly interest in the *Vestiges* controversies beyond the briefer treatments provided in earlier historical surveys of mid-nineteenth-century science. In 2000, John Lynch released a seven-volume set in which he introduced and reprinted chief British reviews of *Vestiges*.[25] That series, however, included no American reactions. As Secord noted, *Vestiges* has frequently been cited by American historians of science and religion, but no comprehensive study of its impact in America has yet been attempted.[26] Referring to the broader historical topic of evolutionary theory in nineteenth-century America generally, Ronald Numbers, who has published more scholarship than any other historian on the nation's long history of debates regarding evolution, found himself admitting in 1998 that historians still lack "satisfactory answers to such basic questions as which American naturalists converted to organic evolution, when they did so, why they did so, . . . and what . . . consequences resulted."[27] It will be argued in this study that historians also lack satisfactory answers to such basic questions as why so many naturalists did *not* convert to organic evolution during the pre-Darwinian era of *Vestiges*. Perhaps a clearer understanding of this issue may also suggest some new ways of approaching the questions Numbers has outlined.

25. John M. Lynch, ed., *"Vestiges" and the Debate before Darwin,* 7 vols. (Bristol: Thoemmes, 2000).

26. Secord, *Victorian Sensation*, 38n66.

27. Ronald L. Numbers, *Darwinism Comes to America* (Cambridge: Harvard University Press, 1998), 25.

The Aim and Organization of This Study

Focusing on Americans' responses to *Vestiges*, this book proposes a model for understanding the relationship between theology and evolutionary theory in nineteenth-century America. The argument will be developed by analyzing how individuals in three academic communities—Princeton, Harvard, and Yale—responded to *Vestiges'* development theory. Drawing from the sociology of religion, as developed by Max Weber and Robert Merton, and an existential model for understanding theology, as proposed by Paul Tillich, three case-study chapters will explore how distinctive theological concerns on each of the academic campuses shaped the manner in which people participated in the natural sciences and evaluated the kind of science proposed in *Vestiges*. As will be further explained in the Methodological Postscript, theology is regarded in this investigation as an existentially grounded system of cultural rationalization that structures the ways in which members of a theologically oriented social group think about and participate in various cultural activities. It will be argued, accordingly, that theology shaped Americans' evaluations of *Vestiges* not merely in terms of how specific doctrinal claims either supported or contradicted statements published in that book, but—much more profoundly—in terms of how the entirety of a community's theology channeled their expectations for what legitimate science ought to be and do, and informed their evaluations of whether or not *Vestiges* could be read in such a way as to meet those expectations.

Chapter 1 will provide an overview of American reactions to *Vestiges* from the first moment that news of its London publication crossed the Atlantic in 1844 until a few years later, when scores of periodicals had reviewed American reprints of both *Vestiges* and its 1845 sequel volume, *Explanations*. Chapters 2 through 4 will apply the methodology outlined below to three case-study communities, each of which had a distinctive theological orientation: Presbyterianism at Princeton, Unitarianism at Harvard, and Congregationalism at Yale. Following these case studies, the Conclusion to this book will highlight how each community's reactions to *Vestiges* compared to the others', and how all of them related to later reactions to Darwin's *Origin of Species*. That final chapter also will suggest some contributions that a similar social-theological analysis might offer to more recent debates concerning evolution.

Three Theologically Distinct Academic Case Studies: Princeton, Harvard, and Yale

The decision to focus upon specific case-study communities reflects, in part, recent studies in the history of science that have emphasized the importance of local context in shaping the development and articulation of people's viewpoints.[28] Secord's study of British reactions to *Vestiges* relied not on published reviews of the book, as had previous scholarship, but on archived correspondence, diaries, lecture notes, pamphlets, publishers' records, and town newspapers. Through a close reading of these sources, Secord uncovered the local social practices through which British readers formed their interpretations of *Vestiges*. Regarding *Vestiges* as a material commodity rather than a transparent conveyor of ideas, Secord identified specific practices of publishing, marketing, purchasing, reading, interpreting, and communicating that together determined individuals' and groups' reactions to *Vestiges*. By revealing many diverse readings of *Vestiges* and situating them within specific community contexts, Secord contributed significantly to the "publishing and reading" turn in the historiography of science that had been advanced by Jonathan Topham and others.[29] Earlier historians' assumption that the meaning of a scientific theory can be found in the text of a book that disseminates that theory has been undermined by Secord's documentation of varied private readership experiences; after Secord, historians of science now must look more than ever to archival evidence of how readers come to their own conclusions when engaging a text such as *Vestiges*.[30]

28. N. Jardine, J. A. Secord, and E. C. Spary, eds., *Cultures of Natural History* (Cambridge: Cambridge University Press, 1996), chaps. 3, 4, 8, 10, 15; Ronald L. Numbers and John Stenhouse, eds., *Disseminating Darwinism: The Role of Place, Race, Religion, and Gender* (Cambridge: Cambridge University Press, 1999), chaps. 1–5, 10.

29. See for example, Jonathan R. Topham, "Beyond the 'Common Context': The Production and Readings of the Bridgewater Treatises," *Isis* 89 (1998): 233–62. For an assessment of recent scholarship by Topham and others, see Adrian Johns, "Science and the Book in Modern Cultural Historiography," *Studies in History and Philosophy of Science* 29, no. 2 (1998): 167–94.

30. The deployment of social constructivism as an historical method does not necessarily mean that there is no objective meaning to the text; it only means that because not all readers recognize that meaning, it can be fruitful for historical investigation to probe the constructed meanings they assigned to the text. Furthermore, the investigator need not apply social constructivism with equal force to all texts. A Christian scholar would, for example, recognize that the text of Holy Scripture has unique properties, such as those identified in Hebrews 4:12, "the word of God is living and powerful, and sharper than any two-edged sword,

This study seeks to apply aspects of Secord's methodology by analyzing a set of unpublished documents that will illuminate day-to-day encounters between the natural sciences and theology in the three academic communities selected for case studies: Princeton, Harvard, and Yale. Professors' lecture notes, students' lecture notebooks, diaries, personal correspondence, and delivery manuscripts for sermons and other public addresses will reveal the "behind the scenes" realities that published writings, often intentionally, obscure. Whereas Secord employed deep archival research to understand how a published book, *Vestiges*, became popular and meaningful through readership practices, this study focuses instead upon the role of three academic communities' distinctive theological identities in shaping how *Vestiges*, arguably the nineteenth century's greatest work of popular science, was interpreted.

Almost overnight *Vestiges* became, in 1845, a part of America's academic life. College professors from Princeton, Yale, Harvard, and Dickinson promptly published reviews of *Vestiges* in the major quarterlies.[31] College presidents and prominent ministers referred to *Vestiges* in public orations at Harvard, Yale, Amherst, and the University of Michigan.[32] Several factors recommend Princeton, Harvard, and Yale, over other academic campuses, to be the case-study focal communities. First, each of them was home to both a liberal arts college that offered courses in the natural sciences and a theological seminary that trained future ministers. These two components of higher education served, respectively, as the *loci* for the social reproduction of men of science (they were

piercing even to the division of soul and spirit, and of joints and marrow, and is a discerner of the thoughts and intents of the heart."

31. [Albert B. Dod (Princeton)], "Vestiges of the Natural History of Creation," *Princeton Review* 17, no. 4 (1845): 505–57; [Francis Bowen (Harvard)], "A Theory of Creation," *North American Review* 60 (Apr. 1845): 426–78; [Asa Gray (Harvard)], "Explanations: A Sequel to 'Vestiges of the Natural History of Creation,'" *North American Review* 62 (Apr. 1846): 465–506; [William H. Allen (Dickinson)], "Vestiges of the Natural History of Creation; Explanations: A Sequel to 'Vestiges of the Natural History of Creation,'" *Methodist Quarterly Review* 3d ser., 6 (Apr. 1846): 292–327.

32. Thomas Hill, *The Annual Address before the Harvard Natural History Society, delivered by Rev. Thomas Hill, Thursday, May 19, 1853* (Cambridge, MA: J. Bartlett, 1853), 16; Theodore Dwight Woolsey, "Inaugural Address," in *Discourses and Addresses at the Ordination of the Rev. Theodore Dwight Woolsey, LL.D., to the Ministry of the Gospel, and His Inauguration as President of Yale College, October 21, 1846* (New Haven: Yale College, 1846), 73–100, at 97–98 (allusion only); Edward Hitchcock, *The Highest Use of Learning: An Address Delivered at His Inauguration to the Presidency of Amherst College*, published by the Trustees (Amherst: J. S. & C. Adams, 1845), 31; Nathaniel West, *The Corruption of Established Truth and Responsibility of Educated Men. An Address before the Alumni of the University of Michigan, June 27, 1856* (Detroit, MI: n.p., 1856), 8, 14, 20.

not yet called "scientists") and men of the cloth. Colleges and seminaries, especially when sharing a common campus, thus were an important venue for contemporary discussions concerning the relations between the natural sciences and theology. Moreover, individuals associated with the colleges and seminaries at Princeton, Harvard, and Yale produced some of America's most widely disseminated responses to *Vestiges,* including lectures, pamphlets, reviews, and books. Not surprisingly, then, when historians have examined either the general relations between science and religion in pre-Darwinian America or else the specific responses of Americans to *Vestiges,* they have tended to focus upon individuals associated with these three institutions. This book, however, seeks to revise those historical narratives which have lumped the three colleges into a common category of representative "American Protestant" views. The case studies identify varieties of American Protestant theology—Presbyterianism, Unitarianism, and Congregationalism—that shaped distinctive reactions to *Vestiges* on those three campuses.

Institutionally, Princeton, Harvard, and Yale were simultaneously religious and academic. Theological concerns justified their existence, sustained their funding, and shaped their educational practices. Theological unity, more than theological diversity, characterized each campus community. College student populations numbered at most a few hundred during this time, and enrollment at the seminaries ranged from a handful to a few dozen students. At Princeton and Yale, students and faculty attended chapel together twice daily; Harvard's morning prayer ritual amounted to little more than attendance-taking, but even students at the Lawrence Scientific School—Harvard's school of natural science—were required to attend Sunday church services.[33] To contemplate the theological implications of *Vestiges* in these three environments did not, therefore, require any special exercise of the intellect; it came automatically, and no one regarded it as optional. Given the prevalent juxtaposition of scientific and religious assessments of *Vestiges* by the principal spokespersons of these colleges, the present study will regard the distinctive theological postures of each community to be integral to, rather than separable from, each community's view of natural science and its response to *Vestiges.* Though Harvard spokespersons

33. On Harvard, see James Turner, *The Liberal Education of Charles Eliot Norton* (Baltimore: The Johns Hopkins University Press, 1999), 42. Details on Princeton and Yale will be provided in chapters 2 and 4.

often claimed their institution was nonsectarian, that campus, no less than those of Princeton and Yale, in actuality served a denominationally specific constituency. The comparative analysis of the three schools' theological orientations will therefore highlight the debates being waged nationally among Presbyterians, Congregationalists, and Unitarians, and suggest that those debates had significance locally for the diverse receptions of *Vestiges* at each institution.

In seeking the particulars of multiple, and sharply distinct, theological communities, this study reinforces a current in postmodern scholarship that replaces generalized "master narratives" with a collection of smaller stories that cannot easily be incorporated into a single summary. Nevertheless, the intended end product is not the "fragmentation of knowledge" at which critics of postmodernism rightfully sound an alarm, but rather a richer integration of human experiences that captures the complexity in which diverse approaches to understanding God, self, and nature have interacted with one another. The three case studies in this project thus find their unity in the historical interactions that took place between them, as people at Yale, for example, shaped their own social and intellectual practices of the sciences based upon their perceptions of what people at Princeton, Harvard, and elsewhere, were doing. In this manner, the narrow focus of each case study enhances, rather than inhibits, the potential for drawing conclusions about "American," rather than merely "Princetonian," "Harvardian," and "Yalensian," intellectual life in the mid nineteenth century. The story of Princeton, Harvard, and Yale is certainly not the story of America, but it is a story with far-reaching implications for American intellectual life, and many of those implications stem from the unique ways in which those institutions integrated theology with the natural sciences.

The lessons to be learned from case studies of Princeton, Harvard, and Yale will be suggestive of a larger cultural picture. These three institutions were widely recognized as leaders in American higher education.[34] The Yale Report of 1828, for example, served as a curriculum standard for numerous other colleges, old and new, during the 1830s and 1840s.[35] Some of the founding members and officers of the American Association for the Advancement of Science were

34. Jon H. Roberts and James Turner, *The Sacred and the Secular University*, introduced by John F. Wilson (Princeton, NJ: Princeton University Press, 2000), chaps. 1, 5.

35. Frederick Rudolph, *Curriculum: A History of the American Undergraduate Course of Study since 1636* (San Francisco: Jossey-Bass, 1977), 72–73.

professors or alumni from these institutions. Over a third of the AAAS membership as of 1859 had graduated from either Harvard or Yale, and although Princeton produced fewer men of science, Princeton's Joseph Henry became the first directing secretary of the Smithsonian Institution and a president of the AAAS.[36] Prominent antebellum intellectual organs, such as the *Princeton Review,* the *American Journal of Science,* and the *North American Review,* had close connections to these colleges, either through direct sponsorship or because a professor from one of those institutions edited or wrote regularly for the journal. From the divinity schools at Harvard and Yale emerged some of the century's most innovative and influential theological movements; Princeton, though more old-fashioned, nonetheless kept pace with the times, as its seminary produced more graduates than either Yale or Harvard, placing them in pulpits, college and seminary classrooms, and administrative posts in the Northeast, South, and West.[37] Professors, editors, publishers, theologians, scientific investigators, and philanthropists formed local networks centered around Princeton, Harvard, and Yale.[38] The faculty at Yale in particular had close ties to Amherst College and Andover Seminary, both in Massachusetts. Thus, however distinctive each local community may have been, none of them was isolated from the broader patterns of nineteenth-century American intellectual life.

A Concept of "Science" during the Era of Vestiges

Before commencing the case studies, it will help present-day readers to reflect for a moment upon what the term "science" meant to Americans who read *Vestiges* during the late 1840s and 1850s. It did not mean what it means today. This difference results in part because much of what comes to mind when one hears of "science" today did not exist during the era of *Vestiges*. Genetics, biochemistry, and nuclear physics remained in an unforeseen future. University research departments were at best just beginning, and government research

36. Kohlstedt, 211–12, 91, 109–10, 158–59.

37. Peter Wallace and Mark Noll, "The Students of Princeton Seminary, 1812–1929: A Research Note," *American Presbyterians* 72, no. 3 (Fall 1994): 203–15; Roberts, *Darwinism and the Divine,* 17.

38. George H. Daniels, *American Science in the Age of Jackson* (New York: Columbia University Press, 1968), esp. chap. 2; Bruce, 38–42, 83–90, and *passim.*

grants generally were to be found on an *ad hoc* basis rather than through agencies established for that purpose. Perhaps these differences can be kept in mind easily enough, but other, more subtle differences pose larger challenges for historians and readers of historical studies.

Most notably, the disciplinary arrangement of scholarship in the mid-nineteenth century differed so profoundly from the university curriculum of the twentieth and early twenty-first centuries that one can easily fail to grasp the fact that "science" from that earlier period (as the German "Wissenschaft" remains today) was as much a precursor to the "humanities" of the more recent period as it was to what today are called "sciences."[39] History, moral philosophy, and philology, for example, were widely considered among the first generation of *Vestiges'* readers to be "sciences" on par with astronomy, geology, and physiology.[40] Harvard's Francis Bowen could issue revised editions of his *Principles of Metaphysical and Ethical Science* during the 1850s without concern that anyone would grimace at the title.[41] Theology, likewise, often was classed among the sciences, as may be inferred from an 1849 appeal by Harvard President Josiah Quincy: "Is no divinity to be taught in Harvard College? Is every other science to be taught there, and the elements and history of religion to be excluded?"[42] Quincy, it may be noted, was a statesman, not a clergyman, but was nonetheless insistent that theology be included among the sciences.[43] When Quincy's successor, the Unitarian minister James Walker, addressed the students in chapel a decade later, he had to apologize for *not* considering religion a

39. Roberts and Turner, esp. 95–105.

40. James Turner, "Le concept de science dans l'Amérique du XIXe siècle," *Annales* 57, no. 3 (May–June 2002): 753–72.

41. Francis Bowen, *The Principles of Metaphysical and Ethical Science Applied to the Evidences of Religion*, new ed., rev. and annotated, for the use of colleges (Boston: Brewer and Tileston, 1855).

42. Josiah Quincy, *Speech of Josiah Qunicy, President of Harvard University, before the Board of Overseers of That Institution, February 25, 1845, on the Minority Report of the Committee of Visitation, Presented to That Board by George Bancroft, Esq., February 6, 1845* (Boston: Charles C. Little and James Brown, 1845), 43.

43. For similar instances at Yale, see Theodore D. Woolsey, "Address to the Professors in the Theological Department," in *Addresses at the Inauguration of the Professors in the Theological Department of Yale College, September 15, 1861* (New Haven: E. Hayes, 1861), 7–15, at 15; Woolsey, "Inaugural Address," 99. As for Princeton, it is well known that seminary professor Charles Hodge declared theology to follow the same scientific methodology as chemistry. See Charles Hodge, *Systematic Theology*, 3 vols. (1871–1872; facsimile rpt., Peabody, MA: Hendrickson, 2001), 1:1–2, 9–15.

science. He tried to explain that religion is a yearning of the heart, whereas theology, "the science of religion," is not religion, but rather the study of it.[44]

As late as 1880, Harvard botanist Asa Gray entitled a lecture not "Science and Religion," but rather "Natural Science and Religion," as if religion, or at least theology, was still assumed to be among the sciences, unless one limited the domain to *natural* sciences.[45] Indeed, one of Gray's main points in those lectures was that natural science and religion both consist of "belief," that is, probabilistic knowledge for which there is no water-tight logical proof (as in geometry). Earlier in Gray's career, many Americans regarded both natural science and theology as capable of stating truths that were as certain as those of geometry; what is significant for the present discussion is not so much that each realm of knowledge became more probabilistic during the mid part of the century, but rather that they changed together, both still retaining a place in the same genre of knowledge, called "science," even for a non-theologian botanist, like Gray. The notion that natural science and theology were fundamentally distinct in method, not merely in subject matter, arose after the era of *Vestiges* and must be put aside in order to recollect what *Vestiges* meant to its early readers.

From these examples, one may see that the now-antiquated term "natural philosophy" was not, as some historians have intimated, a synonym for "science."[46] The latter category was much broader than the former. As Professor Joseph Henry told his Princeton students on the first day of class in 1841, "Natural philosophy is the science which relates to the laws which regulates [*sic*] the phenomena of nature."[47] The professor's language implied that something else could be "the science which relates" to things other than "the laws which

44. James Walker, "Religion not a science, but a want," sermon on Psalm 42:1–2, delivered in the Harvard College Chapel, 3 Apr. 1859 (James Walker Papers, Harvard University, UAI 15.888.5, no. 33).

45. Asa Gray, *Natural Science and Religion: Two Lectures delivered to the Theological School of Yale College* (New York: Charles Scribner's Sons, 1880).

46. For the verbal equation applied to the late seventeenth and early eighteenth centuries, see Michael P. Winship, *Seers of God: Puritan Providentialism in the Restoration and Early Enlightenment* (Baltimore: Johns Hopkins University Press, 1996), 7; likewise for the nineteenth century, see Bozeman, *passim*, and Hovenkamp, *passim*. On other occasions, scholars have noted that "natural philosophy" was an intellectual ancestor of "science," but without explaining the distinction, if any, between contemporaneous historical usages of the two terms. See John Hedley Brooke, *Science and Religion: Some Historical Perspectives* (Cambridge: Cambridge University Press, 1991), 7; Brooke and Cantor, 1.

47. Thomas W. Cattell, Natural Philosophy Lecture Notes, Joseph Henry, 1841–1842 (Lecture Notes Collection, Princeton University, Box 10).

regulate the phenomena of nature." Sure enough, many educators during the nineteenth century regarded history, hermeneutics, and homiletics as sciences.[48] Not only did professors have a capacious view of science, they also passed it on to their pupils. When Daniel F. Gulliver, a Yale student from Boston, Massachusetts, delivered an oration at the college's 1847 Junior Exhibition, he told his classmates and professors that "science is the soul of civilized society" and then expounded upon this claim by discussing the merits of "philosophical, . . . ethical, literary, or historical" contributions to knowledge, that is, to "science."[49] The Yale Rhetorical Society, composed primarily of theological students, regarded theology not just as a science, but a super-science, a "combination of sciences."[50] For clarity, therefore, the term "natural science" will be used in this book to refer to chemistry, botany, and so on, in contrast to "theological science"; when "science" is used without a qualifier, one should construe it in the broad, mid-nineteenth-century sense.

Although *Vestiges* synthesized—or at least attempted to synthesize—numerous sciences, it bore most directly upon theology and those natural sciences categorized under "natural history": cosmology, geology, botany, and zoology. It was in these sciences, as well as in astronomy, philology, and Biblical criticism, that new conceptions of history were being felt most strongly. *Vestiges* attracted attention, and sparked controversy, in part because it accelerated a revolution in natural history that had been slowly taking place for over a century. The very word "history" had formerly meant something like "a comprehensive collection," whether diachronic or synchronic. For example, when Francis Bacon outlined a new methodology for the study of nature in his *Novum Organon* (1620), he called for "a sufficient and suitable *natural and experimental history*"

48. For example, the 1852 tombstone of Moses Stuart, a Yale-trained philologist who made his career at Andover Seminary, hailed him as "The Father of Biblical Science." See John H. Giltner, *Moses Stuart: The Father of Biblical Science in America* (Atlanta: Scholars Press, 1988), 135. It is telling also that Denison Olmsted, an astronomy professor Yale, considered the essential characteristic of "science" to be its concern with "general principles," regardless of the topic to which those principles were to apply. See Denison Olmsted, *The Student's Common-Place Book, on a New Plan; Uniting the Advantages of a Note Book and Universal Reference Book. Applied Alike to the College Student, and to the Professional Man* (New Haven: S. Babcock, 1838), 7.

49. Daniel F. Gulliver, "Fame of Authorship." Junior Exhibition, 27 Apr. 1847, morning (Commencement Orations and Poems, Yale University, Box 1, Folder 10).

50. Secretary's Book, Rhetorical Society of the Theological Department (Records of Clubs, Societies, and Organizations, Yale University, Box 1, Folder 3), entry for 26 Mar. 1845.

to be compiled. To research, for example, the nature of heat, Bacon recommended compiling a catalogue of "fiery meteors," "burning thunderbolts," "natural hot springs," and other "instances" into a "history" of heat. Bacon understood natural history to be a static collection of facts, whether contemporaneous or gathered across the span of centuries by multiple practitioners; history was not a developmental process.[51]

During the period from Bacon to *Vestiges*, the meaning of "natural history," and the goals of the sciences classified under that broad title, shifted dramatically. By the late 1600s, Nicholas Steno and Robert Hooke were championing a new theory that fossils were organic remains of creatures that had lived in the past, rather than *lusi naturae*, or "sports of nature," that had been in those rocks ever since God made the world.[52] A half century later, Georges Buffon, the superintendent of the Jardin du Roi in Paris, began publishing his multi-volume *Histoire naturelle, générale et particulaire* (1749–1779), in which he explored the issues of species fixity and the possibly materialist origins of life. His capstone work, *Les Époques de la nature* (1778), was the Enlightenment's boldest synthesis of natural history and the first full-scale attempt by any modern thinker to connect the origins of the earth, of life, and of diverse species into a comprehensive scheme of natural causes.[53] It did not, however, receive much notice in Britain or the United States, and even today no complete English translation has been published.[54]

In the English-speaking world, it was not Buffon's evolutionary natural history, but the steady-state theory of Scottish geologist James Hutton, that attracted controversy. In 1788, Hutton published an article stating that empirical investigations of the earth could detect "no vestiges of a beginning—no prospect

51. Francis Bacon, *Novum Organon, with Other Parts of the Great Instauration*, trans. and ed. Peter Urbach and John Gibson (Chicago: Open Court, 1994), 144–45 (*Novum Organon*, bk. II, aphorism 11).

52. Martin J. S. Rudwick, *The Meaning of Fossils: Episodes in the History of Palaeontology*, 2d ed. (Chicago: University of Chicago Press, 1985), chap. 2.

53. Rudwick, 93–95; Peter J. Bowler, *Evolution: The History of an Idea*, rev. ed. (Berkeley, CA: University of California Press, 1989), 72–77

54. For a recent Spanish translation, see Georges Louis Leclerc, comte de Buffon, *Las épocas de la naturaleza*, ed. Antonio Beltrán Marí (Madrid: Alianza, 1997). The French orig. has been republished as *Les époques de la nature*, crit. ed., with an introduction and notes by Jacques Roger (Paris: Éditions du Muséum, 1962).

of an end."[55] The concept of an eternal, or even a very ancient, earth conflicted with the 4004 B.C. creation date that Bishop Ussher had calculated in 1650.[56] Those who contemplated the new developments in geology were experiencing what historian Paolo Rossi has termed the "discovery of time," the realization that *"Nature itself has a history* and [fossils] are among the documents of that history."[57] This revolutionary understanding of natural history reached a new plateau when the English lawyer-turned-geologist Charles Lyell published his *Principles of Geology* (3 vols., 1830–1833). In his opening sentence, Lyell summarized the state of the science he was reconstituting: "Geology is the science which investigates the successive changes that have taken place in the organic and inorganic kingdoms of nature."[58]

It was clear by the 1844 publication of *Vestiges* that nature had a past quite different from its present, and that the several sciences of "natural history," such as geology, should explore that past. Developments in philology and biblical criticism reinforced the conviction that the sciences could reveal something about the past, rather than merely the present.[59] Many questions, however, remained unresolved. Lyell, for example, denied that the Bible's account of a worldwide flood had any basis in geology. Many other geologists disagreed. As for the age of the earth, theologians, philologists, and geologists each had a variety of interpretations of both the geological evidence and their King James Bibles. Although Bishop Ussher's calculated creation date of 4004 B.C. had been printed in the margins of many Bibles, practitioners in the three sciences just named

55. Quoted in Bowler, *Evolution*, 46.

56. Ronald Lane Reese, Steven M. Everett, and Edwin D. Craun, "'In the Beginning . . .': The Ussher Chronology and Other Renaissance Ideas Dating the Creation," *Archaeoastronomy* 5, no. 1 (1982): 20–23. For a recent critical edition of Ussher's *magnum opus* of 1650, see James Ussher, *The Annals of the World: James Ussher's Classic Survey of World History*, rev. and updated by Larry and Marion Pierce (Green Forest, AK: Master Books, 2003).

57. Paolo Rossi, *The Dark Abyss of Time: The History of the Earth and the History of Nations from Hooke to Vico*, trans. Lydia G. Cochrane (Chicago: University of Chicago Press, 1984), 4 (history), 36 (time). See also Rhoda Rappaport, *When Geologists Were Historians, 1665–1750* (Ithaca: Cornell University Press, 1997).

58. Charles Lyell, *Principles of Geology*, facsimile rpt. with a new introduction by Martin J. S. Rudwick, 3 vols. (1830–1833; rpt., Chicago: University of Chicago Press, 1990), 1:1. For historical treatments of English geology during Lyell's era, see Nicholaas A. Rupke, *The Great Chain of History: William Buckland and the English School of Geology (1814–1849)* (Oxford: Clarendon Press, 1983); Gillispie, *Genesis and Geology*.

59. Roberts and Turner, 12–13, 95–106.

generally endorsed one or more interpretations of Scripture that allowed for an older earth. A minority refused to accommodate the new historicism and retained Ussher's chronology, but *Vestiges* brought before its readers a more monumental dilemma that powerfully struck young- and old-earth advocates alike. If nature had undergone "successive changes" as Lyell had claimed, then might nebulae have changed into solar systems, nonliving matter into living species, and lower species into higher ones—even humans? If so, what role in that process did God have, and what does this imply for the relationship between God and humanity, and between faith and the sciences?

Broadly speaking, this was the environment in which an anonymous author published what he regarded as "the first attempt to connect the natural sciences into a history of creation."[60] The context will become more refined as chapters 2 through 4 identify distinctive theological situations of Princeton, Harvard, and Yale. Meanwhile, the following chapter offers a preliminary glance at what happened when Americans encountered that "first attempt" in the pages of *Vestiges of the Natural History of Creation.*

60. *Vestiges*, 388.

Study Questions

1. People commonly suppose that Charles Darwin's *Origin of Species* (1859) was the first major work in modern times to bring evolution to the forefront of scientific thinking. However, Americans in fact had been engaging in lively discussions of evolution during the preceding fifteen years, without mentioning Darwin at all. Explain what prompted those earlier discussions.

2. What does the author seek to accomplish by focusing this book on three case study communities—Princeton, Harvard, and Yale?

3. How was "science" defined in nineteenth-century America? Which disciplines were included in "science" back then, but would not be considered part of "science" today?

4. Describe the revolution in thought that had been taking place from the seventeenth through nineteenth centuries concerning the meaning of "natural history." In what sense did Robert Chambers's *Vestiges of the Natural History of Creation* (1844) culminate, rather than initiate, a profound transformation in sciences such as geology?

CHAPTER ONE

America: The *Vestiges of Creation* Crosses the Atlantic

*"It will be a war of tomes and folios,
for a vast deal hinges upon the result."*

An early American review of *Vestiges*, 1845[1]

Introduction

It did not take long for Americans to become familiar with *Vestiges of the Natural History of Creation*. The author himself made little effort to bring his development hypothesis to America, but Americans were accustomed to digesting British works without waiting for an invitation. *Living Age*, a Boston-based anthology of periodical literature edited by Eliakim Littell, quickly reprinted a British review of *Vestiges* that in November 1844 had caught the eye of Alfred Lord Tennyson and enough others to sell out the first London edition within days.[2] It did not matter that when author Robert Chambers had carefully

---

1. Review of *Vestiges of the Natural History of Creation, American Whig Review* 1, no. 2 (Feb. 1845): 215.

2. On Tennyson and early British sales, see James A. Secord, *Victorian Sensation: The Extraordinary Publication, Reception, and Secret Authorship of "Vestiges of the Natural History of Creation"* (Chicago: University of Chicago Press, 2000), 9–10.

orchestrated an anonymous publication in Britain he made no arrangements with American presses. With a second edition appearing in London as early as December 1844 and many American periodicals taking notice early in 1845, New York publisher Wiley & Putnam could be confident of sales when issuing its first pirated edition in January.[3] Chambers may have overlooked Americans when sending complimentary copies of *Vestiges* to selected British gentlemen of science,[4] but no well-read American would overlook *Vestiges*.

In preparation for focused analyses of how specific communities encountered *Vestiges*, this chapter will survey a broad selection of American responses to the work while also providing a basic introduction to the main themes that Americans encountered in the pages of *Vestiges*. The hypothesized "law of development" encompassed the nebular origins of the solar system, the spontaneous generation of life, and the transmutation of lower species into higher ones. As will be indicated here, none of these hypotheses was entirely new to American readers of *Vestiges*, but the combination of them certainly commanded attention, especially considering the boldness with which the author linked all of them under a common "law of development." Although highlighting a few discrepancies among American reviews of *Vestiges*, the overview provided in this chapter will outline evidence suggestive of a broad, nationally shared and denominationally diverse, consensus that emerged within the first two years of Americans' contact with *Vestiges*: the "development hypothesis," it would seem from this initial survey of published reviews, was to be rejected.

As will become progressively apparent, however, especially in the case-study chapters, the American receptions of *Vestiges* are far too complicated to be seen clearly from so generalized a perspective. On the one hand, most published reviewers expressed strong criticism of *Vestiges*, with objections arising from astronomy, comparative anatomy, geology, zoology, and other natural sciences,

3. For a list of British editions of *Vestiges*, see Robert Chambers, *Vestiges of the Natural History of Creation and Other Evolutionary Writings, including facsimile reproductions of the first editions of Vestiges and its sequel Explanations*, ed. with an introduction by James A. Secord (Chicago: Chicago University Press, 1994), appendices A and B. For a comparison of those editions' scientific content, see Marilyn Bailey Ogilvie, "Robert Chambers and the Successive Revisions of the *Vestiges of the Natural History of Creation*," Ph.D. diss., University of Oklahoma, Norman, OK, 1973. For a comparison in terms of culture (including the "material culture" of publishing techniques), see Secord, *Victorian Sensation*. See Appendix B of the present volume for a chart of American printings of *Vestiges* and *Explanations*.

4. Secord, "Introduction," to *Vestiges*, xxvi.

plus theology. On the other hand, reviewers did not always agree with one another regarding, for example, exactly where the geological fallacies of *Vestiges* lay. Moreover, at least a few prominent Americans did not join the majority in dismissing *Vestiges* from the realm of respectable natural history. Finally, opinions concerning the development theory were subject to change over time, as is most clearly apparent with respect to Americans' shifting judgments of the nebular hypothesis (the idea that the solar system originated as a glowing ball of gas). For these reasons, the widespread American rejections of *Vestiges* that will be outlined in this chapter should be understood as constituting a consensus view only in a superficial sense.

In fact, this chapter should be regarded as an attempt to provide a broad survey of American reactions to *Vestiges* from a generalized, and therefore incomplete, "American" perspective. In some respects, this chapter serves not merely as a prelude for, but also as an antithesis to be corrected by, the case-study chapters that follow. Only in chapters 2 through 4 will the significance of the theologically specific frameworks that shaped reactions to *Vestiges* at Princeton, Harvard, and Yale become apparent, and only through those focused analyses will the diversity of American reactions to *Vestiges* begin to become comprehensible. With these caveats in mind, a preliminary inquiry into America's reactions to *Vestiges* may begin.

First Contact

Even before *Vestiges* had been made public in Britain,[5] Edinburgh phrenologist George Combe wrote to American reformer Lucretia Mott to describe the contents of a pre-release copy he had immediately perused upon receipt:

> Yesterday [9 Oct. 1844] I received "from the Author," (who does not give his name either to the public or to me) a book named "Vestiges of the Natural history of Creation," "London, published by John Churchill 1844"; which you would read with interest. It gives a clear & succinct history of the

5. *Vestiges* went on sale in mid October 1844. Chambers already had sent 150 pre-release copies to libraries, editors of periodicals, and other individuals whom he hoped would take interest. See Secord, "Introduction," to *Vestiges*, xxvi.

~~development~~ formation of the Stars, sun & planets so far as Astronomy & Chemistry have revealed their ~~history~~ transitions; next a history of the physical formation of our globe drawn from geological sources; then the geological history of the extinct races; afterwards the physiological characters of the existing inhabitants of the globe; the whole concluding with the "mental constitution of animals,["] & "the purpose & general condition of the animated creation." I have only run thro' it, but I see that [the] author regards the whole Universe as having been formed by a Great intelligent first Cause, who arranged it so perfectly that by the laws which he impressed on its elements it evolved Suns with their Planets, & those with their Satelites [*sic*]; also plants & animals, & finally Man, without his personally interfering, or performing any second, third or subsequent acts of creation. He thinks that a higher species than Man may hereafter be evolved on this globe. He adopts Phrenology as the philosophy of Man; & argues that his views exalt our conceptions of the Deity, & do not dispense with him as many persons would suppose. The work displays great <scientific> learning; it is clearly & calmly written, &, if printed cheaply, would be another battery erected against superstition. The point in which it appears to me to fail is that in which the author advances the hypothesis that organized beings, vegetable & animal, have been evolved out of physical unorganized matter, viz.t carbon, oxygen, hydrogen, & nitrogen; altho' he adduces some facts that go a certain length in sustaining this supposition.[6]

First mentioned only in the privacy of personal correspondence, *Vestiges* became widely publicized in American periodicals early in 1845. The earliest American commentaries came, as in Britain, in the form of brief notices in weeklies and monthlies.[7] By summer the quarterlies were beginning treat the work in twenty-, sometimes fifty-page reviews, revisiting the topic twice, even three times, over the course of a year. From 1845 through 1847, nearly 400 pages of reviews appeared in *Living Age* (four times), the *American Whig Review*

---

6. George Combe to Mrs. Lucretia Mott, 10 Oct. 1844 (National Library of Scotland, Combe Papers, Letter Book, NLS 7388 ff. 768–770, from a handwritten transcription provided by James Secord).

7. Secord, *Victorian Sensation*, 129.

(three times), Francis Bowen's *North American Review* (twice), the *Southern Literary Messenger* (twice), the Presbyterian *Princeton Review*, the Congregationalist *New Englander*, *The Methodist Quarterly Review*, the Unitarian *Christian Examiner*, the *U.S. Catholic Magazine,* and J. D. B. DeBow's *Commercial Review of the South and West*.[8] In January 1846, the *New Englander* folded the discussion back on itself, printing a review of "Vestiges of Creation and Its Reviewers."[9]

Meanwhile, Wiley & Putnam had found a market for its American reprint. The third New York edition of 1845 included a preface by the Rev. George Cheever, a Congregationalist minister and literary critic. In that preface, Cheever warned his audience against reading the very book he was introducing to them. He quickly dismissed as incredulous the author's "Atheistic theory of the creation by the laws of matter" and then mocked his methodology: "The imprint of a duck's foot in the mud, petrified some ages, makes a deeper impression on such a mind [as the author's], than the divine image itself imprinted in the Scriptures."[10] Two reviewers of the Cheever edition complained that Wiley & Putnam should have let *Vestiges* stand or fall on its own.[11] The publisher removed this preface when issuing a new edition later that year, but only to append, following the anonymous author's final chapter, a critical article by David Brewster from the *North British Review*. That article also appeared in three later American editions of *Vestiges*, plus the Boston periodical *Living Age*.[12]

8. See Appendix A for a bibliography of American reviews.

9. [Edward Strong], "Vestiges of Creation and Its Reviewers," *New Englander* 4 (January 1846): 113–27.

10. George B. Cheever, Introduction to *Vestiges of the Natural History of Creation*, 2nd ed. from the 3rd London ed. (New York: Wiley and Putnam, 1845), vi–xxviii, at vii, viii, x–xi. For a biography of Cheever, including a discussion of his previous publications with Wiley and Putnam, and his career goal in offering American readers literature that would reflect noble sentiments and instill religious piety, see Guy R. Woodall, "George Barrell Cheever," in *Dictionary of Literary Biography*, ed. John W. Rathbun and Monica M. Grecu, vol. 59 (Detroit: Bruccoli Clark Layman Book, 1987), 72–79.

11. [Strong], 113–27; J[oseph] H[enry] A[llen], "Vestiges of Creation and Sequel," *Christian Examiner* 4th ser., 5, no. 3 (May 1846): 333–49, at 349.

12. [David Brewster], "Vestiges of the Natural History of Creation," *North British Review* 3 (Aug. 1845): 470–515; rpt., *Living Age* 6, no. 71 (20 Sept. 1845): 564–82; facsimile rpt., in *"Vestiges" and the Debate before Darwin*, ed. John M. Lynch, 7 vols. (Bristol: Thoemmes, 2000), vol. 1 (original pagination retained). For a listing of American editions of *Vestiges* that included Brewster's review, see Appendix B. Subsequent citations are to the pagination in *Living Age*.

Reviewed both from between its own covers, and from without—in a growing number of periodicals—Wiley & Putnam's *Vestiges* looked attractive to New York competitors. W. H. Coyler issued an edition in 1846; Harper & Brothers released two editions in 1847. Beyond New York, Carey & Hart of Philadelphia printed an edition in 1845. By 1852, thirteen American printings had been made, including one in Cincinnati.[13] Often these reprints included the *North British Review* article or else *Explanations*, an anonymous sequel volume Robert Chambers wrote in 1845 in response to his British critics. With or without *Explanations*, all American editions of *Vestiges* were pirated from their London predecessors—a series of revised editions based upon manuscripts penned by Chambers's wife in order to conceal the true author's identity.[14] Whoever was thought to have written it, publishers in America were convinced of one thing: they could profit if only they could undercut their competitors' prices.[15]

As publishers pursued profits, readers' reactions to *Vestiges* proliferated in periodicals, pamphlets, and even books. In 1845, Francis Bowen's fifty-three page refutation from the April *North American Review* appeared in pamphlet form, multiplying its steady criticisms.[16] A Philadelphia press reprinted the Cambridge don William Whewell's *Indications of the Creator* the same year.[17] Though originally written for his *History of the Inductive Sciences* (1837) and *Philosophy of the Inductive Sciences* (1840), the excerpts that Whewell reassembled for *Indications* now supported a new thesis, arguing that *Vestiges* had transgressed the boundaries of true knowledge, whether scientific, religious, or (as it usually was perceived) both. When compiling these excerpts, however, Whewell had chosen not to mention *Vestiges* by name, lest its popularity be increased.[18] In America, however, both the book and its principal New York publisher would be

13. See Appendix B for a chart of American printings of *Vestiges* and *Explanations*.

14. Secord, *Victorian Sensation*, esp. 115–16, 265–79.

15. Ezra Greenspan, *George Palmer Putnam: Representative American Publisher* (University Park, PA: Pennsylvania State University Press, 2000), 141n49.

16. Francis Bowen, *A Theory of Creation* (Boston: n.p., 1845).

17. William Whewell, *Indications of the Creator: Extracts, Bearing upon Theology, from the History and the Philosophy of the Inductive Sciences* (London: W. Parker, 1845; Philadelphia: Carey & Hart, 1845); facsimile rpt. of the 1846 London ed. in *"Vestiges" and the Debate before Darwin*, ed. Lynch, vol. 5.

18. John Hedley Brooke, "Richard Owen, William Whewell, and the *Vestiges*," *British Journal for the History of Science* 10, no. 35 (1977): 132–45.

named, if inaccurately, as in the Philadelphia publication, *Remarks upon a Recent Work Published by Wiley and Putnam of New York, Entitled "The Natural History of the Vestiges of Creation."*[19]

Vestiges and the written responses it occasioned were not only widely published, but also read in earnest. On March 30, 1845, Philadelphia geologist William Parker Foulke read *Vestiges* from cover to cover. Like many readers, he may have sensed some similarity between *Vestiges'* account of naturalistic development and the atheistic theories of Greek atomists, for in April he then read Lucretius' *De rerum natura*. The development theory continued to interest Faulk in the summer. He read Whewell's *Indications of the Creator* and some "Articles in Blackwood [the London-based *Blackwood's Magazine*] on Vestiges of Creation &c" in June.[20] Even without knowing Faulk's own opinion on the matter, it is significant to recognize the manner in which *Vestiges* and its respondents became known to the American reading public. Discussions of the development hypothesis became so ubiquitous that a geologist like Faulk would find *Vestiges* mentioned not only in the *American Journal of Science*, but also in other periodicals he read, such as the Presbyterian *Princeton Review,* a quarterly journal that will be treated at length in chapter 2.

In public addresses, too, Americans invoked "the author of *Vestiges*" with a frequency that presumed audiences to have already heard of this new work. These addresses also popularized *Vestiges* still further, even if usually to criticize it. John Augustine Smith, president of the Lyceum of New York, wrestled with the author of the *Vestiges* in his address of 1846, published that year under a title typically expansive for the Victorian age: *The Mutations of the Earth; Or An Outline of the More Remarkable Physical Changes, of which, in the Progress of Time, This Earth Has Been the Subject, and the Theatre: Including an Examination into the Scientific Errors of the Author of the Vestiges of Creation.*[21] In his 1846 inaugural address for the presidency of Amherst College,

19. *Remarks upon a Recent Work Published by Wiley and Putnam of New York, Entitled The Natural History of the Vestiges of Creation [sic]* (Philadelphia: Carey and Hart, 1846).

20. William Parker Foulke, "Catalogue of Books Read by Me from October A.D. 1834" (Foulke Papers, American Philosophical Society), [35].

21. John Augustine Smith, *The Mutations of the Earth; Or An Outline of the More Remarkable Physical Changes, of which, in the Progress of Time, This Earth Has Been the Subject, and the Theatre: Including an Examination into the Scientific Errors of the Author of the Vestiges of Creation*, 1846 Anniversary Discourse, Lyceum of Natural History of New York (New York: Bartlett and Welford, 1846).

the Rev. Edward Hitchcock, one of America's leading geologists, also spoke of
Vestiges. Like Smith, he criticized its geological errors, but—as will be explored in
chapter 4—he also suggested that its theological implications were perfectly
tolerable. The famed naturalist Louis Agassiz showed less appreciation for
Vestiges' novelties, attacking it in his Boston Lowell Lectures earlier that year.
Agassiz's Harvard colleague Asa Gray had opposed *Vestiges* in that same forum
during the 1845 series. The two professors differed from one another as to what
an organic species is, but both shared a conviction that *Vestiges* had gotten it
wrong. The Rev. Thomas Hill, who later would serve as Harvard president,
echoed Gray's and Agassiz's criticisms of *Vestiges* in his 1853 address before the
Harvard Natural History Society.[22] In the same year Yale geology lecturer
Benjamin Silliman delivered before the New York Mechanical Society a lecture he
had given on four previous occasions: "The Influence of Science and the Arts on
the Moral, Intellectual and Physical Condition of Man Especially in Relation to
Our Own Country." One could hardly give a talk with a title like that during what
was becoming the era of *Vestiges* without addressing the speculations of "the
recent popular author," as Silliman called him in his first lecture in 1846, or "the
author of the Vestiges of the Creation," as he amended his text to read in 1853.[23]

During the late 1840s through 1850s, Silliman, Hitchcock, Agassiz, and
other leading men of science—such as Yale mineralogist James Dwight Dana—
continued on occasion to refer to *Vestiges*. They did so usually to cite an example
of faulty methodology injurious both to natural history and to their particular
Protestant interpretations of the relation between natural science and theology.
Encouraged by the scientific objections presented by these professors, Protestant
apologists from varying backgrounds added the author of *Vestiges* to their
histories of heresy.[24] This anti-*Vestiges* "evidences of Christianity" literature

22. Thomas Hill, *The Annual Address before the Harvard Natural History Society, Delivered by Rev.
Thomas Hill, Thursday, May 19, 1853* (Cambridge, MA: J. Bartlett, 1853), 16.

23. Benjamin Silliman, "The Influence of Science and the Arts on the Moral, Intellectual and Physical
Condition of Man Especially in Relation to Our Own Country," delivery dates spanning 29 Jan. 1845 through
31 Jan. 1853 (Silliman Family Papers, Yale University, Box 45A, Folder 59), 52.

24. See, for example, Thomas Pearson, *Infidelity; Its, Aspects, Causes, and Agencies: Being the Prize
Essay of the British Organization of the Evangelical Alliance* (New York: R. Carter & Brothers, 1854), 97,
109, 111–15, 118, 122, 130, 134, 160, 440; Alonzo Potter, ed., *Lectures on the Evidence of Christianity,
Delivered in Philadelphia, by Clergymen of the Protestant Episcopal Church, in the Fall & Winter of 1853–
4* (Philadelphia: E. H. Butler & Co., 1855), 156–558, 162; Francis Wharton, *A Treatise on Theism, and on the
Modern Skeptical Theories* (Philadelphia: J. B. Lippincott & Co., 1859), 52, 128, 381–94; Alexander

became so pervasive that the New York Mercantile Library listed the work under "Infidelity" in its 1850 catalog, rather than under "Botany and Agriculture," "Civilization," "Cosmogony," "Theology," or "Miscellaneous," as other American libraries were doing.[25]

Catholics also joined the parade, though not without criticizing both *Vestiges* and its Protestant objectors. A reviewer writing for the *United States Catholic Magazine* attributed *Vestiges*, which in its closing pages claimed to draw support from the Bible, as "a natural consequence of the grand Protestant principle, that every one is judge of the sense of the sacred volume." Finding nothing to recommend in the mysterious author's hypothesis, the reviewer stated plainly, "We regret that the book has crossed the Atlantic."[26]

But *Vestiges* had crossed the Atlantic and could not be sent away. For one thing, the book had its American sympathizers, even if not many defenders. In contrast to many Protestants who viewed creation as a static result of a dynamic Creator, Transcendentalists such as Ralph Waldo Emerson and Henry David Thoreau welcomed *Vestiges'* developmental harmony of nature.[27] At a law firm in Springfield, Illinois, attorney Abraham Lincoln received a gift copy of an 1853 printing of *Vestiges* and was intrigued by its revelation of the Deity's "universal law" that drove historical progress—a law of nature's God that resonated with this rising politician's confidence that Providence would preserve the Union.[28] Other

Winchell, *Pamphlets: Education; Philosophy*, 12 pamphlets in 1 vol. (New York: J. Soule and T. Mason, 1858–1889), 67.

25. Mercantile Library Association, *Catalogue of the Mercantile Library in New York* (New York: Baker, Godwin & Co., 1850), 295; similarly, Mercantile Library Association, *Supplement to the Catalogue of the Mercantile Library in New York, Containing the Additions Made to August 1856* (New York: Baker, Godwin & Co., 1856), 154; compared those with: New York State Library, *Catalogue of the New York State Library, January 1, 1850* (Albany: C. Van Benthuysen, 1850), 789 ("Miscellaneous"), 825 ("Theology"); New York State Library, *Catalogue of the New York State Library, 1855: General Library* (Albany: C. Van Benthuysen, 1856), 862 ("Civilization"), 866 ("Cosmogony"); American Institute of the City of New York, *Alphabetical and Analytical Catalogue of the American Institute Library. With Rules and Regulations* (New York: W. L. S. Harrison, 1852), 168 ("Mental and Moral Science"); United States Military Academy Library, *Catalogue of the Library of the U.S. Military Academy, West Point, N.Y.* (New York: J. F. Trow, 1853), 84 ("Botany and Agriculture").

26. "Vestiges of the Natural History of Creation," *United States Catholic Magazine* 6, no. 5 (May 1847): 229–57, quoting 248, 229.

27. Emerson's and Thoreau's views on *Vestiges* will be discussed in chapter 3.

28. William H. Herndon and Jesse W. Weik, *Herndon's Life of Lincoln: The History and Personal Recollections of Abraham Lincoln as Originally Written by William H. Herndon and Jesse W. Weik*, with introduction and notes by Paul M. Angle, 2 vols. (Cleveland and New York: The World Publishing Company,

responses defy easy classification, as when Tayler Lewis, a professor of Biblical languages at Union College in Schenactady, New York, refuted *Vestiges* in his review of 1845 only to be reproved a decade later by James Dwight Dana for supposedly endorsing species transmutation.[29] Initially, however, Dana's own notice in the *American Journal of Science* had "strongly recommend[ed] the work to our readers."[30] The complexities of the Dana–Lewis affair will be explored in detail in chapter 4.

To some American readers, the theory of *Vestiges* was but a continuation of the theory of Jean-Baptiste Lamarck. An assessment of the relation between *Vestiges* and the earlier development of Lamarck's theory, followed by some general portraits of published reviews during 1845–1847, may therefore provide some useful points of departure for those specific case studies.

From Lamarck to Vestiges, *or from* Vestiges *Back to Lamarck?*

Jean-Baptiste Lamarck (1744–1826) served as Professor of Invertebrate Zoology at the National Museum of Natural History in Paris during the late eighteenth and early nineteenth centuries. Lamarck first began promoting his theory of species transformism during his lectures at the museum in 1800. In 1801 he published *Système des animaux sans vertebrès*, which arranged invertebrates in a sequence of structural complexity from simplest to most complex and suggested an historical transformation that connected them along a common ancestral line. The following year, he published *Récherches sur l'organization des corps vivants*, which restated his theory of species transmutation and also included a theory of spontaneous generation, according to which sunlight acted upon subtle fluids of inorganic matter to produce living organisms. Lamarck continued to expand these theories in *Philosophie*

1949), 1:354; William E. Barton, "Vestiges of Creation," chap. 14 in *The Soul of Abraham Lincoln* (New York: George H. Doran Company, 1920), 166–71; Allen C. Guelzo, *Abraham Lincoln: Redeemer President* (Grand Rapids, MI: William B. Eerdmans, 1999), 108.

29. [Tayler Lewis], "Vestiges of the Natural History of Creation," *American Whig Review* 1, no. 5 (May 1845): 525–43; Dana, "Science and the Bible," 80; James Dwight Dana, "Science and the Bible," *Bibliotheca Sacra* 13, no. 49 (Jan. 1856): 80–129, at 90–93.

30. [James Dwight Dana], Review of *Vestiges of the Natural History of Creation*, *American Journal of Science* 48, no. 2 (Jan.-Mar. 1845): 395.

zoologique (1809) and *Histoire naturelle des animaux sans vertebrès* (7 vols., 1815–1822).

The 1815 preface to the latter work presented four laws of evolution: 1) living organisms have a vital force that empowers organic structures to grow; 2) new organs can arise from needs that make themselves continually felt and new movements that arise from those needs; 3) the use of organs reinforces their development; and, 4) any new organs thus acquired may be inherited, thereby producing an enduring change in the species. In short, what caused new species to emerge, according to Lamarck, was *le pouvoir de la vie* (the innate life power that produced a series of plants and animals in ascending complexity) and *le sentiment intérieur* (the inner disposition in higher animals that could strive to meet their *besoins*, or changing needs).[31]

Lamarck's evolutionary works were not translated into English until the early twentieth century.[32] Though some knowledge of Lamarck's theory spread into Britain through the medical lectures of Robert Edmund Grant and others,[33] Anglophone readers became familiar with Lamarck's theory primarily through two British books: Charles Lyell's *Principles of Geology* (2nd volume, 1832), and *Vestiges* itself.

Lyell opposed Lamarck's theory of species change—which at the time was being promoted, in a modified form, by Étienne Geoffroy Saint-Hilaire at the Paris Museum[34]—by emphasizing that no organism had ever been observed to vary beyond specific limits set by nature herself. For Lyell, all phenomena in nature's history had to be explained by causal agents whose action could be presently observed to account for phenomena that were similar in both kind and

31. See Pietro Corsi, *The Age of Lamarck: Evolutionary Theories in France, 1790–1830*, rev. and updated, trans. Jonathan Mandelbaum (Berkeley, CA: University of California Press, 1988), esp. chap. 4.

32. Phillip R. Sloan, "Lamarck in Britain: Transforming Lamarck's Transformism," in *Jean-Baptiste Lamarck, 1744–1826*, ed. Goulven Laurent (Paris: CTHS Publications, 1997), 677–687, at 688.

33. Sloan, 677–84; Adrian Desmond, *Archetypes and Ancestors: Palaeontology in Victorian London, 1850–1875* (Chicago: University of Chicago Press, 1984), 118; Desmond, *The Politics of Evolution: Morphology, Medicine, and Reform in Radical London* (Chicago: University of Chicago Press, 1989), esp. 9, 16, 77–81. Sloan and Desmond each emphasize that Britain's medical anatomists, as compared to natural historians and geologists, deeply explored the transformist biology of Lamarck and Geoffroy during the 1820s and 1830s. To my knowledge no comparable investigation has been done for American medical history.

34. Toby A. Appel, *The Cuvier-Geoffroy Debate: French Biology in the Decades before Darwin* (New York: Oxford University Press, 1987).

degree to the phenomena to be explained. From this methodological standpoint, the transformation of one species into another was as improbable as the biblical account of a global flood, though of course species could exhibit minor variations just as a local rainstorm could flood a valley and thereby reshape a river.[35] Though not everyone shared Lyell's low opinion of the biblical flood narrative, only a few Americans disagreed with his insistence upon species fixity. For example, Kentucky naturalist Constantine Rafinesque, in his *New Flora of North America* (1836), endorsed the idea of species change, and Philadelphia naturalist Samuel Haldeman suggested in January 1844 that Lamarck's theory be revisited. American periodicals, however, seldom took notice of species transmutation[36]—at least not until they carried reviews of *Vestiges*, the book that made both "Lamarck" and "development" household terms.[37]

The author of *Vestiges*, aware that some of his readers would be familiar with Lyell's critique of Lamarck's theory, attempted to distinguish his own theory from Lamarck's, "whose whole notion [of use-inheritance] is obviously so inadequate to account for the rise of the organic kingdoms, that we only can place it with pity among the follies of the wise." Rather than introduce a novel agency, such as Lamarck had done with use-inheritance, the author of *Vestiges*, cleverly

35. Charles Lyell, *Principles of Geology*, facsimile reprint with a new introduction by Martin J. S. Rudwick, 3 vols. (1830–1833; rpt., Chicago: University of Chicago Press, 1990), 1:29–31 (biblical flood), 2:1–17 (Lamarck). Citing Lamarck's *Philosophie Zoologique*, Lyell's discussion of the development theory is sufficiently technical to address the specific claims that the use of an organ can lead to its enhancement, or even the origination of a new organ, and that such enhancements become heritable.

36. The full-text search engine of the Making of America Project, an online database of nineteenth century American literature hosted by the University of Michigan (*http://moa.umdl.umich.edu/*) and Cornell University (*http://cdl.library.cornell.edu/moa/*), locates only two pre–1845 periodical articles referencing "Lamarck," but numerous instances after that year, beginning with the flood of *Vestiges* reviews that were published during 1845–1847. The two earlier references are "The Evils of Unsanctified Literature," *Princeton Review* 15, no. 1 (Jan. 1843): 65–77, at 73 (critical of Lamarck's transmutation theory) and "Philosophy of Natural History," *North American Review* 19, no. 45 (Oct. 1824): 395–411, at 399 (contrasting Lamarck's classification of species with that of Linnaeus, but not mentioning species fixity or change). For another pre-*Vestiges* discussion of Lamarck's transmutation theory, see "The Bible, a Key to the Phenomena of the Natural World" (1829), reprinted in *Essays, Theological and Miscellaneous, reprinted from the Princeton Review, second series, including the contributions of the late Rev. Albert B. Dod, D.D.* (New York and London: Wiley and Putnam, 1847), 1–14, at 3.

37. In addition to the reviews of *Vestiges* discussed in this book, see, for example, J. S. [Allen], Review of *An Essay On the Philosophy of Medical Science*, by Elisha Bartlett, *Southern Literary Messenger* 11, no. 6 (June 1845), 330–340, at 334. For an earlier statement on the significance of *Vestiges* for popularizing Lamarck in America, see Herbert Hovenkamp, *Science and Religion in America, 1800–1860* (Philadelphia: University of Pennsylvania Press, 1978), 194–95.

echoing Lyell's language, sought to "take existing natural means, and shew them to have been capable of producing all the existing organisms, with the simple and easily conceivable aid of a higher generative law." Moreover, *Vestiges* would point men of science back to "the original Divine conception of all the forms of being which these natural laws were only instruments in working out and realizing."[38] The natural law the author had in mind was chiefly that of embryological development. Species change resulted from prolonged gestation, as occasioned by certain environmental conditions, according to the Deity's wise planning.[39]

Despite these attempts at innovation, *Vestiges* often was regarded as merely a rehashing of Lamarck's ideas, long since discredited. Few if any of the reviewers likely had read Lamarck, so their characterization of *Vestiges* as "Lamarckian" should not be understood as a comparison that was more than rhetorical. Writing for Harvard Unitarians, Francis Bowen concluded that Lamarck's hypothesis had been so thoroughly refuted that "the revival of it at this late period seems little more than a harmless exercise of ingenuity, a poetical and scientific dream."[40] Harvard botanist Asa Gray rejected *Vestiges* for "its evident conformity with the old hypothesis of Lamarck," and quoted Lyell's *Principles* against "the Lamarckian view" as a decisive refutation even of the modified elements in the *Vestiges* theory. If anything, Lamarck's theory was for Gray the more probable of the two, for it had new species generating constantly, whereas *Vestiges* had them appearing after irregular intervals.[41] Yale alumnus James Davenport Whelpley suggested that *Vestiges*' errors could be traced not only to Lamarck, but also to the much earlier heresy of Greek atomism.[42] At Princeton, mathematician-theologian Albert Dod likewise traced *Vestiges* back to the Greek atomists and stated that the author's only novel contribution beyond Lamarck's theory is "the diligence with which he has collected and arranged the fragments

38. *Vestiges*, 231.

39. M. J. S. Hodge, "The Universal Gestation of Nature: Chambers' *Vestiges* and *Explanations*," *Journal of the History of Biology* 5, no. 1 (1972): 127–51.

40. [Francis Bowen], "A Theory of Creation," *North American Review* 60 (Apr. 1845): 426–78, at 465.

41. [Asa Gray], "Explanations: A Sequel to 'Vestiges of the Natural History of Creation,'" *North American Review* 62 (Apr. 1846): 465–506, at 491, 492, 494.

42. [James Davenport Whelpley], "A Sequel to 'Vestiges of the Natural History of Creation,'" *American Whig Review* 3 (Apr. 1846): 383–96, at 383.

of various sciences in its apparent support."[43] Similar judgments were expressed beyond the case-study communities, as when Dickinson College's William Henry Allen wrote for the *Methodist Quarterly Review* of the "monstrous doctrines of Lamarck and St. Hilaire" that had been repackaged in *Vestiges*.[44]

Given the paucity of American references to Lamarck prior to the appearance of *Vestiges*, these passages would suggest not that Lamarck *had been* a precursor to *Vestiges* in the sense of exposing a large American readership to the development theory prior to *Vestiges*, but rather that Lamarck *became* a precursor to *Vestiges* through the widespread rhetorical practice by which reviewers instructed their readers that Lamarck had said everything interesting in *Vestiges* already, had said it better, and yet still had been wrong. In fact, one of the earliest American reviews of Lyell's *Principles* to appear outside of scientific periodicals, such as the *American Journal of Science*, was published only after *Vestiges* had become widely known. Unlike the pre-*Vestiges* reviews of Lyell's work, this one placed special emphasis on Lyell's refutation of Lamarck. The reviewer also extrapolated Lyell's earlier analysis to apply now to the more recent *Vestiges*.[45] The important question, then, is not how were the receptions of

43. [Albert Baldwin Dod], "Vestiges of the Natural History of Creation," *Biblical Repertory and Princeton Review* 17, no. 4 (1845): 505–57, at 531.

44. [William H. Allen], "Vestiges of the Natural History of Creation; Explanations: A Sequel to 'Vestiges of the Natural History of Creation,'" *Methodist Quarterly Review* 3rd ser., 6 (Apr. 1846): 292–327, at 317–18; see also [Brewster], 576; Professor Merrick, "The Study of the Bible," *Ladies' Repository* 6, no. 2 (Feb. 1846): 49–51; for later instances, see John Anderson, *Course of Creation: With a Glossary of Scientific Terms* (Cincinnati: Moore, Anderson, Wilstach, and Keys, 1853), 297, 296. On the relation of Allen's review to his career at Dickinson, see Milton C. Sernett, "Allen, William Henry," in *American National Biography*, ed. John A. Garraty and Mark C. Carnes, 24 vols. (New York: Oxford University Press, 1999), 1:349.

Maura Jane Farrelly has attributed the *Methodist Quarterly* article to "William C. Wilson, a professor of natural science at Dickinson College" rather than to Allen. The article was published without any byline, noting simply "Dickinson College, Jan. 31, 1846" on the final page. Secord and I have attributed the unsigned review to William Henry Allen, a professor at Dickinson since 1836 who served as acting president of the college in 1847–1848. Compare Maura Jane Farrelly, "'God Is the Author of Both': Science, Religion, and the Intellectualization of American Methodism," *Church History* 77, no. 3 (Sep. 2008): 659–87, at 672; *Vestiges*, ed. Secord, 228; "William C. Wilson, c. 1855," *archives.dickinson.edu/people/william-c-wilson-c1855*; and, "William Henry Allen (1808–1882)," *archives.dickinson.edu/people/william-henry-allen-1808–1882*. Wilson was an 1850 graduate of Dickinson who served as a professor there later that decade. See Horatio C. King, "Dickinson College," *The American University Magazine* 6, no. 1 (Apr.-May 1897), 4–33, at 12 and 19.

45. C. B. Hayden, "A Resume of Geology," *Southern Literary Messenger* 12, no. 11 (Nov. 1846): 658–671, at 659. For two earlier reviews, neither of which mentioned Lyell's arguments against Lamarck, see Review of *Principles of Geology*, 4th ed., by Charles Lyell, *American Journal of Science* 29, no. 2 (July 1836): 358–59; and, Review of *Principles of Geology*, 5th ed., by Charles Lyell, *American Journal of Science* 32 (July 1837): 210–11.

Lamarck and *Vestiges* related—as if the former receptions not only preceded but also significantly contributed to the latter—but rather, how did the receptions of *Vestiges* (which included retrospective judgments against a hitherto little-known Lamarck) emerge? How, in other words, did the same periodicals that had overlooked Lamarck's theory become almost instantly obsessed with *Vestiges*?

American Reviewers Introduce and Evaluate Vestiges

Vestiges became popular in America in part because the earliest reviews welcomed the new work; only after the initial notices did American periodicals begin to lambast *Vestiges*, and by then their criticisms fueled, rather than smothered, the wildfire of "development." The Boston-based *Living Age*, an anthology of British and American periodical literature, bore on its cover this motto: "These publications of the day should from time to time be winnowed, the wheat carefully preserved, and the chaff thrown away." A November 9, 1844, review of *Vestiges* from the London *Examiner*, a liberal reform weekly, was among the kernels of wheat *Living Age* editor Eliakim Littell preserved for the readers of his January 4, 1845, issue. If this was to be an indication of what would follow, then the author of *Vestiges* would be greatly esteemed among Americans. The review praised from its first sentence "this small and unpretending volume [in which] we have found so many great results of knowledge and reflection, that we cannot too earnestly recommend it to the attention of thoughtful men." The author of *Vestiges* was bold, but not immodest; his thesis startling, but not impious.[46]

After quoting a paragraph in which the author asserted that the efficacy of natural law for the nebular origins of the solar system served as "a powerful argument" for the role of natural law in organic origins, the *Examiner* affirmed that "science supplies facts which bring [this] assumption more nearly home to nature."[47] The *Examiner* also accepted *Vestiges*' presentation of each species' recapitulation, during embryological development, of the adult forms of lower species, to be "truths of physiology, strange as they may seem." Though pointing

46. "Vestiges of the Natural History of Creation," rpt. from the London *Examiner*, *Living Age* 4, no. 34 (4 Jan. 1845): 60.

47. "Vestiges," *Living Age*, 61.

out the difficulty that the "apparently invariable production of like by like" poses for the transmutation hypothesis, the *Examiner* presented as satisfactory the author's notion that the law of "like-production" is subordinate to a higher law of development, by which new species occasionally are produced.[48]

Turning to the final chapter of *Vestiges,* entitled "Purpose and General Condition of the Animated Creation," the *Examiner* admitted that "its views may in certain points seem too material," but then immediately cautioned that these "not be hastily judged." The reviewer next quoted an extended passage in which the author of *Vestiges* affirmed the benevolence of the Deity by dismissing any misfortunes in this life as but a step of progress toward a more fully evolved system of nature. In the reviewer's assessment, this theodicy was precisely what "the religion of Christ has assured to its humble and undoubting followers—that this present existence is the preparation for a better." The author of *Vestiges* had concluded the chapter with an exhortation to "be of good cheer"; this first review of his work to appear in America applied that lesson to the author as well, saying that he, by hiding from criticism behind a cloak of anonymity, "underrates the aptness of the time for an inquiry conducted with so much modesty and so much knowledge . . . and a reverent admission of the Goodness and Mercy of God."[49]

As in Britain, several reviewers in America soon would present alternative conceptions of the methods of scientific inquiry and the goodness of God, corresponding to a less favorable estimation of *Vestiges of Creation* than that expressed by the *Examiner.*[50] But criticisms, however strong in the months to come, did not arise readily. In this first winnowing of new works, *Littell's Living Age* had preserved *Vestiges* among the wheat rather than casting it aside with the chaff.

In February, the *United States Democratic Review* announced that New York publisher Wiley & Putnam "had just printed a work of striking character" treating "an interesting group of topics connected with the primal condition of physical and animal creation." *Vestiges'* author reportedly had "evident ability, evincing a thorough acquaintance with scientific lore, and a pleasing facility of

---

48. "Vestiges," *Living Age,* 62.

49. "Vestiges," *Living Age,* 64.

50. Milton Milhauser, *Just Before Darwin: Robert Chambers and Vestiges* (Middletown, CT: Wesleyan University Press, 1959), 119–39; Secord, *Victorian Sensation,* esp. 310.

popularizing the same."[51] *Vestiges* was still wheat, but perhaps being tossed out with the literary chaff was not the only misfortune the work could experience in America. Though the *Democratic Review* made no specific criticisms in its brief notice, it classified *Vestiges* as a popular work of "scientific lore," distancing the book from reputable scientific scholarship. In the same notice, for example, the *Democratic Review* announced Wiley & Putnam's release of "the renowned Dr. [Johannes] Müller's 'Chemistry of Vegetable and Animal Physiology,' edited by Professor Silliman." Whereas the author of *Vestiges* had concealed his identity in order to protect his name from readers' animosity to the development hypothesis, Dr. Müller's name—in "high repute"—brought esteem to his "learned work," a book that "no scholar will neglect."[52]

Also appearing in February 1845, the first of four articles concerning *Vestiges* in the *American Whig Review* faulted Wiley & Putnam's edition with "heterodoxy." The unsigned reviewer accused the author of "join[ing] issue with grave and revered doctrines which lie at the very core of the existing Christian theory of things." The author of *Vestiges* had replaced "Creation by special exercise" with "Creation by Law," from which "subversive" consequences "imperiously" follow. Such consequences were not specified in this one-page article, which concluded with a promise that the *American Whig Review* would "take it up again" in a future issue. Lest brevity and procrastination be mistaken as indications of apathy, the reviewer affirmed that *Vestiges* must be "refuted fully. . . . It will be a war of tomes and folios, for a vast deal hinges upon the result."[53]

By March, as reported in the *New York Journal of Medicine and the Collateral Sciences*, Wiley & Putnam's edition had gained "wide circulation in this country." For the *Whig* reviewer, this would require "tomes and folios" to counteract *Vestiges'* heresies, but the *New York Journal of Medicine* encouraged readers "to purchase the work," promising it would "furnish food for contemplation of the noblest and most gratifying kind," despite "many inaccuracies" as to "purely scientific and professional details." Its author—said to be "a distinguished member of the British Parliament, who is an *amateur* in

51. "Monthly Literary Bulletin," *U.S. Democratic Review* 16, no. 80 (Feb. 1845): 202–7, at 204.

52. "Monthly Literary Bulletin," 203.

53. Review of *Vestiges of the Natural History of Creation, American Whig Review* 1, no. 2 (Feb. 1845): 215.

scientific matters"—had commendably "written with great elegance, and an apparent comprehension of the whole subjects of Astronomy, Geology, Physiology, &c. &c."[54]

Sir Richard Vyvyan was the Member of Parliament whom British readers had pegged as the author of *Vestiges*; during some six months of journalistic flurry and parlor fury, his was "the most canvassed name" in that country concerning the new natural history. A devoted reader of Lamarck's transformist biology, Vyvyan had worked out a pre-*Vestiges* development theory in his privately circulated *Harmony of the Comprehensible World* (1841), a work in which he embraced geological progress, spontaneous generation, species transmutation, and phrenology.[55] As for authoring *Vestiges,* he in fact played no role, but New York publishers Wiley & Putnam helped spread the Vyvyan rumor to America when they advertised their second edition of 1845 as Vyvyan's work.[56]

No one could know for sure who had written *Vestiges*. The *Southern Literary Messenger* referred to Vyvyan as "the supposed author" in May 1845.[57] In February 1846 the *American Whig Review* ran a twelve-page philological argument that pinned the authorship on Isaac Taylor, whose *Physical Theory of Another Life* (1836) bore many similarities to *Vestiges*, the two being "but parts of one book." Americans also suspected Francis Newman, Lord Sackville, Sir Philip Francis, and of course John Nichol, whose *Views of the Architecture of the Heavens* (1838) brought popular currency to the nebular hypothesis.[58]

54. "Vestiges of the Natural History of Creation." *New York Journal of Medicine and the Collateral Sciences* (Mar. 1845): 269.

55. Secord, *Victorian Sensation,* 180. See also Boyd Hilton, "The Politics of Anatomy and an Anatomy of Politics c. 1825–1850," in *History, Religion, and Culture: British Intellectual History 1750–1950,* ed. Stefan Collini, Richard Whatmore, and Brian Young (Cambridge: Cambridge University Press, 2000), 179–97.

56. For a reproduction and discussion of the advertisement, see Secord, *Victorian Sensation,* 380–82.

57. "European Correspondence," *Southern Literary Messenger* 11, no. 5 (May 1845): 323–26, at 324.

58. "The Author of the Vestiges of the Natural History of Creation," *American Review: A Whig Journal of Politics, Literature, Art and Science* 3 (Feb. 1846): 168–79, quoting 173; at 170 other suspected authors are listed: "Lee, Burke, Lord Sackville, and Sir Philip Francis." An attribution to John Nichol is regarded as "probably a mere guess" in "Vestiges of the Natural History of Creation," *Southern Literary Messenger* 11, no. 4 (Apr. 1845): 255. For suggestions of "Vivyan [sic], Nichol, Newman, and lastly Taylor," see [Benjamin Blake Minor], "Notice of New Works," *Southern Literary Messenger* 12, no. 3 (Mar. 1846): 189–92, at 191. Regarding Nichol's contribution to the popularizaiton of the nebular hypothesis in America, see Ronald Numbers, *Creation by Natural Law: Laplace's Nebular Hypothesis in American Thought* (Seattle: University of Washington Press, 1977), 21.

Privately in Britain, Nichol wrote to phrenologist George Combe that Scottish publisher Robert Chambers had plagiarized from *Views of the Architecture* when writing *Vestiges*,[59] but on the American side of the ocean, Chambers was not yet widely suspected, at least not in print. An 1848 reviewer deemed an encyclopedia edited by Chambers's brother and publishing partner William to be "worthy of the brilliant genius of the author of the Vestiges of Creation," but this was at best an oblique suggestion.[60] Less easily deflected, a *North American Review* article in 1849, deploying a philological approach for determining authorship, underscored similarities between *Vestiges* and Robert Chambers's *Ancient Sea-Margins* (1849).[61] But no one had the final say. In 1858, the editor of the *North American Review* apologized for an article that had recently, and "believed to be erroneous[ly]," attributed *Vestiges* to "an Oxford Professor."[62] Meanwhile, in 1853, a writer for *Debow's Review* referred to "the theory of the still unknown, and probably ever-to-be-unknown author of the 'Vestiges of Creation.'"[63] Even for decades to come, commentators on the development hypothesis usually would refer to an unspecified "author of *Vestiges*," whose book became a sort of individual in its own right.

Whoever the author of *Vestiges* was, the *Southern Literary Messenger* announced in April 1845 that the book "has created a sensation in England, and will no doubt attract almost universal attention in the United States, in spite of certain features in it which may make it regarded by the unphilosophical as irreligious in its tendencies." As was the case in the *American Whig Review*, these tendencies were not expressly indicated, but the editor added a parenthetical note that the work "will probably receive farther notice in our next

---

59. Simon Schaffer, "The Nebular Hypothesis and the Science of Progress," in *History, Humanity, and Evolution*, ed. James R. Moore (Cambridge: Cambridge University Press, 1989), 131–64, at 154–5 and 163n70; James A. Secord, "Behind the Veil: Robert Chambers and *Vestiges*," in *History, Humanity, and Evolution*, ed. Moore, 165–94, at 192n40.

60. Review of *Chambers' Miscellany of Useful and Entertaining Knowledge*, ed. by William Chambers, *Debow's Review* 5, no. 4 (Apr. 1848): 399.

61. See the discussion in chapter 3.

62. "Errata," *North American Review* 87, no. 181 (Oct. 1858): 572.

63. "Fish and Fishermen," *Debow's Review* 15, no. 2 (Aug. 1853): 143–60, at 143.

[issue]." In the meantime, booksellers "Drinker & Morris have the work"—reprinted by Wiley & Putnam for sale to the *Southern Literary*'s readers.[64]

In May the *American Whig Review* made good on its promise to inspect the work more carefully. This and subsequent articles gave reviewers the opportunity to articulate over the course of several, sometimes scores of, pages their judgments concerning what they considered the most significant claims in *Vestiges of the Natural History of Creation*. As William Henry Allen, a professor of chemistry and natural history at Dickinson College in Carlisle, Pennsylvania, concluded in his article for the *Methodist Quarterly Review*, "not one of the three hypotheses [the nebular hypothesis, spontaneous generation, and species transmutation], by which the scheme of law-creation is upheld, has any foundation in nature and truth."[65] Nevertheless, *Vestiges*' reviewers did not themselves share a common "foundation" of "nature and truth," for, as the case-study chapters will reveal, their theological orientations differed radically, even within Protestantism.

Most American respondents did agree, however, that *Vestiges* stood for "development," a universal development that dared to encompass the origin of everything (except the Deity himself). Reviewers generally identified three distinct development hypotheses within *Vestiges*: the nebular origins of the solar system, the spontaneous origin of life, and the transmutation of living beings from one species into another. A fourth stage in the developmental scheme posited the progressive development of human consciousness. This stage received less attention than the other three, perhaps because many critics thought that their arguments against the first three automatically undermined the author's materialist theory of future development. That civilization had a history and that humanity had dispersed into diverse races and language groups with industrialized Victorian Britain at the pinnacle of progress, were perhaps the least controversial points presented in *Vestiges*; the most controversial point was the climax of the third stage of development: rational, morally responsible organisms called *Homo sapiens* were the natural offspring of animals in a world where God was distant from his creation.

---

64. "Vestiges of the Natural History of Creation," *Southern Literary Messenger* 11, no. 4 (Apr. 1845): 255.

65. [William H. Allen], 327.

The Nebular Hypothesis: Did It All Begin As a Fire-Mist?

Vestiges began, like the world whose history it narrated, with a fire-mist, glowing and thinly dispersed throughout the universe. The universal law of gravitation pulled this fiery substance into clusters that, according to the same law of gravitation, began to orbit a central body of fiery material, the sun. These clusters cooled into liquid and then solid spheres, forming the planets and their several satellites. Some Americans already had read of similar "nebular hypotheses" proposed during the late eighteenth century by William Herschel and Pierre Simon de La Place, both of whom received due credit in *Vestiges*. Historian Ronald Numbers has demonstrated that Americans of diverse scientific and theological allegiances voiced increasing approval of these pre-*Vestiges* nebular hypotheses during the 1830s and early 1840s. The version promoted by *Vestiges* differed as to the finer details, but its most significant distinction was a bold generalization that linked nebular development to the origin of life, even human life, under a law of nature as universal as Newton's law of gravitation.

Already in 1822—over twenty years before *Vestiges*—the *American Journal of Science* had printed an article by the Rev. Isaac Orr, a Yale alumnus, endorsing several concepts Laplace had suggested a generation earlier. Orr himself had been unaware of Laplace's *Exposition du système du monde* (1796) and *Traité de mécanique céleste* (1798), but an 1830 article in the *American Quarterly Review* introduced Laplace to American readers and defended his developmental cosmogony from charges of atheism. During the 1830s and early 1840s, numerous works on natural science and natural theology, including the Bridgewater Treatises, either affirmed or at least expressed an openness to the nebular origins of the universe as God's mode of creation by natural law. By the early 1840s, the nebular hypothesis had become a staple in American college curricula and also was taught in public lectures, such as those delivered by Cincinnati astronomer Ormsby MacKnight Mitchel. Joseph Henry at Princeton and Benjamin Peirce at Harvard each promoted an openness to, if not an outright endorsement of, the nebular hypothesis during those years.[66] It would seem that the time had ripened for the author of *Vestiges* to present a development theory founded upon the nebular hypothesis, but in fact *Vestiges'* author was about to

---

66. Numbers, *Creation by Natural Law*, 16–27, 36–38.

become a victim of poor timing, to be branded by many as a perpetrator of conjectural reasoning.

American attitudes toward the nebular hypothesis shifted suddenly in 1845—the year of *Vestiges'* first New York printing—when Lord Rosse, Earl of Parsonstown, Ireland, announced what he had seen through his new 72-inch aperture reflecting telescope.[67] His observations one clear night in March 1845 revealed that several nebulae that previously had been thought to be solar systems-in-the-making were in fact clusters of fully formed stars. By implication, concluded several American reviewers of *Vestiges*, perhaps all nebulous glimmers in the sky were likewise distant clusters of stars too far away, or viewed with too weak a telescope, to be recognized for their true nature. With a telescope that would remain the world's largest reflector until 1917, Rosse caused what one reviewer of *Vestiges* called "sad havoc among the old nebulae on which the hypothesis of Herschel was based."[68] If no real nebulae exist, argued several reviewers, then to posit that the present solar system once was a nebula is to imagine something both beyond experience and beyond analogy to known facts.[69]

David Brewster's review in the *North British Review*, the one reprinted in America both as an article in *Living Age* and as an appendix to several American editions of *Vestiges*, went a step further than some reviews by asserting that even before the discoveries of Lord Rosse:

> there was every reason to believe, from analogy as well as from observation, that nebulae are mere collections of stars, deriving their general lustre, or the lustre of their individual parts, from the brightness and the number of the stars of which they are composed, and often exhibiting the appearance of [nebulous] globes or discs, from the inability of our telescopes to detect [their true nature].[70]

Could "one inflexible law," asked Princeton's Albert Dod, explain the varying rotational patterns observed among the planets?[71] The nebular

---

67. Numbers, *Creation by Natural Law*, 47–49. For further background on Rosse, see Michael J. Crowe, *Modern Theories of the Universe from Herschel to Hubble* (New York: Dover, 1994), 166–68.

68. [William H. Allen], 307. This and other reviews discussed here are listed in Appendix A.

69. See [Bowen], 441; [Brewster], 566; [Dod], 537, and [Strong], 116, 119, 120.

70. [Brewster], 568.

71. [Dod], 537.

hypothesis, emphasized Allen and Bowen, failed also to account for the anomalous orbits of comets, the irregular periods of Jupiter's moons, and the tilt of planetary axes away from the plane of their orbits. In short, nebular development was at best a suggestive hypothesis, and at worst a wild conjecture; either way it was a far cry from a law of nature.[72] Reviews published in America, whether coming from Britain or a local source, thus presented a strong consensus of opposition to the first tenet of the bold developmental theory advanced in *Vestiges.*

The nebular hypothesis was problematic enough on its own terms, but *Vestiges* further complicated matters by linking it with two other controversial topics that American men of science had never endorsed as a community: the spontaneous organization of nonliving material into a living cell, and the transmutation of lower life forms into higher forms. Both were united to the nebular hypothesis by a common law of development:

> The fact of the cosmical arrangements being an effect of natural law, is a powerful argument for the organic arrangements being so likewise, for how can we suppose that the august Being who brought all these countless worlds into form by the simple establishment of a natural principle flowing from his mind, was to interfere personally and specially on every occasion when a new shell-fish or reptile was to be ushered into existence on *one* of these worlds? Surely this idea is too ridiculous to be for a moment entertained.[73]

Undermining the author's objective in reasoning from nebular to organic development, reviewers for the *New Englander, Debow's Commercial Review of the South,* and the *Princeton Review* noted that their refutations of the author's nebular foundation discredited the remainder of his book. They also, however, acquiesced rhetorically with the nebular hypothesis for the sake of exploring whether the other stages of development postulated in *Vestiges* had any factual basis.[74] The author himself, when issuing revised editions of *Vestiges,* continued

72. [William H. Allen], 298–89; [Bowen], 444–45.

73. [Chambers], *Vestiges,* 154.

74. [Strong], 116; [George Taylor], "Theories of Creation and the Universe," review of *Vestiges* and *Explanations, Debow's Review* 4, no. 2 (Apr. 1847): 177–94, at 183–84; [Dod], 550–51.

to endorse a nebular hypothesis amid similar criticisms encountered in British reviews, but he also clarified in his sequel volume that his theories of spontaneous generation and transmutation did not require that the nebular hypothesis be confirmed. The real foundation of his theory was not the nebular hypothesis, but the more general claim of development by natural law.[75] Reviewers, therefore, would have to consider each phase of development independently.

Spontaneous Generation: Whence the First Living Cell?

Vestiges' reviewers focused the debate concerning spontaneous generation upon the author's reliance on recent (1836–1844) experiments conducted by Andrew Crosse and confirmed by William Henry Weekes. The results of these experiments, favorable to the theory that life can arise spontaneously from nonliving matter, appeared in both *Vestiges* and *Explanations*.[76] British reviews of *Vestiges* had contradicted its author severely on this point, prompting him to include as an appendix to *Explanations* two September 1845 letters from Weekes that recounted the rigorous method by which he had demonstrated an occurrence of spontaneous generation. This addition, however, did not mitigate American reviewers' denials of the *quod demonstratum*, and the claim was explicitly derided by Harvard botanist Asa Gray in the *North American Review*:

> The terrestrial races [says the author] all crawled forth from the water, like frogs of the second plague of Egypt; and in the famous Acarus-breeding experiments of Mr. Weekes, triumphantly detailed in the appendix to put down all unbelief, we come near to having the third Egyptian plague realized [in which the dust of the earth turned into lice, Exodus 8:16–19].[77]

75. *Explanations*, 4–8.

76. For a discussion of these experiments and their relation to the *Vestiges* controversies, see James A. Secord, "Extraordinary Experiment: Electricity and the Creation of Life in Victorian England," in *The Uses of Experiment: Studies of Experimentation in the Natural Sciences*, ed. David Gooding, Trevor Pinch, and Simon Schaffer (Cambridge: Cambridge University Press, 1989), 337–83. A broader survey of debates concerning spontaneous generation (including distinctions between abiogenesis and heterogenesis) may be found in John Farley, "The Spontaneous Generation Controversy (1700–1860): The Origin of Parasitic Worms," *Journal of the History of Biology* 5 (1972): 95–125.

77. [Gray] (see citation at note 41), 488–89.

Reviewers quite unanimously portrayed Crosse's and Weekes's laboratory work as sloppy. William Henry Allen, a chemist writing for the *Methodist Quarterly Review*, lamented that readers "unacquainted with chemical manipulations might regard [the work of Crosse and Weekes], as our author does, an *experimentum crucis*." To the contrary, "the practical chemist" would see immediately that the "defect was in his valve." The animalcules that began growing in the jar had seeped in from the outside, rather than arising spontaneously from its original, inorganic contents. "The marvel," concluded the reviewer with a smirk of irony, "is not that ova got into the jar, but that they did not get in sooner."[78] And so it is, asserted the *American Whig Review*, that only a "drunken, irresponsible fancy may create chimeras and generate animalcules in the mud of a convenient ocean; the scientific chemist, in full possession of his wits, is unable to do so."[79]

As if to corroborate the claim of this proverbial "scientific chemist," other American reviewers cited four scientific authorities who had spoken against Crosse and Weekes: Franz Ferdinand Schulze (a Berlin professor of chemistry), Christian Ehrenberg (a German naturalist and microscopist), Johannes Müller (a University of Berlin professor of anatomy and physiology), and the British comparative anatomist Richard Owen.[80] In contrast to these "men of science,"[81] reviewers portrayed Weekes, whose results no one had replicated, as "a name known as yet to science only through this dubious experiment."[82] Owen's lectures on Schulze's "carefully conducted experiments," which experiments the *North British Review* portrayed as decisively having "set this question to rest," were presented as sufficient reason to disbelieve any theory founded on spontaneous generation.[83] The facts stood firm: scientists had found "no authentic instance of

78. [William H. Allen], 313. See also [Dod], 519–20, 546; [James Dwight Dana], "Sequel to the Vestiges of Creation," *American Journal of Science and Arts* 2nd ser., 1, no. 2 (Mar. 1846): 250–54, at 252; [Whelpley], 383; "Explanations: A Sequel to 'Vestiges of the Natural History of Creation,'" *Living Age* 8 (7 Mar. 1846): 442–45, at 443.

79. [Whelpley], 384.

80. [Bowen], 454–56; [Strong], 124; [William H. Allen], 312, 315.

81. [Whelpley], 384.

82. [Dod], 520.

83. [Brewster], 575.

the generation of an animal out of dead substances."[84] By contrast, scientific denials of spontaneous generation rested securely on a "law of nature" derived from "an induction comprising innumerable instances, deciding that the fixed law of organic production is *'omne vivum ex ovo.'*"[85]

Concluding spontaneous generation to be "utterly untenable," a commentator in the *North American Review* moved to the "next point in his system,—the most chimerical of all,—the gradual development of the higher orders of being out of those next beneath them in the scale."[86] This chimera the reviewers called "transmutation," and it could be viewed as the author's answer to a challenge posed in the *North American Review*: even conceding the spontaneous generation of "animalcules, or beings very small in size," how could one "pretend, that a dog, a horse, or a man can thus be created"?[87]

Fossils, Facts, and Opinions:
Debating Progress while Denying Transmutation

American reviewers shared a strong consensus that the author of *Vestiges* had no evidence in his favor for spontaneous generation, but their assessments of the fossil evidence relating to species fixity were more varied. Though most reviewers discredited the species transmutation theory of *Vestiges*, they did so by portraying contemporary evidence—especially the fossil record—in contradictory ways. Unfortunately for the author of *Vestiges*, geology was not so firmly established as to provide a solid foundation for his theory. Neither British nor American geologists could agree amongst themselves concerning issues such as: whether the same strata sequences are found across the globe, or vary by location; whether major fossil kinds (classes, orders) are distributed sequentially within the strata; whether minor fossil kinds (families, genera, species) are distributed sequentially within the strata containing their classes and orders; and, whether any of those sequences indicate progress from simpler to more

---

84. [Whelpley], 393.

85. [Dod], 520, 547.

86. [Bowen], 457.

87. [Bowen], 434.

complex forms.[88] These debates, which had intensified during the 1830s and early 1840s, left American reviewers of *Vestiges* with variant readings of the fossil record against which to judge the development theory. Their somewhat broad range of opinion was, nonetheless, sufficiently narrow to exclude *Vestiges*.

The author had read from the fossil record a history of naturalistic transmutations through "an obvious graduation" (rather than saltation), "from the simplest lichen and animalcule respectively up to the highest order of dicotyledonous trees and mammalia."[89] Countering this claim, the *New Englander*, *North British*, and *Methodist* reviewers employed fossil evidence to deny that fossil types were distributed in a progressive sequence of geological strata.[90] The *New Englander*, for example, portrayed the fossil record as evidence for a relatively static history of the earth's life forms: "in the identical rocks where you find the lowest order of fishes you also find *all*, even the highest that his *law of development* has ever evolved."[91] Far from displaying an increase in complexity, noted that reviewer, "any change appear[ing] . . . is a change the wrong way, i.e. retrograde."[92] After mentioning that the Swiss naturalist Louis Agassiz "denies most unequivocally . . . transformation," the commentator quoted from the Rev. William Buckland's Bridgewater Treatise, *Geology and Mineralogy* (1836), that "although the sauroid fishes occupy a higher place in the scale of organization than the ordinary forms of bony fishes, yet they are found in abundant numbers in the carboniferous and secondary formations." Lyell also was cited, for he had concluded "after a full and candid examination" that the "attributes and organizations" by which species are distinguished, are immutable.[93] Indeed, as *Debow's Review* emphasized, Lyell had invalidated

88. For a general background, see Peter J. Bowler, *Fossils and Progress: Paleontology and the Idea of Progressive Evolution in the Nineteenth Century* (New York: Science History Publications, 1976), esp. chaps 3–5. Specific controversies are addressed in Martin J. S. Rudwick, *The Great Devonian Controversy: The Shaping of Scientific Knowledge among Gentlemanly Specialists* (Chicago: University of Chicago Press, 1985); and James A. Secord, *Controversy in Victorian Geology: The Cambrian-Silurian Dispute* (Princeton, NJ: Princeton University Press, 1986).

89. *Vestiges*, 191.

90. Peter Bowler has termed this position "anti-progressionism." See Bowler, *Fossils and Progress*, 67.

91. [Strong], 122.

92. [Strong], 125.

93. [Strong], 122.

progressionist readings of the fossil record by discovering that "vertebrated animals, true fishes, are found among the earliest types of organism."[94]

The *North British* reviewer, citing Buckland's *Geology and Mineralogy*, Richard Owen's 1842 report on British mammalian fossils to the British Association for the Advancement of Science, and Agassiz's *Recherches sur les poissons fossiles* (1833–1843), concluded that the distribution of fossils in no way supports *Vestiges'* claims: "geologists of all classes repudiate them as incorrect and unfounded."[95] The reviewer detailed "six Palaeozoic systems of rocks, [which] satisfied the intelligent reader that the forms of organic life which they successively display, have not been the result of progressive development." Confidently, the "chronicle of geological events is now so complete, that we do know what has happened in preceding ages," leaving *Vestiges* with "not one single fact in its favor."[96]

Similarly, the *Methodist Quarterly* reviewer cited an April 1845 report from the *American Journal of Science* regarding a recent discovery of birds and reptiles "in the carboniferous series of Pennsylvania." This was one among many examples showing that:

> the author of Vestiges is not only incompetent to translate the hieroglyphics of nature's fossil archives, but that he plays the interpolator with the translations of men who are competent, [for] . . . the labors of geologists are constantly revealing new proofs of the antiquity of some of the higher forms of the vertebrata.[97]

Other reviewers, though equally critical of *Vestiges*, portrayed the fossil record in a different light. Asa Gray, for example, entertained the possibility of a progressive fossil sequence, since he was confident that invertebrate forms appeared prior to vertebrate forms. But, despite this strata sequence and the temporal sequence it represented, Gray insisted against *Vestiges* that "it is not probable that there was any historic progress *through* the invertebrate series, or even that there was *[with]in* each order a regular sequence from the lowest to the

94. Lyell, quoted in Taylor, 185.

95. [Brewster], 572.

96. [Brewster], 573 and 575, but also see 572.

97. [William H. Allen], 307.

highest forms."[98] Thus, even though Gray was more willing than Brewster to identify historical progress *among* major groupings in the fossil record, he shared the British reviewer's judgment that the scale of nature's complexity indicates "only the order of procedure which the Almighty chose to pursue."[99]

James Davenport Whelpley, a Yale-trained geologist, agreed with Gray that the fossil record contained progressive sequences among major kinds, and like Gray he also emphasized that geological succession does not in itself entail biological progress, much less transmutation, because "it seems more agreeable to observation, if we conclude that species are something quite fixed and immutable; and that no growth of one will improve it into another."[100] After all, 3000-year-old mummies unearthed from Egyptian tombs had revealed no significant morphological differences from nineteenth-century humans.[101] Moreover, "there is no transitional group between the ape and man which has one-half the whole *moral* quality proper to man."[102]

The overwhelming conclusion of these reviewers—as compared to the initially positive brief notices of *Vestiges*—was that the transmutation of one species into another had never been witnessed, nor had it ever occurred.[103] With his "unquestionable authority," Owen had opposed it, and was supported (said various reviewers) by Agassiz, Cuvier, Henry Thomas De la Beche (a British geologist), Justus von Liebig (a German chemist), Charles Lyell, and of course Moses, the earth's divinely appointed historian whose *magnum opus* had quoted the very words of God: "Let us make man in our own image."[104] *Vestiges* had provided no viable objection to the "the celebrated German physiologist"

---

98. [Gray], 479–80 (emphasis added).

99. [Brewster], 570; [Gray], 483.

100. [Whelpley], 388–89, 394, and quoting from 391.

101. [Whelpley], 385, 387, 389; also see 383. For historical background concerning interpretations of the mummy discoveries, see Appel, *The Cuvier-Geoffroy Debate*, 72–82.

102. [Whelpley], 386, 387. Whelpley's saltational view of geological history fits an approach that Bowler has called "discontinuous progression"; *Vestiges*, by contrast, endorsed "continuous progression," in which one species literally gave birth to another. See Bowler, *Fossils and Progress*, 67.

103. See especially [Whelpley], 394.

104. Quoted in [Bowen], 461. For the other scientific endorsements, see [William H. Allen], 326; [Strong], 122; Taylor, 186–88.

Johannes Müller's respected opinion that *"there is not a remote possibility* that one species has been produced from another."[105]

Embryology and Macleay's Circles of Life:
Living Evidence for Transmutation?

If the arguments in *Vestiges* and *Explanations* lacked scientific credibility, it was not for the mysterious author's lack of trying. Fossils, which the reviewers made speak against him, even if somewhat cacophonously, were not his only foundation for transmutation. He also could point to the solid, empirical evidence of embryology. Although multiple reviewers rejected the author's claim that simple forms had transmuted into more complex ones, nevertheless men of science generally agreed by 1844 that primordial cells do develop into the complex adult forms of sundry species. Not only that, but some naturalists had claimed that the process of embryological development displays a series of transformations in the individual that progressively mirror the forms of lower species, culminating in the form of the species to which that individual belongs. Wrote the author of *Vestiges*:

> Physiologists have observed that each animal passes, in the course of its germinal history, through a series of changes resembling the *permanent forms* of the various orders of animals inferior to it in the scale. Thus, for instance, an insect, standing at the head of the articulated animals, is, in the larva state, a true annelid, or worm, the annelida being the lowest in the same class. . . . The frog, for some time after its birth, is a fish with external gills, and other organs fitting it for aquatic life, all of which are changed as it advances to maturity, and becomes a land animal. . . . Nor is man himself exempt from this law. . . . His organization gradually passes through conditions generally resembling a fish, a reptile, a bird, and the lower mammalia, before it attains its specific maturity.[106]

105. [Brewster], 574; Müller is quoted identically in [Taylor], 188. This is not surprising considering that Taylor was reviewing an American edition of *Vestiges* to which the publishers had appended Brewster's review.

106. *Vestiges*, 198–99.

According to the development theory of *Vestiges*, this recapitulation of lower forms was neither coincidental nor merely an ornament manifesting the Creator's design in nature. Gestation was the process by which a seed developed into an adult form, and in the same manner a new species could develop from "a new stage of progress in gestation, an event as simply natural, and attended as little by any circumstances of a wonderful or startling kind, as the silent advance of an ordinary mother from one week to another of her pregnancy."[107]

When American reviewers addressed *Vestiges'* recapitulation argument, they interpreted the facts of embryology differently than the author had desired. Princeton's Dod argued that the resemblance of embryological stages with the mature forms of lower species indicated a typological, not genealogical, "unity of organic nature."[108] Yale alumnus Whelpley, likewise, concluded that "it is contrary to all science, to assert that a species of quadruped *is*, at any foetal stage, either an animalcule, a mollusk, or a fish; though . . . it may *resemble* those forms."[109] At Harvard, Gray faulted the author "with forgetting that analogy is very different from identity," and Bowen emphasized that "the essential internal organization" determines species, not the external similarities that manifest themselves during various stages of development.[110]

Bowen and Gray seem to have been aware of something that Whelpley did not notice: *Vestiges* was inconsistent in the way it incorporated two different kinds of recapitulation theories.[111] One version of recapitulation, developed by Johann Friedrich Meckel, Friedrich Tiedemann, and Étienne Reynaud Serres during the 1820s and 1830s, held that the embryos of higher species pass through the adult forms of each lower species. This understanding of recapitulation strongly informed the first edition of *Vestiges*, but already it was outdated science. In 1828 and 1837, the German embryologist Karl Ernst von Baer had put forth evidence showing that embryonic development is a process of differentiation: the embryos of all organisms look identical at the first stage, but as time passes they branch out into diverse forms. Insofar as higher species

107. *Vestiges*, 223.

108. [Dod], 547; cf. 548.

109. [Whelpley], 392 (emphasis added).

110. [Gray], 498; [Bowen], 458–59.

111. Compare [Bowen], 462, and [Gray], 505, with [Whelpley], 392.

recapitulate the forms of lower species, argued von Baer, they resemble only the embryonic forms (not the adult forms) of only those lower species that share a common branch within the embryological tree of life (not of all lower species). By the early 1840s, "von Baer's law" had become well-known in Britain. In fact, Chambers even included aspects of it in his first edition of *Vestiges*, and he revised subsequent editions in a manner that put increasing emphasis on von Baer's branching model of embryological analysis.[112] But by then it was too late. No matter which version of recapitulation the author of *Vestiges* would endorse, American reviewers, like many writers in Britain, were not convinced that embryological resemblances were evidence of genealogical relationships.

The author of *Vestiges* had further supported his transmutation theory with William Sharp Macleay's "quinirian" system of classification. Macleay, a British entomologist, had published this system in his two-volume work *Horae entomologicae* (1819–1821). Macleay arranged natural kinds of animals in circles, with their proximity to one another along the circle's perimeter being proportionate to their anatomical affinities. As the author of *Vestiges* explained, the very fact that specimens could be so arranged testified to the naturalness of this circular arrangement. Moreover, within each circle, five smaller circles could be made. Thus, for example, the circle of the animal kingdom consists of five smaller circles for mammals, reptiles, fish, amphibians, and birds, and within the mammal subkingdom, five more circles can be made, one for each mammalian order.[113] It is worth noting that circular taxonomic systems were regarded seriously in Britain during the 1840s by such scientific authorities as Richard Owen, William Carpenter, and Robert B. Todd.[114] *Vestiges* pushed the limits not by using a circular system *per se*, but by suggesting that species' affinities indicated a genealogical connection between them. Macleay's quinirian system, argued the author of *Vestiges*, supplies *"a powerful additional proof of the hypothesis of organic progress by virtue of law."*[115]

112. Secord, "Behind the Veil," 180–82. For a history of the various schools of recapitulation, see Robert J. Richards, *The Meaning of Evolution: The Morphological Construction and Ideological Reconstruction of Darwin's Theory* (Chicago: The University of Chicago Press, 1992), 42–61.

113. *Vestiges*, 238–39.

114. Secord, "Introduction," to *Vestiges*, [ix]-[xlv], at p. [xv].

115. *Vestiges*, 250 (emphasis original).

William Allen found the "celebrated hypothesis of circular groups" to be sloppily applied in *Vestiges*, since the author had ignored that "nature has set up impassable barriers to prevent confusion of species."[116] The reviewer for the *New Englander* dismissed the quirinian system as "the most objectionable feature of his work, as originally published."[117] That last phrase hinted at a fact emphasized also in the *North British Review* article that was coupled to many American reprints of *Vestiges*: "He has actually omitted from his 4th edition the whole section on the Macleay system."[118] Even then, however, the *New Englander*'s commentator remained discontent, and wondered why the author had not jettisoned the rest of his theory together with Macleay.[119] Arranged on quirinian circles or not, concurred Bowen in the *North American Review*, "the interval between two species most nearly allied to each other seems to be quite as impassable as the broadest gulf of separation."[120] A consensus thus had emerged that there was little if any reason to accord *Vestiges* a respectable place in the literature of natural science.

Interpreting the Theology of Vestiges

The chief reviewers of *Vestiges* also felt theologically agitated by the work's tone. It was not simply a matter of Protestants being troubled by an irreligious theory of natural development; rather, *Vestiges* undeniably had a religious tone, but it was a tone that did not harmonize with many reviewers' own religious views. "That God created animated beings," wrote the author of *Vestiges*, "is a fact so powerfully evidenced, and so universally received, that I at once take it for granted."[121] Though not denying God's existence or His power, *Vestiges* nevertheless collided with the lived experience that many American Protestants had with their God. The author of *Vestiges* thus sparked controversy when writing, "Some other idea must then be come to with regard to *the mode* in which

116. [William H. Allen], 320, 326.

117. [Strong], *New Englander*, 116.

118. [Brewster], 576.

119. [Strong], 123.

120. [Bowen], 458.

121. *Vestiges*, 152.

the Divine Author proceeded in the organic creation. . . . What is to hinder our supposing that the organic creation is . . . a result of natural laws, which in like manner are an expression of his will?"[122] The ascription of God's creative work to natural law, even if those laws were themselves but the expression of God's will, troubled many Protestant readers who preferred a different understanding of both God and the laws of nature. As Allen wrote for the *Methodist Quarterly Review*, "Why then should God exist for no purpose but to cause the existence of [a law of development] which could have existed eternally as easily as himself?"[123]

As the next three chapters will indicate, concerns about God's relation to nature and humanity were not idle speech within the three case-study communities. Writing as a Presbyterian at Princeton, Albert Dod admitted that *Vestiges* acknowledged some sort of Creator, but he nonetheless found the work tending toward atheism.[124] Yale alumnus James Davenport Whelpley took a more congenial approach, suggesting that *Vestiges* had been written with the intention of refuting atheism rather than supporting it.[125] On the surface, Gray and Bowen at Harvard sounded similar to Dod in accusing the author of a tendency toward atheism and materialism,[126] but they in fact were speaking from within a tradition that was critical of Dod's own understanding of Christian theism. As chapters 2 through 4 will illustrate, what was at stake for *Vestiges'* respondents at Princeton, Harvard, and Yale, was not merely the distinction between their theology and that of *Vestiges*, but also—and perhaps more especially—the distinctions among their three rival theologies, which had entered into sharp conflicts during the years just before *Vestiges* was published.

One could, of course, as other historians have done, write an account of *Vestiges'* theological receptions in American by emphasizing pervasive pan-Protestant themes, such as the personal intervention of a providential God in the course of natural events.[127] The case studies that follow adopt the challenge of showing that such a generalized portrait misidentifies what was really at stake for

122. *Vestiges,* 153–54.

123. [William H. Allen], 295.

124. [Dod], 533.

125. [Whelpley], 383.

126. [Bowen], 474, 475, 477; [Gray], 468, 485, 501, 502.

127. The pertinent literature will be referenced in the case studies as applicable.

each theological community. "What must I do with *Vestiges*?" was intertwined with "What must I do to be saved?" Disagreements regarding that latter question were to shape distinctive approaches to the former question. For this reason, even the far-reaching national consensus, as outlined in this chapter, that *Vestiges* had fallen short of the standards of natural science, must now be reexamined in terms of how each of the case-study communities, living out its own theological concerns, participated in the natural sciences to the glory of their God.

Study Questions

1. If neither the Scottish author Robert Chambers nor his British publishers arranged for *Vestiges* to be reprinted in America, then how did it come about that so many American editions appeared in so short a time?

2. Identify the four laws of evolution proposed by Jean-Baptiste Lamarck in the early 1800s. To what extent was Lamarck's evolutionary theory known in America prior to the publication of *Vestiges*?

3. Which persons did American readers suspect to have authored *Vestiges*, and how did these suppositions shape the reputation of the book's development theory?

4. Summarize each of the three major components of the development theory presented in *Vestiges*: the nebular hypothesis, spontaneous generation, and species transmutation.

5. The author of *Vestiges* had hoped that his development theory could ride atop the previously favorable reception of the nebular hypothesis, but most men of science instead began to reject both the nebular hypothesis in general and his book in particular. What led to this shift in opinion?

6. Summarize the scientific consensus in the 1840s concerning the hypothesis of spontaneous generation.

7. Although most geologists rejected the theory of species change set forth in *Vestiges*, they did not themselves agree as to exactly what the fossil record indicated. Describe the range of interpretations that these men of science had concerning the history of now-fossilized life forms.

CHAPTER TWO

Princeton: *Vestiges* Meets the Scientific Sovereignty of God

"The Vestiges has been all the rage for a time."

a Princeton undergraduate, reviewing *Vestiges*, 1847[1]

Introduction[2]

In November 1845, one year after the first edition of *Vestiges* had appeared in England, an early death took Princeton College's chair of mathematics from this world, but not before he had written a fifty-three page critique of the mysterious author's development hypothesis. Albert Baldwin Dod (1805–1845) was no small contender in Americans' debates concerning *Vestiges*. He had held Princeton's chair of mathematics since 1830 and established a reputation as a brilliant teacher, speaker, and writer, committed to both scholarship and his Lord Jesus Christ. The United States Military Academy at

---

1. Review of *Explanations: A Sequel to the Vestiges of the Natural History of Creation*, by the author of that work, *Nassau Monthly* 6, no. 7 (May 1847): 217–28, at 217.

2. Portions of this chapter have been delivered before the Intellectual History Seminar of the University of Notre Dame's Dept. of History (Notre Dame, IN, 5 Apr. 2002) and the History of Science Society Annual Meeting (Milwaukee, WI, 8 Nov. 2002). A modified version of this chapter was published as Ryan C. MacPherson, "Natural and Theological Science at Princeton, 1845–1859: *Vestiges of Creation* Meets the Scientific Sovereignty of God," *Princeton University Library Chronicle* 65, no. 2 (2004): 184–235.

West Point once offered him the Chaplaincy, with a professorship, but this Princeton doctor of divinity remained at his alma mater. There Dod earned Princeton Theological Seminary professor Charles Hodge's praise as "one of the ablest men New Jersey has ever produced." He received the highest salary of any professor at the college. Dod also contributed several important articles to the *Princeton Review*, which Hodge edited. During the 1850s, his biographers would recall especially his April 1842 defense of capital punishment and his October 1845 review of *Vestiges*. The former was adopted wholesale by a committee in the New York legislature; the latter became a public document of another sort, as it added to Americans' ongoing war of words concerning the implications of what Dod called "a mechanical theory of the universe."[3]

Historians previously have understood Dod's review in terms of broadly shared, national and interdenominational understandings of how religion and natural science should interrelate. Herbert Hovenkamp, for example, cited the review as a typical Protestant deployment of the design argument against evolutionary naturalism.[4] Theodore Bozeman classified Dod's argument against *Vestiges* as typical of "Baconianism," a broadly dispersed synthesis of natural science with natural and revealed theology that emphasized empirical facts against hypothetical speculations. To Protestants who felt threatened by the secular tendencies of the Enlightenment, Baconian science provided reassurance that nature clearly revealed God's wisdom, goodness, and power.[5] Two doctoral dissertations exploring science and theology in nineteenth-century America—one of them surveying a broad spectrum of literature and the other focused upon Princeton Presbyterians—also emphasized broadly shared Protestant themes as characteristic of Princeton thought. In these accounts, it was not so much the

3. Henry Clay Fish, *History and Repository of Pulpit Eloquence*, 2 vols. in 1 (New York: Dodd and Mead, 1850), 560–61, quoting 560 (Hodge); *The Papers of Joseph Henry*, ed. Nathan Reingold (vols. 1–5) and Marc Rothenberg (vols. 6–) et al., 8 vols. to date (Washington, D.C.: Smithsonian Institution Press, 1972–1998), 6:120n6 (salary); John Lewis Blake, *A Biographical Dictionary* [etc.] (Philadelphia: H. Cowperthwait and Company, 1859), 370–71; [Albert Baldwin Dod], "Vestiges of the Natural History of Creation," *Biblical Repertory and Princeton Review* 17, no. 4 (1845): 505–57, quoting 505 (theory). See also the various clippings in the Albert Dod Faculty File at the Seeley G. Mudd Manuscript Library, Princeton University.

4. Herbert Hovenkamp, *Science and Religion in America, 1800–1860* (Philadelphia: University of Pennsylvania Press, 1978), 198–200.

5. Theodore Dwight Bozeman, *Protestants in an Age of Science: The Baconian Ideal and Antebellum American Religious Thought* (Chapel Hill: University of North Carolina Press, 1977), 107, 200n24.

design argument or Baconianism that mattered to Dod and his companions, but more especially the doctrine of a fatherly God who superintended the universe. Princetonians, according to this view, rejected *Vestiges* primarily because its vision of a distant deity denied the widely shared Protestant belief in God's providence.[6]

In contrast to these portrayals of Dod as a typical American Protestant of his time, this chapter will establish a more localized and more denominationally specific context in which Dod's review of *Vestiges* can be more adequately interpreted. Following the approach previewed in the Introduction to this book and explained more fully in the Methodological Postscript, this case study will seek first to identify the distinctive systematic theology that was cherished by the Old School Presbyterians who taught, preached, and enrolled for classes at Princeton College and Princeton Theological Seminary. At the center of their theology was something that Presbyterians called the "sovereignty of God," which was their answer to the existential question, "What must I do to be saved?" All Christians, of course, professed a belief that salvation comes from God, but Old School Presbyterians actively distinguished themselves from other Christians (and at times even from other Presbyterians) by insisting that God alone, according to His sovereign will, predestines and elects certain people to salvation and others to damnation, and orders human lives accordingly. Princetonians prized, as the central organizing principle of their theology, God's sovereign election of their souls for salvation. It was here that no new system, including that of *Vestiges*, could trespass.

Careful attention to the institutional setting of Princeton will reveal the manner in which this understanding of God's sovereignty patterned the ways in which faculty, students, and ministers participated in the sciences—both natural and theological. Through extensive analysis of unpublished manuscript sources, this chapter will then uncover reactions to *Vestiges* at Princeton as taught in college and seminary lecture halls, as discussed by undergraduates outside of class, and as preached from Presbyterian pulpits by members of the Princeton community. Admittedly, some of those reactions will appear at first glance to be

---

6. John F. McElligott, "Before Darwin: Religion and Science as Presented in American Magazines, 1830–1860" (Ph.D. diss., New York University, 1973), esp. 8, 49; Bradley John Gundlach, "The Evolution Question at Princeton, 1845–1929" (Ph.D. diss., University of Rochester, New York, 1995), esp. 1–7. See also the recently published Bradley J. Gundlach, *Process and Providence: The Evolution Question at Princeton, 1845–1929* (Grand Rapids: William B. Eerdmans, 2013).

typical of American Protestants generally. It is, therefore, little wonder that earlier studies have lumped Princeton's Albert Dod together with non-Presbyterians into a generic category of American Protestantism. Nevertheless, the close analysis employed in this case study will suggest that Princetonians' reactions to *Vestiges* are best understood as expressions of distinctively Presbyterian—even distinctively Old School Presbyterian—concerns about the relationship between God and humanity. The subsequent case-study chapters concerning Harvard and Yale, by offering analyses of Unitarianism and Congregationalism for comparison, will reveal even more clearly how significant God's sovereignty was at Princeton when *Vestiges* came to campus.

The Presbyterian Setting:
Princeton College and Princeton Theological Seminary

The College of New Jersey at Princeton, informally called "Princeton College" until it officially became Princeton University in 1896, was unmistakably a Presbyterian college. Its charter of 1746 had been granted by the Colony of New Jersey, which had no established church, with the express statement that the college would serve all denominations, but in fact those who directed the college's affairs from its founding through the early twentieth century were Presbyterians who saw to it that Princeton would serve not only the public at large but also the Presbyterian Church in particular.[7] During the 1840s, all of the faculty and most of the students had strong ties to the Presbyterian Church. However non-denominational the charter, most constituents took for granted that Princeton's president should be an ordained minister of the Presbyterian Church.[8] Students were required to attend the campus Presbyterian chapel twice daily, with unexcused absences being placed on their permanent record.[9]

---

7. Glenn T. Miller, *Piety and Intellect: The Aims and Purposes of Ante-Bellum Theological Education* (Atlanta, GA: Scholars Press, 1990), 88–89; Thomas Jefferson Wertenbaker, *Princeton, 1746–1896* (Princeton, NJ: Princeton University Press, 1946), 28, 232, 239.

8. This generalization is reported in John Maclean to Charles Hodge, 24 Aug. 1853 (Charles Hodge Papers, Department of Rare Books and Special Collections, Princeton University Library [hereafter, PUL], Box 17, Folder 28).

9. See, for example, the report cards pasted in George W. Ketcham, Scrapbook, 1859 (Scrapbook Collection, PUL).

Looking back in 1877, former college president John Maclean, himself an ordained Presbyterian minister, regarded the College of New Jersey as a "Presbyterian College" that should be graced architecturally with "the proper form of a Presbyterian Chapel."[10] Amid his relation of significant events on campus, however, he noted that the college had once hired a Roman Catholic foreign language instructor (who resigned in 1836), and that the faculty admitted "two Greek youths from Athens" (presumably Eastern Orthodox) in 1840. Moreover, when eighteen students joined Presbyterian congregations following a religious awakening that swept the campus in early 1850, six other students joined an Episcopal church, and one became a Methodist.[11] Nevertheless, students and faculty alike recognized a Presbyterian ambiance that shaped campus life, including the intellectual milieu discussed here.[12]

The thickly Presbyterian atmosphere at Princeton was certainly savored by members of that denomination, but the "insider-outsider" perspective of a non-Presbyterian student at Princeton may provide a more illuminating entrance into the college's religious environment. Sixteen-year-old Theodore Dwight Tallmadge, a Methodist from Lancaster, Ohio, took the train from Wheeling, Virginia (now West Virginia), to begin studies at Princeton College in 1843. Tallmadge's early letters to his family expressed his dismay that the town of Princeton had no Methodist church. Otherwise pleased with his roommate and professors, he wrote that this was "the only thing that I regret exceedingly, for I cling to the Methodists."[13] And cling he did. By the spring semester, Tallmadge was able to write home with joy, for he and a Methodist classmate had found a Methodist church midway between Princeton and Trenton. By special arrangement with Dr. John Mclean, the college vice president and head

10. John Maclean, *History of the College of New Jersey, from Its Origin in 1746 to the Commencement of 1854*, 2 vols. (Phialdelphia: J. B. Lippincott and Co., 1877), 2:317.

11. Maclean, 2:300 (Roman Catholic), 304 (Athenians), 239 (awakening).

12. Edward Shippen, a 1845 graduate of the college, later recalled Princeton as a somewhat suffocating "Presbyterian Preparatory School," referring immediately to poor dormitory conditions, but throughout his memoir also recalling aspects of campus religious life. Edward Shippen, "Some Notes about Princeton," ed. J. Jefferson Looney, *Princeton University Library Chronicle* 59 (Autumn 1997): 15–57, at 31. His negative attitude may be attributed to his poor academic performance, which he apparently blamed on the college. J. Jefferson Looney, "'An Awfully Poor Place': Edward Shippen's Memoir of the College of New Jersey in the 1840s," *Princeton University Library Chronicle* 59 (Autumn 1997): 9–14, at 14.

13. Theodore Tallmadge to Miss Elmira Cattin (his cousin in Ohio), 12 Nov. 1843 (Tallmadge Papers, PUL, Box 1, Folder 3).

disciplinarian, the two boys were excused from the requisite Presbyterian Sunday services to worship with their Methodist brethren.[14] The music there was "new and fresh, not like the select, stiff chain, with their formal tunes, or the 'fal, de, rae' [fa, do, re], of the aristocratic Episcopal, but all shouted out with the praise, making a noise that had it been in Princeton it would of [sic] shaken its very foundation."[15] Among a student body of some 300, Tallmadge and his Methodist friends numbered but "two or three"; Princeton's Presbyterian foundation could not be shaken by them or their music.[16]

Princeton College's Presbyterian commitments also were apparent to prospective students beyond the campus. Fifteen-year-old Thomas W. Hall, Jr., wrote from Baltimore in 1848 to his father, who at the time was traveling in New England. He scribbled out an urgent plea that his father would "make some inquiries about Trinity College. . . . For you know, I would much prefer to go to a respectable & ably managed Episcopal College [than] to any Presbyterian or Puritan concern in this country." He then related a rumor that Princeton's Vice President John Maclean, "a dull, common-sense Scotch Presbyterian, of little or no genius," would soon replace President James Carnaham. His plea resumed: "if there are good colleges holding our own faith, I see no reason for following after & patronizing the schools of the Presbyterians[,] Unitarians or any other sectaries." The insistent lad's father must have been unsuccessful in finding a suitable Episcopal college, but at least Hall did not have to attend Princeton against his will. He enrolled instead at the University of Virginia, founded on a Jeffersonian commitment to personal religious liberty—a safe haven for someone uncomfortable with Presbyterian education.[17]

Thomas Hall missed out on a golden age of Princeton education in the nineteenth-century sciences. Under Vice President Maclean (who was the real engine behind President Carnaham and who himself became president in 1853), Princeton had revived itself from the poorly funded college of declining

14. Theodore Dwight Tallmadge, *Remiscences of T. W. Tallmadge, Class of 1846, While a Student at the College of New Jersey, 1843–1846* (Bound Ms. Collection, PUL), 6.

15. Theodore Tallmadge to Sarah A. Tallmadge (mother), 20 May 1844 (Tallmadge Papers, PUL, Box 1, Folder 3).

16. Tallmadge, *Reminiscences*, 6.

17. Thomas W. Hall, Jr., to Thomas W. Hall, 24 July 1848, typed transcription with biographical annotations (Office of the President Records, PUL, Box 6, Folder 5).

enrollment that faltered during the 1820s. Enrollment blossomed from 70 in 1829 to 270 in 1839 as Profs. Albert Dod (mathematics), John Torrey (chemistry), Joseph Addison Alexander (ancient languages), Joseph Henry (natural philosophy), and Stephen Alexander (astronomy) came on staff.[18] By the 1840s, Princeton had completed its long transition from a school of statesmenship under President John Witherspoon (during whose tenure James Madison, the architect of the United States Constitution, graduated) to a school offering strong curricula in all the sciences—natural, philological, and political.[19]

All of this stood in contrast to the first three decades of the nineteenth century, when the college lacked stability. Student disruptions were not uncommon in American colleges during the early decades of the nineteenth century, but Princeton's constituents took special alarm at the record number of discipline cases (for offenses ranging from intoxication to arson) in 1807. When the faculty rejected a petition of appeal signed by 160 students in regard to recent disciplinary findings, several students retaliated by breaking windows and doors. The trustees faulted President Samuel Stanhope Smith's leadership, whose reputation was further injured by additional uprisings in 1809 and 1812.[20] Smith's efforts to establish a divinity department for training future ministers at the college showed little promise. Consequently, in 1812, the General Assembly of the Presbyterian Church established as a separate institution the Theological Seminary of the Presbyterian Church at Princeton, or "Princeton Theological Seminary," as it was familiarly known.

Even more so than Princeton College, Princeton Theological Seminary was thoroughly committed to Presbyterianism. Whereas final authority over Princeton College rested with its board of trustees, which consisted primarily of Presbyterian ministers plus the New Jersey governor as president *ex officio*, Princeton Theological Seminary had to answer, via its trustees and directors, to the General Assembly itself. Faculty members were required, by the seminary's charter, to take an oath to teach the Presbyterian theology embodied in the

18. Maclean, 2:278, 308, 284.

19. David B. Calhoun, *Princeton Seminary. Vol. 1: Faith and Learning, 1812–1868; vol. 2: The Majestic Testimony: 1869–1929* (Carlisle, PA: The Banner of Truth Trust, 1994), 1:168.

20. Mark A. Noll, *Princeton and the Republic: The Search for a Christian Enlightenment in the Era of Samuel Stanhope Smith* (Princeton, NJ: Princeton University Press, 1989), 6, 168–69, 227–30, 245–46; Miller, 96.

Westminster Confession of 1646. Trustees had to renew their own oaths annually.[21] Thirty-one years after the seminary's founding, pressure was still strong to remain faithful to the Westminster standards. When college trustee James Lenox donated his library to Princeton Theological Seminary in 1843, he stipulated in the deed of property transfer that the books must be returned if ever the seminary deviated from teaching the Calvinist doctrines codified at Westminster. Specifically, Lenox listed universal and total depravity; election; atonement; imputation of Adam's sin; imputation of Christ's righteousness; human inability; and regeneration, conversion, and sanctification by the Holy Spirit.[22] Such tightly monitored political and theological control goes some distance to explain why, in the view of several historians, "Princeton was the only American theological institution to teach a consistent theological position for more than two generations."[23]

The doctrines Lenox specified formed a cohesive system bound by the central dogma of Old School Presbyterianism, namely, that God wills and acts in sovereignty over each person's soul.[24] During the 1840s and 1850s, Princeton

---

21. Miller, 105–6.

22. Calhoun, 1:270–71; William K. Selden, *Princeton Theological Seminary: A Narrative History, 1812–1992* (Princeton, NJ: Princeton University Press, 1992), 39.

23. Miller, *Piety and Intellect*, 107. For a similar statement, though with a concession that Harvard Divinity also held to a fairly consistent teaching during the nineteenth-century, see Mark A. Noll, "The Princeton Theology," chap. 1 in *Reformed Theology in America: A History of Its Modern Development*, ed. David F. Wells (Grand Rapids, MI: Baker, 1997), 15–35, at 29. Hodge himself claimed, in 1872, that no new doctrine had been introduced at the seminary during his long tenure. For an interpretation that does not question Hodge's claim, but does move under its surface in order to understand how such a self-perception functioned to construct the notion of a seminary as a community of unified believers, see E. Brooks Holifield, "Hodge, the Seminary, and the American Theological Context," in *Charles Hodge Revisited: A Critical Appraisal of His Life and Work*, ed. John W. Stewart and James H. Moorhead (Grand Rapids, MI: William B. Eerdmans, 2002), 103–28.

24. Pertinent selections from the Westminster Confession of Faith (1646), to which Old School Presbyterians subscribed, include chap. V, art. I (God's sovereignty throughout the universe); chap. IX, art. III (total depravity); chap. X, art. I (God's sovereignty in predestination, salvation, and conversion); and chap. X, art. II (God's sovereignty in conversion). The text may be found in *The Confession of Faith; the Larger and Shorter Catechisms, with the Scripture-Proofs at Large, together with the Sum of Saving Knowledge* (Edinburgh: D. Hunter Blair, 1845), 19–105. For a Princeton commentary, see Arhibald Alexander Hodge, *The Confession of Faith: A Handbook of Christian Doctrine Expounding the Westminster Confession* (1869; rpt., London: Billing and Sons, 1978). John C. Vander Stelt, *Philosophy and Scripture: A Study of Old Princeton and Westminster Theology* (Marlton, NJ: Mack Publishing Company, 1978), 123, identified scientific induction and biblical inspiration as the two themes that dominated Princeton theology. John W. Stewart, *Mediating the Center: Charles Hodge on American Science, Language, Literature, and Politics*, Studies in Reformed Theology and History series, vol. 3, no. 1 (Princeton, NJ: Princeton Theological

Theological Seminary and the adjacent Princeton College served as the flagship institutions for the Old School wing of the Presbyterian Church, which had expelled the New School contingent from the General Assembly in 1837. The New School had appropriated elements of the "New Haven theology" that Congregationalist divinity professor Nathaniel W. Taylor was teaching at Yale College. Taylor's emphasis on personal agency in matters of moral responsibility and spiritual rebirth seemed to suggest that God's sovereignty had limits— people's own initiative somehow contributed to their salvation. Concerned over this Arminian tendency in the New Haven–influenced theology of New School Presbyterians, Old Schoolers regarded themselves as the last bastion of Presbyterian orthodoxy in America.[25] They committed themselves to the Calvinist doctrines of total depravity and predestination as preserved in the seventeenth-century Westminster Confession of Faith, which affirmed the absolute sovereignty of God. Applied to the cosmos, this meant that it was the will of God alone that caused the planets to travel in their orbs; regarding the human soul, God's sovereignty determined who would be saved and graciously saw to it that those people were converted to faith in Christ. In neither case was the result dependent upon the nature of the planet or of the response of a person's soul.[26]

Both at and beyond Princeton, Old School Presbyterians championed God's sovereignty above all other concerns. In 1854, a writer for the *Presbyterian Magazine*, a journal edited by Princeton Theological Seminary alumnus and fundraiser Cortland Van Ransselaer, argued that God's sovereignty should be

Seminary, Winter 1995), 22, identified three components in the "Princeton paradigm": 1) Scottish Common Sense Realist epistemology; 2) doxological uses of science; and, 3) the pursuit of theology following the model of natural science. Without questioning the importance of these elements at Princeton, I maintain that these each functioned for Princetonians as tools for promoting what was in fact the central organizing principle of Old School Presbyterianism, namely, the sovereignty of the God confessed at Westminster.

25. George Marsden identified six areas of doctrinal dispute leading to the 1837 split: 1) degree of allegiance to the Westminster Confession; 2) polity; 3) Presbyterian involvement in interdenominational voluntary organizations; 4) revivalism; 5) statements concerning the nature of unregenerate man (the Old School accusing the New School of Arminian theology with tendencies toward Finneyan disorder); and, 6) slavery. Paul Conkin has underscored the role of the Nathaniel Taylor's New Haven theology (which was intended to refute Arminianism, but did so in a manner that looked suspiciously Arminian) in the division between the Old and New Schools. See George M. Marsden, *The Evangelical Mind and the New School Presbyterian Experience: A Case Study of Thought and Theology in Nineteenth-Century America* (New Haven: Yale University Press, 1970), 67–84; se also Paul K. Conkin, *The Uneasy Center: Reformed Christianity in Antebellum America* (Chapel Hill: University of North Carolina Press, 1995), 212, 262–68. Taylor's own perspective on the debate will be addressed in chapter 4 of the present volume.

26. See the portions of the Westminster Confession cited in note 24, above.

preferred over God's love as the foundation for the doctrine of atonement. In the writer's view, God's sovereignty underpinned the Calvinist doctrine that Christ died to redeem only the elect (limited atonement). In keeping with the Westminster Confession, God's sovereignty in this "economy of grace" was explicitly analogous to His sovereignty in the "economy of nature"; in both realms, the Creator's will, not creation's response, determined God's action.[27]

Princeton College and Seminary were charged with the responsibility of preserving this Presbyterian orthodoxy both in the present generation and for those to come. Beyond the banner of Presbyterianism, Princetonians were convinced there lay even greater dangers: the blatantly Arminian exhortations for personal contributions to conversion experiences among Methodists and Baptists, the denial among Unitarians of Christ's divinity, the idolatrous practices of Roman Catholics, the pantheism of Transcendentalists, and the skepticism of deists. Princeton's Old School Presbyterians defended God's sovereignty over the soul's conversion, the planets' orbits, and the American political order as different aspects of a single truth. The New School Presbyterians' seemingly Arminian emphasis on individual conversion experiences, for example, was feared among Old School Princetonians to atomize religious as well as social and political authority, for it fragmented the structured freedom embodied in both Presbyterian polity and the United States Constitution.[28]

Princeton's theologians aimed to articulate and defend in a systematic way their Old School Presbyterian theology, with its attendant vision for American society, as it related to any intellectual movement of the day. The seminary's charter required that each student be well-versed in "the principal arguments and writings relative to what has been called the deistical controversy" and learn "carefully and correctly Natural Theology" in order "to become a defender of the Christian faith."[29] The *Princeton Review*, the quarterly journal in which Prof. Albert Dod's analysis of *Vestiges* appeared, served as a means for seminary and college faculty to evaluate any new book or idea. Classroom lectures, which followed a consistent outline from year to year at both the college and the

27. "The Extent of the Atonement," 2 pts., *Presbyterian Magazine* 4, no. 11 (Nov. 1854): 496–502, 4, no. 12 (Dec. 1854): 533–40; and, Calhoun, 1:260 (on Rensselaer's support of PTS during the 1840s).

28. James H. Moorhead, "The 'Restless Spirit of Radicalism': Old School Fears and the Schism of 1837," *Journal of Presbyterian History* 78, no. 1 (Spring 2000): 19–33.

29. Calhoun, 1:423–24.

seminary, provided a forum for training the next generation in the kind of natural and revealed theology, natural and moral philosophy, and natural and sacred history—in short, the kind of "science," or systematized knowledge—that was constitutive of Old School Presbyterian intellectual life.[30]

Officially, the seminary at Princeton had closer ties to the Presbyterian Church than did the college, but historians have on occasion over-emphasized this distinction. For example, some scholars have portrayed the seminary's first professor, Archibald Alexander, as a conservative theologian who opposed the strong concentration of natural sciences in the college curriculum. Accordingly, Alexander's proposal before the General Assembly in 1812 to form an independent seminary, rather than pursue Smith's plan of a divinity department at the college, has been understood as a conflict between Presbyterian theologians and Princeton College's advocates of science.[31] Such a polarization, however, was unknown at Princeton. Alexander's academic concerns for ministerial students stemmed not from a tension between theology and natural science—as if anyone at Princeton could coherently have conceived of one of these apart from the other—but rather from a preferential emphasis upon theological and exegetical sciences over natural sciences in the training of future clergy. In 1829, Alexander joined with Profs. Charles Hodge and Samuel Miller at the seminary and Profs. Albert Dod, John Maclean, and President James Carnaham at the college in an "association of gentleman" who conducted the *Biblical Repertory, a Journal of Biblical Literature and Theological Science*.[32] (For simplicity, this periodical has been named throughout this book by its later title, *Princeton Review*). After Joseph Henry was hired as the college's professor of natural philosophy in 1832, Archibald Alexander could be found in Philosophical Hall participating in Henry's electromagnetism experiments.[33] It is doubtful that Alexander and Henry ever felt much tension between their respective scientific specialties (theology and physics); for one thing, Henry often introduced his classroom experiments with words like, "Young gentlemen, we are

---

30. See the discussion under the subheading "A Concept of 'Science' during the Era of *Vestiges*," in the introduction to the present volume.

31. Hovenkamp, 17; Miller, 103–4.

32. Calhoun, 1:193.

33. Stewart, *Mediating the Center*, 20; Bozeman, 41.

about to ask God a question."[34] Natural and revealed knowledge at Princeton were two parallel paths to the same divine facts, and Alexander both encouraged Henry's natural science and promoted natural theology for his own part.[35] Indeed, the seminary's charter required him to do so.[36]

Though the seminary had been founded largely because the General Assembly had lost confidence in the college, by the 1840s the two institutions were both faring well, and doing so arm-in-arm. They were linked, above all, by their common "Princeton Theology," a tie that remained strong throughout much of the century.[37] Each institution formally had its own board of trustees, but the two boards overlapped significantly, in terms of interests and also personnel. Professors from one institution served as trustees for the other, and some trustees served simultaneously on both boards.[38] College professor Joseph Henry, for example, served as a trustee of the seminary during 1844–1851.[39]

Family ties further linked Princetonians into a single community. Henry's brother-in-law (and cousin) Stephen Alexander taught astronomy at the college. Archibald Alexander's protégé Charles Hodge graduated from both the college (1815) and the seminary (1819) and became the nineteenth-century's most prominent professor at the seminary, instructing some 3,000 students during his half century of service. Hodge and Alexander each served on the college board of trustees for over a quarter of a century. Perhaps no student was shaped by these two men more than Hodge's son, tellingly named Archibald Alexander Hodge. The young Hodge also deeply admired Joseph Henry and nearly pursued a professorship in natural science, but ultimately he followed his father's footsteps

34. Calhoun, 1:167.

35. Mark A. Noll, "Science, Theology, and Society: From Cotton Mather to William Jennings Bryan," chap. 4 in *Evangelicals and Science in Historical Perspective*, ed. David N. Livingstone, D. G. Hart, and Mark A. Noll (New York and Oxford: Oxford University Press, 1999), 99–199, at 113; Archibald Alexander, *A Brief Outline of the Evidences of the Christian Religion* (Philadelphia: American Sunday School Union, 1825).

36. Calhoun, 1:423–24.

37. Mark A. Noll, "Introduction," in *The Princeton Theology, 1812–1921: Scripture, Science, and Theological Method from Archibald Alexander to Benjamin Breckinridge Warfield*, ed. Mark A. Noll (Grand Rapids, MI: Baker Book House, 1983), 11–48, at 20.

38. Gundlach, ix; Noll, "Introduction," 20–21; Selden, 57.

39. *Catalogue of the Officers and Students of the Theological Seminary at Princeton, New Jersey, 1844–1845* through *Catalogue . . . 1850–1851* (Princeton: John T. Robinson, 1845–1851), available in the reference room at the Mudd Manuscript Library, Princeton University.

in the paths of theological science, becoming a professor at the seminary in the 1850s.

Given the numerous ties linking the college and seminary, it likely astonished no one that, beginning in 1855, students at the seminary were permitted to attend classes at the college with tuition waived. More interesting for this study, seminary students had since 1834 been permitted to attend three particular courses at the college free of charge: natural philosophy, chemistry, and natural history.[40] For seminary students, these courses were expected to provide a learned foundation for the ministry of the gospel. The truths concerning God's world that the natural sciences catalogued were truths that ministers would marshal to refute six "isms" that attacked the church from without: deism, atheism, materialism, pantheism, idealism, and skepticism.[41] Archibald Alexander had himself first become interested in theology when reading a book of "Christian evidences" garnered from natural theology.[42] Generations of his students studied—in addition to their first-year courses in Hebrew, Biblical Exegesis, and Biblical History—a course in Mental and Moral Science and another entitled "The Evidences of Natural and Revealed Religion." The former course, drawn largely from the Scottish Realist philosophy of Thomas Reid, as popularized by Dugald Stewart, impressed upon students the reliability of the human mind in perceiving facts about the observable world, and the latter course arranged those facts into arguments supporting theistic belief.[43]

A fifteen-minute walk north from the seminary, Princeton College provided its students with an undergraduate-level exposure to such proofs. Juniors were required to study Archibald Alexander's *A Brief Outline of the Evidences of the Christian Religion* (1825) in the fall semester and William Paley's *Natural Theology* (1801) in the spring.[44] The college and seminary curricula were thus coordinated to present theology as the queen of the sciences

40. *Catalogue of the Officers and Students of the Theological Seminary at Princeton, New Jersey, 1834–1835* through *Catalogue . . . 1855–1856* (Princeton: John T. Robinson, 1834–1855), available in the reference section of Speer Library, Princeton Theological Seminary.

41. Thomas S. Malcom, Notes on Lectures on Polemic and Pastoral Theology and Sermons, by Archibald Alexander, 1840–1841 (Archibald Alexander Collection, Department of Special Collections, Henry Luce III Library, Princeton Theological Seminary [hereafter, PTS Library], Box 25, Folder 5).

42. Miller, 110.

43. Calhoun, 1:87–88; Allen H. Brown, Lecture Notes on the Evidences of the Christian Religion, by Archibald Alexander, 1840 (Archibald Alexander Collection, PTS, Box 4, Folder 2).

to which all other sciences must bow, presenting their gifts humbly at her feet for service to the church.

Princeton theologians were so convinced that natural and revealed knowledge were two sides of a single, God-ordained truth that they exercised patience whenever natural and theological science seemed in conflict. For them, no true conclusions of one science (such as geology) would be irreconcilable with those of another (such as theology). "Skeptics," "rationalists," and other foes of the church, against whom Princeton's students were being trained to be on guard, had ridiculed the Mosaic account of creation in Genesis chapters 1–2 because it speaks of God creating in six days, some 6,000 years ago, a world that geologists had concluded to have been formed successively during innumerable eons in the distant past. Archibald Alexander, during the 1830s and early 1840s, and Charles Hodge, during the late 1840s and 1850s, taught a generation of seminary students in their "Didactic Theology" course that when it comes to Genesis and geology, "The truth has nothing to fear from truth." Perhaps geologists are correct that the earth is extremely ancient, they willingly conceded. "If Geologists prove their theory," explained Alexander, "then we as theologians will give a different interpretation to Gen. 1." Indeed, noted these professors, some theologians already had suggested new interpretations: the term "day" in Genesis might be mean "eon," or else a gap of innumerable ages might have been passed over in the Mosaic narrative between Genesis 1:2 and 1:3. Assuming such a gap, the six days could refer to a literal six-day period some 6,000 years ago during which God renewed a small portion of the globe as Eden and there made Adam and Eve.[45]

---

44. Tallmadge, *Reminiscences*, 7; Maclean, 2:285; *Catalogue of the College of New Jersey, for 1858–1859* (Princeton: John T. Robinson, 1859), 19; James Buchanan Henry and Christian Henry Scharff, *College As It Is: Or, The Collegian's Manual in 1853*, ed. with an introduction by J. Jefferson Looney (Princeton, NJ: Princeton University Libraries, 1996), 196. Looney, at 196n, suggests that what Henry and Scharff (class of 1853) referred to, in identical language as the college catalogue, as "Alexander's Evidences of Christianity" was Archibald Alexander's *Evidences of the Authenticity, Inspiration, and Canonical Authority of the Holy Scriptures* (7th ed., 1836), rather than *A Brief Outline of the Evidences of the Christian Religion* (1825). Alexander's *Brief Outline* originated in a college sermon aimed at checking some skeptical tendencies among the students. See Calhoun, 1:87. Whichever book was used for the natural theology curriculum, it gave Princeton a means to supplement the more generic theistic apologetics of William Paley with one of their own Presbyterian minister's contributions to the "evidences" literature.

45. Peter Lesley, Questions on Theology, by Archibald Alexander, ca. 1842–1845 (Archibald Alexander Collection, PTS, Box 3, Folder 1), 121; Allen H. Brown, Lecture Notes on Didactic Theology, by Archibald Alexander, Sep. 1843 (Archibald Alexander Collection, PTS, Box 4, Folder 4); Henry Van Vleck Rankin, Notes from Hodge, A. Alexander, Miller, and J. A. Alexander, 1845–1847 (Charles Hodge Collection, PTS, Box 36, Folder 1).

Princeton Theological Seminary's exegetes-in-training were encouraged to exercise similar flexibility regarding the interpretation of Noah's Flood: if the earth is young, then the Noachian Deluge transformed it to look old; if it is old, then fossils are the remains of pre-Adamite animals and the Deluge was perhaps a local event.[46] In any case, one should patiently trust that once geologists have gathered more evidence, theologians will know exactly what kind of reconciliation will be required in their interpretations and then proceed accordingly. "We fear not the light and results of sound science," Alexander told his students in 1843.[47]

Princeton's interpretative flexibility regarding the Bible's accounts of creation and the flood were by no means indicative of a general liberality in Biblical exegesis. Quite to the contrary, Princeton's professors were convinced that what Alexander called "sound science" would support a particularly Presbyterian understanding of Scripture; indeed, they regarded their theology as *the* scientific interpretation of the Bible. From the founding of Princeton Theological Seminary in 1812 until the publication of Charles Hodge's *Systematic Theology* in 1871–1872, Archibald Alexander, Charles Hodge, and Archibald Alexander Hodge each in turn taught courses in Didactic Theology from Francis Turretin's *Institutio theologiae elencticae* (1679–1685). This work systematized the Calvinist Reformed doctrines to which the Presbyterian Church pledged itself at Westminster in 1646. By the early nineteenth century, American Presbyterians had fused this tradition with elements of the eighteenth-century Scottish Enlightenment, resulting in a conviction that the Bible was a book of theological facts, that the human mind was competent to identify and properly arrange those facts, and that the task of theology was to do precisely those things in order that the gospel may be known and proclaimed to all. In their theology classes, seminary students learned that non-Calvinist interpretations of Scripture resulted either from denial of the facts or else a failure to arrange the facts in the most coherent and consistent manner.[48]

Variants within the Christian tradition (such as Arminianism) and pagan philosophies (such as pantheism) were alike regarded as "systems" that were

46. Lesley, Questions on Theology, 121; Brown, Lecture Notes on Didactic Theology.

47. Brown, Lecture Notes on Didactic Theology.

48. Calhoun, 1:90–91, 262; Miller, 94, 111.

demonstrably false for their failure to make sense of all the facts known from nature and from Scripture. "The Arminian system," explained Charles Hodge, "proceeds upon the false assumption that <u>faith</u> is the believer[']s own act, that it is man & not God that decides the question whether a given person accepts the gospel or not."[49] His son preserved the family theological tradition by teaching that the Calvinist system, which maintains that conversion is a sovereign act of God alone, is the only coherent presentation of the facts.[50] As with the soul, so also with the cosmos: God must be sovereign. "What <u>systems stand opposed</u> to the doc[trine] of a Creation ex nihilo?," Alexander asked his theology class in 1842. Prior to *Vestiges*, Alexander listed six: 1) the Stoic and Epicurean view that the world is eternal; 2) "the old dualistic theory" that the world emerged from a primordial polarity of good and evil; 3) hylozoism (the theory that matter has an inherent and divine life principle); 4) gnosticism (which holds that the world emanated from the divine being); 5) pantheism (which holds that God and nature are one); and, 6) the "modern German philosophy" that held that since "Creator" is an essential part of God's being, God must be eternally creating, so creation, like God, must be eternal.[51] In contrast to these, Princeton's theologians presented the "Calvinist system" as a faithful catalogue of all of the facts of nature and of Scripture to demonstrate that God is distinct from and sovereign over his creation, just as He acts in sovereignty over the conversion of a person's soul to faith in Christ Jesus.[52]

The oft-repeated lesson that the Presbyterian Church's Westminster Calvinism, with its central emphasis upon the sovereignty of God, was the only adequate systematization of the facts gathered from both natural and revealed science, spread from the seminary to the pulpit. In their sermons Princeton Seminary graduates drew from their acquaintance with natural philosophy to defend the teachings of their church. In 1846, the Rev. Eli Field Cooley, who

49. Armstrong, Notes from Lectures on Didactic Theology.

50. A. A. Hodge, A Comparison of the Main Distinguishing Positions of the Three Rival Systems of Pelagianism, Semipelagianism, & Augustinianism, ca. 1862 (Archibald Alexander Hodge Collection, PTS, Box 3, Folder 4).

51. Lesley, Questions on Theology.

52. Archibald Alexander, Creation, ca. 1846 (Archibald Alexander Collection, PTS, Box 9, Folder 34); Charles Hodge, Creation, ca. 1847 (Charles Hodge Collection, PTS, Box 1, Folder 22), Providence, 12 Dec. 1846 (Box 1, Folder 25), and Election, 13 Dec. 1847 (Box 1, Folder 20); Armstrong, Notes from Lectures on Didactic Theology.

served as a seminary and college trustee for some thirty years, preached concerning God's use of natural laws to preserve the universe from degenerating into chaos.[53] In 1852 he emphasized the parallel between God's providential sovereignty over the universe and God's redemptive sovereignty over the souls of the elect: in each case, God is ever-present as an active caretaker. Without God, only chaos and self-destruction would result.[54] Such sermons formed an enduring tradition. For example, the Rev. John T. Duffield, professor of mathematics, preached a similar message in the college chapel in 1862, affirming "the sovereignty of God over men, both in the sphere of grace as well as of providence."[55]

Because God's providential care over all creation and his redemptive plan for human souls were understood as parallel instances of God's sovereign governance, natural and theological science enjoyed a long and stable partnership at Princeton. Occasionally, some new idea—often recognized by Princetonians as a resuscitated old heresy—would arise to threaten that precious marriage between natural and revealed knowledge. Such was the case, for example, with something known as "the mechanical theory of the universe." Prof. Hodge quizzed his seminary students in preparation for their high calling: "What is the doc[trine] of preservation as held by Deists & other advocates of the Mechanical theory of the univ[erse]?" Again: "What is the mechanical theory of the nature of providential government? What are objections to it?"[56] The student notebooks in which these study questions appear do not indicate any answers. Prof. Albert Dod, however, was grounded in the same Princeton heritage as seminary professor Hodge, and when he identified *Vestiges* in his review as "a mechanical theory of the universe," he made a point to expound upon what that should mean for Princeton Presbyterians.[57]

53. Eli Field Cooley, Sermon on Psalm 97:1–2, 26 Nov. 1846 (Eli Field Cooley Papers, PUL, Box 3, Folder 2).

54. Cooley, Sermon on Psalm 19:7, 26 Dec. 1852 (Eli Field Cooley Papers, PUL, Box 3, Folder 3).

55. John T. Duffield, Sermon on Ephesians 1:3–6, College Chapel, 1862 (Duffield Family Papers, PUL, Box 4, Folder 3).

56. Charles B. Scott, Questions in Systematic Theology, by Charles Hodge, 1846 (Charles Hodge Collection, PTS, Box 36, Folder 2); Henry Clay Cameron, Notes on Courses, v. 14: Lectures on Nature of Theology and Sources for Our Religion, 1853, Lecture 20 (Cameron Family Papers, PUL, Box 6).

57. [Dod], 505.

The Princeton Review:
Defending Presbyterian Science against Vestiges

Dod structured his review of *Vestiges* in two parts, each extending about twenty-five pages. First, he described the work as a "mechanical theory," which he defined to mean a theory proposing that the entire universe and especially life itself, "in all its forms and with all its endowments, [had] evolved through the action of mechanical and chemical causes."[58] Along the way Dod indicated numerous errors in fact and in reasoning that together, even if not severally, sufficed in his mind to refute the author's theory. He concluded that the entire system tended—despite its author's occasional doxologies to a Creative Intelligence—toward atheism. *Vestiges* had merely "embellished by modern science" the "sty of Epicurus," which Anaximander had anticipated and Gassendi, Hobbes, the French encylopedists, Erasmus Darwin, and Lamarck had defended in more recent times. To distance the Creator from his creation, to ascribe its development and subsistence to innate tendencies and natural laws is tantamount to atheism and "has been so recognized in all ages," stated Dod about midway through his article.[59] In other words, since *Vestiges* had tried to subvert God's sovereignty, the book's adherents were on the ancient path toward godlessness.

One might suppose that Dod could have concluded his review right there. Instead he continued, explaining, "But we propose to make a further examination of this system upon its merits as a scientific hypothesis."[60] In the second half of his article, Dod repeated many of the scientific objections he already had made in the preceding pages: the eighteenth-century French astronomer Pierre Simon Laplace had responsibly offered his nebular hypothesis of the solar system's origin as merely a conjecture, whereas the author of *Vestiges* rashly advanced an inferior version of that hypothesis as something more certain; despite the well-established inductions that all life comes from the living and that like produces like, the author built upon ungrounded speculations of spontaneous generation and species transmutation; Andrew Crosse's spontaneous generation experiments, presented as decisive evidence in *Vestiges*, had not been

58. [Dod], 515–16.

59. [Dod], 530–31.

60. [Dod], 534.

corroborated by reputable men of science. From page to page, *Vestiges* was littered with mistaken facts and unproven assertions. Because the author's misapplied analogies failed to acknowledge how fundamentally "the plant is differenced from the stone," Dod concluded that the plausibility of Laplace's nebular hypothesis did not warrant the speculative conclusion in *Vestiges* that plants, animals, and even "man extends thus in a direct line back to the original nebulous matter of which the universe was composed."[61] Thus far the first and second halves of Dod's review agreed.

It is not readily apparent why Dod chose to subject the chief elements of the development theory—its nebular hypothesis, spontaneous generation, species transformation, and the animal (indeed, the nebular) origin of human beings—to two waves of criticism within the same review, especially considering that some statements were repeated almost verbatim. Matthew Boyd Hope, who made arrangements for the printing and distribution of the October 1845 *Princeton Review*, wrote to editor Hodge from Philadelphia with a concern about Dod's article: "I do not think it is in his best style by any means."[62]

Back in Princeton, Dod had good reason not to be at his best. In mid September, while he was preparing the article, he had been troubled by much more than the development hypothesis. When a few students skipped morning recitations to go hunting, one of them was fatally wounded by an accidental gunshot. It was Dod who carried the young Richard Stockton Boudinot from the woods—"his brains oozing out" of the head wound, as a classmate later recalled. Dod kept the injured boy in his own home for about a month, until he was moved to a guest house with his parents where he died in early November. Philadelphia physicians were fascinated by the unprecedented case: a wound through which the division between the left and right hemispheres of the brain could be observed. Dod, who earlier in the year had been experiencing an accelerated pulse and irregular heartbeat, was noticeably shaken over the matter, as it left him with nightmares that intensified his ongoing bout with a "nervous disease."[63]

61. [Dod], 550 (quotation); 516 (quotation).

62. Matthew B. Hope to Charles Hodge, 4 Oct. 1845 (Charles Hodge Papers, PUL, Box 16, Folder 49).

63. Theodore Tallmadge to Sarah A. Tallmadge (his mother), 28 Sep. 1845 (Tallmadge Papers, PUL, Box 1, Folder 3); Charles Hodge to Hugh Lenox Hodge, 16 Apr. 1845, 16 Sep. 1845, 29 Sep. 1845, 16 Oct. 1845, 29 Oct. 1845, 30 Oct. 1845, 4 Nov. 1845 (Charles Hodge Papers, PUL, Box 11, Folder 3); George William Doane to the Editor of the Burlington Gazette, 27 Nov. 1845, in Bethuel L. Dodd and John R. Burnet, *Genealogies of the Male Descendents of Daniel Dod, of Branford, Conn., a Native of England*

Had circumstances been different, Dod's review of *Vestiges* might have been more polished, less rambling, and less redundant.

As Dod's review stood, the mass of scientific objections in the two sections led to slightly different aspects of Dod's principal conclusion. During the first pass, Dod traced the author's thesis to what he considered its unavoidable religious implication: atheism. The second time, he emphasized a closely related philosophical implication: materialism. Either way, *Vestiges* would deprive that boy's family and friends of the comfort that a sovereign God had watched over him during the fatal accident that removed him from this world to the next.

"With most of our readers," wrote Dod, "we trust it would be deemed an ample refutation of any system to show clearly that it was atheistic in its essential character."[64] *Vestiges,* however, posed a special challenge. True, Dod felt he had "shown that the author of this work has failed at every point, in establishing his different positions, but"—and this "but" seems to have been Dod's impetus for passing over the evidence a second time—"we have not shown that some other explorer in the same direction may not be more successful."[65] After all, Dod himself twice acknowledged that Laplace's version of the nebular hypothesis might withstand some of the criticisms he was laying against the "system" in *Vestiges,* and he made a similar concession, regarding species transmutation, to Jean-Baptiste Lamarck's supporters.[66] Dod then admitted that the only way to stop "some other explorer in the same direction" from being "more successful" was to appeal to "man's intellectual and moral nature."[67]

Rhetorically, at least, Dod was "perfectly willing now to yield every position which we have taken against this author's theory," while still maintaining that humans had been created in the image of God, not evolved from animals.[68] He appealed to "the bar of human consciousness," bringing testimony that matter ("that which *appears,*" producing "phenomena") and mind ("that

(Newark, NJ: Daily Advertiser Office, 1864), 206–8, esp. 207.

64. [Dod], 534.

65. [Dod], 552–53.

66. [Dod], 506, 513, 549.

67. [Dod], 549.

68. [Dod], 550–51.

which *perceives*" those phenomena) are fundamentally distinct.[69] This contradicted a central tenet of materialism—namely, that all interactions in the universe are material and none require the direction of an immaterial mind, soul, or deity—but it did not show that humans could not have been evolved from animals. The crucial plank in Dod's argument came next: "Man . . . is free from all law except that which is self-imposed." Free will separated man, according to Dod, "not in degree only, but in kind, from any thing that is found in the brute creation."[70] Free will, wrote Dod, was an impassible division between man and brute that not only undermined any theory of transmutation between animals and humans, but also required that *Vestiges*' "line of descent from below" be redrawn, for humans at least, as a more orthodox "line of descent from above."[71] Man's direct creation in the image of his heavenly Creator could thus be established, Dod insisted, even with "the sacred scriptures" excluded from the discussion.[72]

Dod's confidence rested on a rigid mind-matter distinction that formed the foundation of natural philosophy as taught at Princeton. As Robert Lenox Banks, class of 1848, recorded in his notebook of Prof. Joseph Henry's 1847 lectures on natural philosophy, "Science is separated into two great divisions viz Material & Mental . . . Mind is that which thinks and is capable of Moral Emotions [but] Matter is inert."[73] The author of *Vestiges* had blurred this distinction between mind and matter when claiming that the Deity had endowed matter with the capacity to transform itself into living, feeling, even reasoning organisms. The problem ran deeper than the prospect of inert, nonliving matter acquiring mental properties; even *Vestiges*' claim that nebulae could, by some law of development, become consolidated into stars, transgressed the barrier between mind and matter.

69. [Dod], 557 and 553.

70. [Dod], 554. It is important to note that the kind of free will that distinguished humans from animals was nevertheless impotent in matters of salvation, and thus in no way impinged upon God's sovereign activity of converting the elect unto salvation. See Westminster Confession, chap. IX.

71. [Dod], 551.

72. [Dod], 550.

73. Robert Lenox Banks, Natural Philosophy Lecture Notes, Joseph Henry, 1847 (Lecture Notes Collection, PUL, Box 24), 3.

Consistent with a belief in the transcendence of God, many Christians favored a sharp distinction between the immaterial and material worlds. At Princeton, however, the mind-matter distinction also resonated powerfully with the Presbyterian conceptions of a sovereign God and a universe obedient to his will. According to Prof. Henry, no law of matter could be the true cause of any transformation. Laws were merely descriptions of the observable patterns in which, wrote one of Henry's students in May 1846, the "volition of the Deity" acts as the "intelligent cause" throughout the universe.[74] Properly speaking, recorded another of Henry's students in his notebook for the same course, the term "cause" should be employed exclusively for "the power and volition of an intelligent being."[75] God has created and is sustaining the universe, but matter itself can neither be the cause nor self-sufficiently contain the cause of anything.

Still, one might wonder whether this mind-matter distinction, regarded by Old School Presbyterians as a corollary to the sovereignty of God, left at least a small opportunity for *Vestiges* to find accommodation at Princeton. Despite Dod's claim that the system tended toward atheism, the author of *Vestiges* had in fact ascribed the law of matter's development to the mind of God. At the close of his first chapter he affirmed "a First Cause to which all others are secondary, and ministrative, a primitive almighty will, of which these laws are merely the mandates."[76] The author similarly underscored in *Explanations,* the sequel volume, that "natural law" does not refer to "a system independent or exclusive of Deity, but one which only proposes a *certain mode of his working.*"[77] One might suppose that Princeton's faculty would delight in *Vestiges*' subordination of natural law to divine will. After all, Prof. Henry in the 1840s, and Profs. Stephen Alexander and Arnold Guyot in the 1850s, taught that a "law of nature" means simply "the mode in which divine wisdom operates in producing the changes of

74. Henry Clay Cameron, Natural Philosophy Lecture Notes, Joseph Henry, 1846–1847 (Lecture Notes Collection, PUL, Box 24), 13.

75. James Finley Davison, Natural Philosophy Lecture Notes, Joseph Henry, 1846 (Lecture Notes Collection, PUL, Box 23), 5.

76. [Chambers], *Vestiges,* 26, in *Vestiges of the Natural History of Creation and Other Evolutionary Writings, Including Facsimile Reproductions of the First Editions of "Vestiges" and Its Sequel "Explanations,"* ed. with an introduction by James A. Secord (Chicago: Chicago University Press, 1994).

77. [Chambers], *Explanations,* 3, in *Vestiges,* ed. Secord (emphasis original).

nature."[78] When studying the greatest achievement in natural philosophy—Newton's laws of motion—Princeton undergraduates were led to conclude that "no other reason can be given [for these laws of motion] . . . but the will of the Deity."[79] What prevented Dod from permitting the author of *Vestiges* to apply a similar statement to the "law of development"?

The difference arose from the details of how Presbyterians viewed the constancy of God's workings in nature. In ascribing Newton's laws to divine will, Princeton's professors were strengthening a marriage bond, called "natural theology," between natural philosophy and Presbyterian theology in order to confirm the empirically discernable sovereignty of God over creation. The constancy of natural law was at the same time a foundation for science and a pious exploration of Providence: God actively sustains his world in predictable patterns, which, when discovered as "laws," noted Henry's student, would provide "proofs of the power of the Creator."[80] No laws fulfilled this duty more faithfully than Newton's three laws of motion, which operate in the same way in all places and at all times as constant means by which God preserves a universal order.

The vestigian law of development, by contrast, required progressive stages of action in nature: first causing the solar system to form, later causing life to emerge, and finally causing diverse organic species to evolve. These transformations, each departing from the common pattern of observed events, sounded miraculous to Presbyterian ears, yet the author of *Vestiges* insisted they had occurred in nature without any special interventions by God. This dynamic vision of nature's laws sounded thoroughly unscientific in the stalwart halls of Princeton, where God's predestinarian sovereignty was preached daily in chapel and each spring the junior class recorded that "Science assumes as a fundamental principle that the laws of Nature are constant. Without this assumption there

78. Caspar W. Hodge, Natural Philosophy Lecture Notes, Joseph Henry, 1847 (Lecture Notes Collection, PUL, Box 25), [3]. See also Banks, 1; T. Pickney Huger, Geology Lecture Notes, Arnold Guyot, 1859–1860 (Lecture Notes Collection, PUL, Box 17), 2; Henry E. Hale, Phyiscal Geography Lecture Notes, Arnold Guyot, 1858–1859 (Lecture Notes Collection, PUL, Box 18), 82; John Howard Wurtz [sic: Wurts], Lectures on natural history and natural philosophy, 1858 (American Philosophical Society Library, 504/W95), 8.

79. Cameron, Natural Philosophy Lecture Notes, 11.

80. Theodore Tallmadge, Natural Philosophy Lecture Notes, Joseph Henry, 1846 (Lecture Notes Collection, PUL, Box 24), 3.

would be no Science."[81] Thus, Dod spoke for the college (Henry later called him "our representative and mouthpiece"[82]) when he refuted *Vestiges* by pointing to an obvious truth about God's world: "There is no law of nature more firmly established than that like produces like."[83] Such was Princeton's empirical knowledge of God's sovereignty over nature, and *Vestiges* suffered ridicule for daring to contradict a fact so obvious, and so intimately connected to the central organizing principle of the Presbyterian faith, namely, God's sovereignty over each soul's salvation.

Student Life after Vestiges*:*
The Development Theory Becomes an Undergraduate Pastime

By drawing into question the constancy of natural law and attributing to matter the dynamism that only mind could possess, *Vestiges* had transgressed the boundaries of science as demarcated at Princeton. The book outlived its Princeton reviewer, but its life on this American campus was subjected to a role that its British author had never desired. *Vestiges* at Princeton became an example of unphilosophic speculation, an example from which half a generation of students would learn what to avoid if they desired to become men of science. The development of *Vestiges'* role at Princeton may be traced from Dod's review of 1845 to the faculty's lectures during the 1850s, but before professors brought *Vestiges* into the lecture halls, students already had taken the initiative to learn about its "law of development."

John Robert Buhler, a senior when Dod reviewed *Vestiges*, recorded his curiosity about the development hypothesis in a pencil diary entitled *My Microscope*. Buhler held Dod in high regard, remarking after class one day that his lecturing style "is so gorgeous & fluid. His language is all incarnate drapery or the vivified colours of a fine painting." However attentive in class, Buhler was the sort of student who often overslept, missing both breakfast and chapel. His diary entry for Sunday, November 9, 1845, stands out for its mention of a chapel service that, for once, he attended. As Buhler's classmate Theodore Tallmadge—

81. Banks, 3; see also Davison, 3; Tallmadge, Natural Philosophy Lecture Notes, 1.

82. Joseph Henry to [William Leslie Harris], Aug. 1846, in *Papers of Joseph Henry*, 6:490.

83. [Dod], 549.

the Methodist who by now had adjusted to Princeton—indicated in a letter to one of Dod's friends, Prof. Dod preached that morning for a memorial service as the campus community reflected upon the tragic death of the boy who had skipped class to go hunting two months earlier.[84] Buhler recalled that Dod emphasized "the Inevitability of Death—its Immediate Consequences & the Necessity of Preparing for it." This sermon so impressed Buhler—"Never heard such a sermon in my life . . . so solemnly awful"—that he desired the college to publish it. Eleven days later that sermon suddenly took a personal meaning. Buhler recorded, "It's a shame when Dod is dying," and later added, "He died in peace this aft at a quarter past 4 o'clock. What a tremendous blow to the College . . . what a sad affair."[85] Charles Hodge soon issued a tract to assure the community that the Rev. Prof. Albert Dod had, as his sermon admonished, prepared himself for a Christian death when pleurisy so quickly overtook him.[86] Meanwhile, the significance of *Vestiges* at Princeton lived on, as Buhler's diary reveals.

During the spring of 1846, Buhler continued to miss morning prayer, but found time to explore some issues raised by *Vestiges* and its critics. For example, he played a practical joke on his friends by pretending that the faculty had asked him to collect ten dollars from each student to finance a new review journal to be edited by "Mr. LORD."[87] Most likely he was referring to Eleazar Lord, a New York businessmen and former trustee of Princeton Seminary who had made an unstable reputation as a scriptural geologist and millenarian. His 1843 book, *Geological and Scriptural Cosmogony,* had claimed that the earth was not older than six millennia and that Christ would return to rule on the earth for the seventh and final millennium.[88] In a later work Lord explicitly rejected the

84. Theodore Tallmadge to the Rev. William Cox, 22 Nov. 1845 (Tallmadge Papers, PUL, Box 1, Folder 8).

85. John R. Buhler, *My Microscope* (General Manuscript Collection, PUL), 7, 9, 17, and 20 Nov. [1845].

86. Charles Hodge, *A Brief Account of the Last Hours of Albert B. Dod: Nov. 20th, 1845* (Princeton: John T. Robinson, [1845?]).

87. Buhler, *My Microscope*, 16, 18, 26, and 27 Feb. 1846.

88. Attempts to locate a copy of this book have failed. The characterization given above is based upon personal communication from Robert Whalen, who saw a copy of the book some years ago. Whalen has published a series of articles that suggest a possible context for understanding Buhler's joke: Robert Whalen, "Calvinism and Chiliasm: The Sociology of Nineteenth Century American Millenarianism," *American Presbyterians* 70, no. 3 (Fall 1992): 163–72; "Eleazar Lord and the Reformed Tradition: Christian Capitalist in the Age of Jackson," *American Presbyterians* 72, no. 4 (Winter 1994): 219–28; and, "Genesis, Geology, and Jews: The New York Millenarians of the Antebellum Era," *American Presbyterians* 73, no. 1 (Spring

natural history in *Vestiges* for inexcusably violating his principles of biblical interpretation, but this Old Schooler's young-earth geology and millenarian eschatology found little solace at Princeton, where faculty taught a vast antiquity of the earth and followed Augustine's amillenialist reading of Revelation 20.[89] Far from insisting upon a literalist Mosaic cosmology, Prof. Henry had begun his 1841 geology course by lecturing favorably upon the nebular hypothesis (a safe topic at Princeton before it became linked in *Vestiges* to a denial of God's continual and sovereign engagement with nature).[90]

In the year preceding Buhler's "fund-raiser" for Lord's scriptural geology, his classmate Theodore Tallmadge recorded from Henry's lectures that the "Bible was not instituted to teach Physical truths—these are only indirectly mentioned."[91] Similarly, Dod wrote for the *Princeton Review* that *Vestiges* erred not in contradicting the "Mosaic record," but in denying the "historical fact" that God created the first pair of humans directly. To distinguish Princetonians from scriptural geologists, Dod wrote, "it is undoubtedly true that the scriptures were not given to teach us natural philosophy."[92] Understandably, then, Buhler burst into "incontrollible [*sic*] laughter" when his classmates, falling for his joke,

1995): 9–22. As for scriptural geology, Lord's contemporary Old School Presbyterians "c[ould] not concur in the details of his logic." See Review of *The Epoch of Creation; The Scripture Doctrine Contrasted with the Geological Theory*, by Eleazer Lord, *Presbyterian Magazine* 2, no. 1 (Jan. 1852): 39–40. Nevertheless, Lord's standing as a former trustee and "an elder in the church" sufficed to prompt Charles Hodge to have *Epoch of Creation* (1851) reviewed in the *Princeton Review*. See Charles Hodge to Joseph Henry, [1851] (Papers of Joseph Henry, Smithsonian Institution Archives). On Lord's service as a seminary trustee during 1823–1826, see Joseph H. Dulles, ed., *Princeton Theological Seminary Biographical Catalogue, 1909* (Trenton, NJ: MacCrellish and Quigley Printers, 1909), 7.

89. Eleazar Lord, *The Epoch of Creation; The Scripture Doctrine Contrasted with the Geological Theory* (New York: Scribner, 1851), esp. ix–x. See also the companion volume written by his brother and business partner, David N. Lord, *Geognosy; or, The Facts and Principles of Geography Against Theories* (New York: Franklin Knight, 1855), esp. 276–77. For a discussion of these works, which were equally hostile to mainstream British and American old-earth geological theories as to *Vestiges*, see Rodney L. Stiling, "Scriptural Geology in America," chap. 7 in *Evangelicals and Science in Historical Perspective*, ed. Livinstone, Hart, and Noll, 177–92, at 181–83. On Princetonians' views of the earth's age, see the discussion above; on eschatology, see Thomas R. Markham, Notes from Lectures on Didactic Theology, by Charles Hodge, 1850–1851, vol. 3 (Charles Hodge Collection, PTS, Box 39, Folder 2); Henry Clay Cameron, Notes on Courses, v. 1: Lectures on the Apocaplyse by Dr. J. A. Alexander, 1854–1855 (Cameron Family Papers, Box 3). For a broader historical discussion of Hodge's views on the millennium, see Conkin, 221–24.

90. Ronald Numbers, *Creation by Natural Law: Laplace's Nebular Hypothesis in American Thought* (Seattle: University of Washington Press, 1977), 25–26, 28.

91. Tallmadge, n.p.

92. [Dod], 535.

"began to disclaim with eloquent indignation against such an imposition" as financing the likes of "Mr. LORD."[93]

Buhler's diary also contains frequent references to Prof. Henry's natural philosophy course, apparently ranked as the diarist's favorite. On February 17, 1846, Buhler noted, "Prof. HENRY gave us an acct of the Creation of a Louse [or spontaneously generated animacule] by Mr. CROSS[E] of Eng'd." Like Dod, Henry was skeptical of Crosse's spontaneous generation experiments that the author of *Vestiges* had presented as evidence in favor of the development hypothesis.[94] But Buhler's cousin John Trumbull Scott, a recent Princeton graduate, felt compelled by a certain "un-get-over-ability of the *Theory of Development*." The tension between these opinions left Buhler eager to learn more. With ample punctuation he gleefully wrote on Tuesday, March 3: "Had a long conversation with Prof. HENRY at the Bookstore this morning on *Physics & Metaphysics !!! !!!*" Henry recommended that he purchase a copy of German naturalist Alexander von Humboldt's *Kosmos* and share it with Scott. Humboldt's contribution to scientific literature was regarded by Old School Presbyterians as an admirable "attempt at reducing and generalizing scientific data, so as to bring the multiform and perplexing details into the simplicity and unity of nature."[95] Whereas the author of *Vestiges* "understands the mere Literature of Science," Henry told Buhler, "Baron HUMBOLDT comprehends the Science of Science!" As for "the un-get-over-ability of the *Theory of Development*," Henry refuted *Vestiges* that morning by saying, "it lacked even originality."[96]

Original or not, *Vestiges* fascinated Princeton's students. With Dod no longer alive to review the book's sequel, *Explanations*, the undergraduates assumed the duty for themselves. In 1841 they had established their own version of the *Princeton Review*, which they called the *Nassau Monthly* (named after Nassau Hall, Princeton's oldest academic building). The writing quality was impressive for boys in their teens; but, of course, these were no ordinary boys. They had passed examinations in Greek and Latin just to get admitted to

93. Buhler, *My Microscope*, 27 Feb. 1846.

94. *Papers of Joseph Henry*, 3, no. 320–22, 4, no. 167.

95. Buhler, 3 Mar. 1846; Review of *The Lives of the Brothers Humboldt, Alexander and William*, by Juliette Bauer, *Princeton Review* 25, no. 2 (Apr. 1853): 324–26, at 325.

96. Buhler, 17 Feb. and 3 Mar. 1846.

college.[97] Then, after two years of studying the classics, such as Xenophon's *Anabasis* and Virgil's *Aeneid*, in their original languages, Princeton undergraduates performed the festive charade of "Sophomore Commencement," at which they delivered mock orations for one another's entertainment. Finally, as juniors and seniors they could study natural philosophy under Prof. Henry and natural theology under President Carnaham.[98] The students learned these upper-division lessons eagerly, and learned them well. "Science," wrote one for the April 1847 number of the *Nassau Monthly,* "when she is the true and faithful interpreter of nature, cannot disagree with the right understanding and appreciation of the truths contained in the Inspired Volume."[99] Immediately the discussion focused upon creation. Then the writer drew a principled conclusion:

> While, then, the relations of all sound philosophy to revealed religion are thus most intimate, they are also as peaceful and harmonious as they are intimate. It is only when science forsakes her true vocation and becomes unmindful of the interests humanity has entrusted to her keeping that these relations are disturbed. When she becomes proud and swelled with conceit, strutting with no unsandelled [*sic*] feet on holy ground; and regardless of God's glory vainly endeavors to substitute second causes for the great first Cause—it is then that she introduces discord into the kingdom of knowledge.[100]

During the late 1840s, one could hardly read, or write, such a passage without thinking of *Vestiges,* the most ridiculed instance lately of mishapen science introducing "discord into the kingdom of knowledge" by conflating secondary causes with the First Cause. In the May issue, the students became more explicit: "The *Vestiges* has been all the rage for a time." A student review of *Explanations* filled twelve pages that month, appearing as the lead article. "The centre table was deprived of an ornament so long as this book did not grace it; and many an admiring Sophomore, or the spectacled Misses Blimbers of our boarding schools have talked of it in high terms to their friends, so that one

97. Henry and Scharff, 144.

98. Tallmadge, *Reminiscences,* 7, 9, 18; *Catalogue of the College of New Jersey, for 1840–41* through *Catalogue . . . 1856–57* (Princeton: John T. Robinson, 1840–1856); Henry and Scharff, 144.

99. "Agreement of Science and Revelation," *Nassau Monthly* 6, no. 6 (Apr. 1847): 181–85.

100. "Agreement of Science and Revelation," 182–83.

would think it was just the volume to make us wise."[101] Precisely because of its popularity, *Vestiges* had to be addressed.

Ostensibly, the student article was not a "review," but merely sought to "present a few thoughts suggested to our own minds in the perusal of the Vestiges."[102] The students' chief objection to an otherwise fascinating book was "the melancholy intelligence that *God is Law*." The writer admitted that the author of *Vestiges* "speaks reverently of the Deity," but considered this all a thin veneer over "the philosophy of the Vestiges," namely, that the laws endowed by the Deity "assumed *de facto*, the creative powers of Deity, and put him so far off that we almost forget such a Being exists."[103] How could a God so far removed convert a soul unto salvation?

The *Nassau Monthly* faulted *Vestiges* methodologically for mistaking "analogy for identity—appearance for reality" when it came to "the successive progressions in the animate creation."[104] Though nature presented abundant evidence as to the fundamental similarity linking all life forms, the students found even more compelling evidence in nature that different life forms are distinct, especially humans from animals with regard to moral and intellectual traits. The following year a student writing on the "Value of History" concluded that natural history has established the facts in opposition to the development hypothesis: "What minerals, vegetables, and irrational animals are now, with little variation, they always have been."[105]

In November 1848, three years after Prof. Dod's death, Princeton undergraduates continued to contemplate the development hypothesis. A report in the *Nassau Monthly*, gathered by "mesmerism," related the proceedings of a recent meeting of the "Mosquito Geologico-Phlebotomist Society." At that meeting, a certain "Prof. Gallynipper" presented evidence that watermelons were formed by the same nebular development that produced the earth. For example, just as the earth's core is hot, as a vestige of its origin as a hot vapor, so also the

---

101. Review of *Explanations: a Sequel to the Vestiges of the Natural History of Creation*, by the author of that work, *Nassau Monthly* 6, no. 7 (May 1847): 217–28, at 217.

102. Review of *Explanations*, 218.

103. Review of *Explanations*, 219.

104. Review of *Explanations*, 222.

105. "The Value of History," *Nassau Monthly* 8, no. 1 (Sep. 1848): 12–14, at 13.

core of a watermelon is red, indicating its previously hot state. Prof. Gallynipper extended these conclusions:

> We have thus shown briefly, how erroneous is the common sectarian doctrine as to the formation of water-melons. Geology, true, scientific geology, gives a very different account of their origin. We hope to show hereafter that the common, bigoted opinions about other so-called vegetables are equally erroneous, and that their origin is similar in its main features to that of water-melons. We believe also that the germs of all species of being, from which according to the author of the "Vestiges of Creation," all living things have been in the course of ages developed, are volcanic globules effused from the interior of the earth. ... Give us then the nebular hypothesis, to start with, and we can then go on and easily account for all the phenomena of the universe, without that unscientific sectarian idea of a creative God.[106]

This student satire resulted as much from the students' own ingenuity and curiosity about the development hypothesis as from their professors' instruction concerning the standing of *Vestiges* in relation to proper scientific methodology. Two things especially should be noted. First, Prof. Gallynipper proceeded, like the author of *Vestiges*, according to the method of analogy: watermelons and the earth are similar, therefore they must have originated by an identical process. As the next section indicates, Princeton's faculty saw to it that their students understood the difference between true analogies and misapplied analogies, using *Vestiges* to exemplify the latter. Second, Prof. Gallynipper contrasted "true, scientific geology" with "bigoted opinions" drawn from the "unscientific sectarian idea of a creative God." To be satirical, one had to say that, for at Princeton hardly anything could be more backwards. Students had learned that the Presbyterian idea of a creative, sovereign God laid a solid foundation for all the sciences, including geology, so faith had nothing to fear from any of the sciences if only these could be protected from charlatans like the author of *Vestiges*.

106. "Geological Paper," *Nassau Monthly* 8, no. 3 (Nov. 1848): 81–88, at 86.

The Curriculum:
Preparing Students to Protect Science from Vestiges

Conversing in the bookstore, Prof. Henry had told John Buhler that *Vestiges* "lacked even originality," but this was not the only weakness of *Vestiges* that Henry's authoritative opinion exposed at Princeton.[107] Just as Dod had assailed the author of *Vestiges* for ignoring that "the plant is differenced from the stone" when subsuming organic and inorganic development to a common law of nature,[108] so also Henry instructed his natural philosophy students that true science is built upon true analogies, not spurious speculations about superficial similarities—like those littering the pages of *Vestiges*. The same outline for the natural philosophy lectures that one of Henry's students recorded during the 1840s was followed in the 1850s by Henry's successor and brother-in-law, Stephen Alexander. Henry and Alexander each taught that "When one system of facts is similar to another and when we therefore infer that the Laws of the one are similar to the Laws of the other we are said to reason from Analogy."[109] A full decade after Henry had left Princeton to serve as the first directing secretary of the Smithsonian Institution in Washington, D.C., his brother-in-law continued to repeat Princetonians' most obvious example of the method of analogy improperly applied: *Vestiges of Creation*. As Alexander's student John Howard Wurts recorded in his lecture notebook for 1857:

> We must not be led away by what Prof. Henry calls "rhetorical analogy." The supposition for example, of the author of the "Vestiges of Creation," that the growth of the crystal and that of a plant were similar, is a mere rhetorical analogy. For there is in the plant a principle of life according to wh. it is formed and wh. issues its existence; this the crystal entirely wants. If you take away from the plant this principle assimilation, accretion and the whole process of growth will be immediately stopped. In the crystal, on the other hand, there is no such principle and its growth is incessant, and dependent upon no

---

107. Buhler, 3 Mar. 1846.

108. [Dod], 550.

109. From "Heads of Lectures on Natural Philosophy Delivered by Prof. Joseph Henry LL.D.," in an unidentified student's Natural Philosophy Lecture Notes, Prof. Elias Loomis, 1848–1849 (Lecture Notes Collection, PUL, Box 30).

internal organism. The resemblance then is only fancied, that is rhetorical.[110]

Wurts's classmate George Gray also recorded Alexander's repetition of Henry's warning about the "mere <u>Rhetorical Analogy</u>" that the author of *Vestiges* had attempted to pass off as legitimate natural philosophy. Regarding the vital principle that distinguishes the organic from the inorganic, Gray wrote, "When an animal is struck dead by electricity, the arrangement of the parts is just as nice as before, but the difference is that it is dead. There is a principle of life in the vegetable, and in the animal, in addition to the principle of <u>growth</u> in the Crystal."[111]

Vestiges at Princeton was no passing fad. The following year, Alexander was still mentioning *Vestiges* in his third introductory lecture to the natural philosophy course required of all juniors at Princeton. George Ketcham, class of 1859, wrote "Vestiges of Creation" in the margin of his notebook beside a paragraph that read:

> If facts are similar in given instances, we are led to believe that the laws are similar. In this way we reason from analogy, a process based upon the simplicity of the laws of Nature. "But," says Prof. Henry, "we must be careful not to be misled by a mere Rhetorical analogy.["] The author of "Vestiges of Creation" traces an analogy between the cells in crystallization and the cells in vegetables, and therefore concludes that the same principles hold in each case; whereas the one has life, and the other has not.[112]

Students learned about "rhetorical analogies" in other classes, too. Prof. Matthew Boyd Hope taught in his rhetoric course for juniors that a "rhetorical analogy" is an abuse of figurative language. "It gives rise to sophistry . . . introducing what is untrue as the basis of an allegory." Sir Isaac Newton had employed a "logical analogy" and tested it by experiment when identifying the fundamental similarity—the law of gravitation—that links an apple's fall and the moon's orbit. "Rhetorical analogy," emphasized Hope, "is not adequate for force

---

110. Wurts, 18–19.

111. George Gray, Natural Philosophy Lecture Notes, Stephen Alexander, 1857–1858 (Lecture Notes Collection, PUL, Box 2), [19].

112. George W. Ketcham, Natural Philosophy Lecture Notes, Stephen Alexander, 1858–59 (Lecture Notes Collection, PUL, Box 2) , 16.

& argument, but beauty."[113] Yet *Vestiges* did not even have beauty in the eyes of Princeton beholders. As George Ketcham learned in Prof. Lyman Atwater's intellectual philosophy course, human and animal intelligence differ by discrete, not continuous, grades. "This distinction completely annihilates the 'Development Theory' contained in a book called the 'Vestiges of Creation.'" Methodologically, *Vestiges* had confused continuous for discontinuous gradation and thus misapplied the method of analogy.[114]

Albert Dod would have been pleased had he lived to see Ketcham write up a clean copy of his natural philosophy lecture notes in preparation for his June 1859 graduation.[115] Ketcham, trained in the method of analogy by Hope, Atwater, and Alexander, dutifully recorded Alexander's distinction between a hypothesis and a theory: "An hypothesis verified rises to the dignity of a theory –: therefore a veritable theory is a perfect hypothesis."[116] The development hypothesis of *Vestiges of Creation*, though it had been known to Princetonians for nearly fifteen years, remained unverified—a mere hypothesis, and merely a "rhetorical" one at that.

To those observing from a theological distance, Princetonians' own analogy between the development hypothesis and atheism may appear to be just as "rhetorical" as the analogies in *Vestiges* that linked humans and animals. But Princeton theologians were neither sloppy nor insincere in labeling *Vestiges* an atheistic work. For them, anything that opposed the system of Westminster Calvinism, as defended by the Old School Presbyterian Church, ultimately would lead people away from the faith and thus, out of love for humanity, it should be

113. David Ebenezer Smith, Notes on Rhetoric, by Matthew B. Hope, 1847 (Lecture Notes Collection, PUL, Box 26), 38. For a discussion of the method of analogy in antebellum American science, see George H. Daniels, *American Science in the Age of Jackson* (New York: Columbia University Press, 1968), chap. 8. Daniels suggests that analogies served as a methodological compromise between raw inductivism and hypothesis framing, the former seeming inadequate if one wanted not only to describe but also to explain nature, and the latter seeming too speculative if one wanted to remain confident about one's account of nature.

114. George W. Ketcham, Intellectual Philosophy Lecture Notes, Lyman Atwater, 1858–1859 (Lecture Notes Collection, PUL, Box 3), 17.

115. During the spring term of their senior year, Princeton students made a presentation copy of their lecture notes from the preceding spring and fall terms, submitting these for correction by the professor, as evidenced by interlineations by a second hand. See, for example, George W. Ketcham, Natural Philosophy Lecture Notes, Stephen Alexander, 1858–59 (Lecture Notes Collection, PUL, Box 2) .

116. George W. Ketcham, Natural Philosophy Lecture Notes, 16.

exposed for its atheistic tendencies. The Rev. Prof. Dod quite expectedly argued in November 1845 that *Vestiges* presented an atheistic system by drawing an analogy between the kind of "deity" that the author of *Vestiges* claimed to revere and the godless, pagan theories of Greek atomists.[117]

Vestiges had attributed to natural law certain transformations that broke the empirically discerned pattern of events in nature, yet did not consider these transformations to be miraculous acts of God. It was daring enough that *Vestiges* did not allow for God to have agency in ordinary natural events, but to exclude God also from never-been-observed evolutionary transformations that sounded (if they were to be believed at all) like miracles, this was atheism even bolder still. A writer for the *Princeton Review* confronted a similar proposition in 1846 when addressing a theory that the resurrection of the dead can be explained by a natural law. "This is the same absurd and atheistic principle of development," concluded the reviewer, "which has recently been pushed out to such a ridiculous extent in the work entitled 'Vestiges of Creation.'"[118]

The threat of atheism was not abstract, but existential. Though Princetonians may have often seemed heady, truth at Princeton was not only propositional but also heartfelt.[119] The *Princeton Review*, written primarily by and for ministers, dealt with intellectual debates precisely because of the pastoral relevance of contemporary thought. Atheism could at times be felt on a very personal level. One morning in January 1845, the Rev. Ashbel Green, a former president of Princeton College and a long-standing seminary trustee, who frequently met with students to nurture them in their commitment to the Westminster Confession,[120] had himself felt besieged by "infidelity and atheism" while trying to say his morning prayers. Not until the afternoon was he able to repent of his spiritual doubts and receive comfort through prayer.[121]

About a year later, Archibald Alexander drafted some notes for his seminary lecture concerning "Creation." He scribbled a reference to Dod's review and two others—one being David Brewster's *North British Review* article that

---

117. [Dod], 530–31.

118. Review of *The Soul*, by George Bush, *Princeton Review* 18, no. 2 (Apr. 1846): 219–60, at 259.

119. Noll, "Introduction," 33–34.

120. Calhoun, 1:185–86.

121. Ashbel Green, Diary (transcript), 27 Jan. 1845 (Ashbel Green Collection, PUL, Box 16).

was reprinted as an appendix to many American editions of *Vestiges*, and the other being Cambridge geology lecturer Adam Sedgwick's article in the *Edinburgh Review*, which had prompted Robert Chambers to issue *Explanations* as an anonymous sequel volume to his *Vestiges*. All three reviews claimed that *Vestiges* was essentially materialist or atheist in its tendency, an accusation to which Chambers objected vehemently in *Explanations*.[122] Bolstering the charge of atheism before his theological students, Alexander decried "the folly of learned atheism" as exemplified by Lamarck, who—according to Alexander—had claimed that "the stimulus of desire" could cause nonliving matter to become living until "in the lapse of ages a monad becomes a man!!!" Alexander admitted that Lamarck "sometimes speak[s] of God. But His God is nature . . . [and] can possess no efficacy." He continued:

> The same theory is developed in a late ingenious & popular work intitlted [sic] 'Vestiges of Creation.' The anonymous author professes to be a theist, but his principles lead to atheism. He renders a designing cause unnecessary by the hypothesis that organized animal bodies may be found [sic: formed?] by the laws of nature, as the particles of matter arrange themselves in singular uniform parts, so by certain laws of chemical affinity animal bodies are formed from wh. organization life proceeds.[123]

The good news for Princetonians was that atheism, even when subtly packaged as theism, had a cure: natural theology. At Princeton Theological Seminary, the faculty trained the church's future ministers how to deploy natural theology when preaching against *Vestiges*. Prof. Joseph Addison Alexander, a son of Archibald Alexander who beginning in the 1830s taught ancient languages at both Princeton College and Princeton Theological Seminary, discussed *Vestiges*

122. [Adam Sedgwick], Review of *Vestiges of the Natural History of Creation*, *Edinburgh Review* 82 (July 1845): 1–85, at 3, 12, 60, 62–64; facsimile rpt. in *"Vestiges" and the Debate before Darwin*, ed. John M. Lynch, 7 vols. (Bristol: Thoemmes, 2000), vol. 1 (original pagination retained); and, [David Brewster], Review of *Vestiges of the Natural History of Creation*, *North British Review* 3 (Aug. 1845): 470–515, at 503, 506; facsimile rpt. in *"Vestiges and the Debate before Darwin*, ed. Lynch, vol. 1 (original pagination retained). Brewster's review was appended to at least four American printings of *Vestiges of the Natural History of Creation* (New York: Wiley and Putnam, 1845, 1847, 1848; New York: Coyler, 1848) and also was reprinted in *Littel's Living Age* 6, no. 71 (20 Sep. 1845): 564–82. For Robert Chambers's anonymous rebuttal to Sedgwick's and Brewster's claims of an irreligious tendency in *Vestiges*, see [Chambers], *Explanations*, esp. 3, 23, 123–25, 169–88.

123. Archibald Alexander, Creation, ca. 1845 (Archibald Alexander Collection, PTS, Box 9, Folder 34).

in his homiletics, or "Sacred Rhetoric," course at the seminary during the spring of 1851. The first several lectures in that course dealt with argumentation, gesturing, and other practical skills. By Lecture 25, the professor began to suggest how natural theology may be applied in the pulpit. Henry Clay Cameron's notebook for that lecture contains a reference to the Bridgewater Treatises, a series of British works written about two decades earlier by learned men who adduced evidences of God's wisdom, power, and goodness from recent researches in astronomy, geology, physiology, and other sciences.[124] In Lecture 26, Prof. J. A. Alexander sketched out some arguments by which God's eternality, independence, and immutability could be demonstrated "from Reason," even apart from Scripture. Cameron, who as a student at the college had learned from Joseph Henry that the laws of nature all point back to God, their Author, was now learning at the seminary that the natural sciences could be applied in the pulpit in order to convince those who would not listen to Scripture. The "Vestgs of Crtn," Cameron abbreviated in his notebook, had theorized the "eq[ui]v[o]c[a]l g[e]n[e]r[tio]n of pl[a]nts[,] B[ut this had] n[e]v[er been] d[i]sc[o]v[e]r[e]d"; all observations indicated that plants came "only fr[o]m s[ee]d." Whereas empirical evidence favoring the development hypothesis was lacking, Cameron learned to recognize abundant evidence that nature had been designed: food was adapted to the appetite, human organs were adapted to their uses, and so on. Design requires a mind, and mind is independent of matter, therefore a Creator-God must exist who is wise, benevolent, and independent.[125]

This argument was generally in keeping with the commonplace Anglo-American natural theology of the early Victorian era, but at Princeton the conclusion that God was "independent" received greater emphasis than God's other naturally revealed attributes, such as wisdom, power, and benevolence, because God's independence, or sovereignty, was central to the Presbyterian faith. "The Scriptures plainly teach," said Archibald Alexander Hodge in his 1860 course on Didactic Theology, "that God is a being self existing, and independent, and perfectly sovereign." This doctrine requires, Hodge explained, that God's act

124. For a recent revisionist treatment of these, see Jonathan Topham, "The Bridgewater Treatises and British Natural Theology in the 1830s," Ph.D. diss., Lancaster University, 1993. Departing from consensus histories, Topham argues that different segments of the British reading public understood the Bridgewater Treatises in unique ways.

125. Cameron, Natural Philosophy Lecture Notes; Cameron, Notes on Courses, v. 7: Lectures on Sacred Rhetoric, by Dr. James [sic: Joseph] A. Alexander, 1850–1851 (Cameron Family Papers, PUL, Box 4).

of creation be *ex nihilo*, or "from nothing," for otherwise He would have been limited by "the necessary existence of matter." For Princeton's theological students, this understanding of creation resonated with the doctrine of election, according to which God's "effectual call" unto salvation accomplishes its end *ex nihilo*—that is, "not from any thing at all foreseen in man, who is altogether passive therein," but "of God's free and special grace alone."[126] In their theology courses, seminarians learned that all other Christian denominations erred precisely on that point, for they each ascribed to individuals some role, however small, in a salvation that Presbyterian orthodoxy insisted was accomplished by God alone.[127]

The sciences, as proclaimed from Presbyterian pulpits, were to establish God's sovereignty over nature, which then could serve as a foundation for proclaiming God's sovereignty over the eternal fate of each person's soul. Among the most frequently preached texts, "The Lord reigneth" (Psalm 97:1) provided Princeton's ministers an opportunity to proclaim God's sovereignty in the parallel realms of creation and redemption.[128] Sermons on other texts also emphasized these themes, providing occasion for *Vestiges* to be implicitly, even if not always explicitly, refuted. As John T. Duffield preached at the college chapel in 1862, "our conception of the Gospel plan of salvation—and indeed of the whole system of truth which God has revealed in his word, will be determined, in a greater or

126. Westminster Confession, chap. X, art. II.

127. Armstrong, Notes from Lectures on Didactic Theology. The distinguishing mark of Presbyterian orthodoxy was razor sharp, identifying not only gross deviations from the doctrine of election, such as Pelagianism, but also more subtle distinctions. Hodge noted, in the lecture just cited, that the Lutherans' *Formula of Concord* (1577) quite emphatically insisted that God alone converts the sinner to faith, but warned that this Calvinist-sounding doctrine would not hold among Lutherans, since they rather asymmetrically taught that unbelievers actively reject God. As Hodge here described the *Formula*, "The attempt was to refer the salvation of man entirely to grace but to get rid of the consequence that the sovereignty of God must in that case be concerned in the perdition of the ungodly." The only consistent and orthodox view, in Hodge's mind, was that God was equally sovereign, and man equally passive, in the case of believers and unbelievers (Calvinist double predestination, as compared to the Lutheran *Formula*'s single predestination). See also A. A. Hodge, A Comparison of the Main Distinguishing Positions of the Three Rival Systems of Pelagianism, Semipelagianism, & Augustinianism, ca. 1862 (Archibald Alexander Hodge Collection, PTS, Box 3, Folder 4).

128. This text recurs numerous times in the sermons by Princeton-affiliated ministers during the 1830s through 1850s. See, for example, Cooley, Sermon on Psalm 97:1–2, 26 Nov. 1846 (Eli Field Cooley Papers, PUL, Box 3, Folder 2) and Sermon on Psalm 97:1, 14 July 1850 (Box 5, Folder 5); John Maclean, Jr., Sermon at the College on Ps. 97:1, 12 May 1861 (Office of the President Records, PUL, Box 33, Folder 11). For favorable commentary on an earlier sermon by Maclean on the same text, see Henry Clay Cameron, Diary, 1 Jan.–31 Dec. 1859 (Cameron Family Papers, Princeton University Libraries, Box 8), 14 Aug. 1859.

less [*sic*] degree, by our opinion[,] belief[,] views, in regard to the doctrine of election." The Rev. Duffield then quoted from a section of the Westminster Confession that presents the Presbyterian doctrine known as "God's Eternal Decree." That doctrine holds that God has foreordained both the course of events in nature as well as the eternal destiny of every person, some of whom he elects to salvation and the remainder of whom he predestines to damnation. "The basis of the whole doctrine of election," explained Duffield, is "that God dispenses His favors to men according to His sovereign good pleasure" and "withholds from others," whom he predestines to damnation, according to the same "absolute sovereignty of God."[129]

From Princeton to the American Association for the Advancement of Science

The Princeton commitment to God's sovereignty can be detected also among the writings of a notable layman who served on the seminary's board of trustees and carried Princeton's vision beyond both Princeton and Presbyterianism. As indicated above, trustees affirmed by oath that they would preserve the theology of the Westminster Confession. Joseph Henry, who had taught natural sciences at the Princeton College prior to becoming the first directing secretary of the Smithsonian Institution in 1846, served as a Princeton Theological Seminary trustee from 1844 to 1851. He also served as president of the recently formed American Association for the Advancement of Science (AAAS) during 1849–1850. When acting in this capacity, Henry did not leave his Princeton Presbyterianism behind, though this fact has been obscured by a tendency among historians to overemphasize the role of the AAAS in separating science from religion.

When writing about the formative years of the AAAS, historian Sally Gregory Kohlstedt concluded that men of science had only peripheral interest in religion. She suggested that although many of the founding members of the AAAS held strong personal religious convictions, their professional interests seldom related to religious issues beyond the social necessity of assuring theologians and

---

129. Sermon on Ephesians 1:3–6, College Chapel 1862 (Duffield Family Papers, PUL, Box 4, Folder 3), with explicit reference to the Westminster Confession of Faith, chap. III, arts. III and V.

laypersons that the advancement of science would not lead to a decline in religion. When arguing for these conclusions, Kohlstedt admitted that speakers at AAAS meetings occasionally referred to the design in nature that bore testimony to the Designer, but she emphasized that the AAAS would not permit such religious expressions to encroach upon the formulation of scientific methodology. Regarding Henry's leadership as exercised through the customary presidential address in 1850, Kohlstedt reported simply, "Henry spoke on the need to defend science's public reputation by protecting it from charlatans and illustrating its moral functions."[130]

Henry's 1850 presidential address in fact had a much stronger religious element, and it likely stemmed from his ties to Princeton Presbyterianism. Toward the end of his lecture, Henry alluded to some accusations that the AAAS promoted atheism since it members "referr[ed] all the phenomena of nature to the operation of physical laws."[131] Henry's own brethren at Princeton had made a similar accusation against the author of *Vestiges*. As *Vestiges* continued to be reprinted and discussed in America in 1850, it was important for members of the AAAS to ensure that they could have both their laws of nature and their theistic respectability, too. Otherwise, the newly formed AAAS would be castigated along with the author of *Vestiges*.

Henry himself had reached a solution to this dilemma in 1846, during his final semester as a full-time Princeton professor. His student Henry Clay Cameron recalls that day in class vividly:

> He [Prof. Henry] was walking to and fro [*sic*], and had just dictated: "We all explain a fact when we refer it to *law*;" and then it occurred to him to express the corresponding idea in a similar form: "We explain a *law* when we refer it to the *will* of God." He stopped, and exclaiming, "Yes! that is it!" he repeated the expression.[132]

130. Sally Gregory Kohlstedt, *The Formation of the American Scientific Community: The American Association for the Advancement of Science 1848–60* (Urbana: University of Illinois Press, 1976), 114 (reassuring theologians; prohibiting methodological encroachments), 74–75 (occasional doxologies), 109–10 (quotation).

131. *Papers of Joseph Henry*, 8:99.

132. Henry C. Cameron, *Reminiscences of Joseph Henry* (1878), 8. A copy is available in Faculty Files, Joseph Henry, folder 2, PUL.

As noted earlier, Henry and his successors at Princeton expanded upon this idea by teaching that a "law of nature" means simply "the mode in which divine wisdom operates in producing the changes of nature."[133] Similarly, when speaking before the AAAS, Henry asserted, "The essential characteristic of matter is inertness, or inability to change its state without extraneous force, and the proper higher meaning of the term <u>law of nature is our conception of the mode in which divine wisdom invariably operates in producing the phenomena of nature</u>."[134] Henry emphasized the qualifier "invariably" by stating that the "laws of change" in nature are themselves "as immutable as the purposes of <u>Him</u> who knows no change."[135]

The inertness of matter and the constancy of natural law—two scientific obstacles that Princetonians had placed in the path of the author of *Vestiges*—had a solid Presbyterian foundation, even if Henry hoped that non-Presbyterian men of science also would employ these standards to protect natural science from atheism. As discussed earlier, the inertness of matter resonated with the passivity of man under God's sovereign predestination. The idea that nature's laws were immutable could be found in the Westminster Confession's statement that "God from all eternity did by the most wise and holy counsel of his own will, freely and unchangeably ordain whatsoever comes to pass."[136] A member of the New York Avenue Presbyterian Church in Washington, D.C., and a trustee of Princeton Theological Seminary,[137] Henry originally had intended to refer to this, but in an apparent effort to make his Presbyterian science more palatable to an interfaith scientific gathering, he struck a direct reference to "the Westminster confession of faith" from his draft manuscript. He chose simply to summarize the doctrine for his AAAS audience without any sectarian attribution: "God hath foreordained everything which cometh to pass."[138]

133. Caspar W. Hodge, Natural Philosophy Lecture Notes, [3]. See also Banks, 1; Huger, 2; Hale, 82; Wurtz, 8.

134. *Papers of Joseph Henry*, 8:99.

135. *Papers of Joseph Henry*, 8:99.

136. Westminster Confession, chap. III, art. I.

137. Samuel S. Mitchell, *Joseph Henry. In Memoriam. Funeral Address, by His Pastor, Rev. Samuel S. Mitchell, D.D., May 16, 1878* (Washington, DC: Thomas McGill and Co., 1878), dedication page.

138. *Papers of Joseph Henry*, 8:99.

The Limits of God's Scientific Sovereignty at Princeton

Thus far, this chapter has emphasized the pervasive concord among faculty, students, and ministers associated with both Princeton College and Princeton Theological Seminary as to the central organizing principle of their Old School systematic theology, namely, God's sovereignty. The archival evidence firmly demonstrates that God's sovereignty, more than anything else, shaped Princetonians' reactions to *Vestiges*. In order to understand more fully how that theology was situated socially at Princeton, and, in turn, how it shaped reactions to *Vestiges*, it will help to consider briefly some exceptions to the consensus perspective documented above.

One would be mistaken to conclude that all who passed through Princeton College and Princeton Theological Seminary held to a completely uniform acceptance of the Old School doctrine of God's sovereignty and its implications for synthesizing the natural with the theological sciences. In some cases, seminary students' interest in natural science carried them away from their commitments to Westminster theology. The Rev. Peter Lesley, who studied at Princeton Theological Seminary during the early 1840s, became so interested in geology that he resigned from the ministry in 1848 to pursue a career as a state geological surveyor. His geological work, in turn, brought him into close contact with "unorthodox"—specifically, Unitarian and Transcendentalist—scholars in the Boston area. Lesley married a woman who was troubled by the "grim spectre of Calvinism,"[139] and the two them ultimately became Unitarians.[140] It seems neither of them could cope with the predestinarian sovereignty of God, as presented in this poem by Princeton Seminary's Joseph Addison Alexander, which Peter Lesley copied into his private notebook:

There is a time we know not when
A point we know not where,
That marks the destiny of men
To glory or despair . . .[141]

139. [Lydia] Maria [Child] to Susan Lyman [Lesley], 4 July 1847 (J. Peter Lesley Papers, American Philosophical Society, B/L56).

140. Martha B. Kendall, "Lesley, J. Peter," *Dictionary of Scientific Biography,* ed. Charles Coulston Gillispie, vol. 8 (New York: Charles Scribner's Sons, 1973), 260–61.

141. "Reprobation," by Rev. J. Ad. Alexander, Princeton, n.d. (J. Peter Lesley Papers, American Philosophical Society, HS Film 12, reel 11).

Nevertheless, the exceptionality of Lesley's trajectory in faith and scholarship reinforces, rather than counteracts, the thesis that Princetonians' ultimate concern was

God's sovereignty. Lesley did, after all, leave both Princeton and Presbyterianism. Similarly, when Yale alumnus Elias Loomis, who desired that natural sciences be made independent of theology, replaced Henry on Princeton's faculty in 1848, he soon found that he did not fit the social *milieu* in which the natural sciences were to serve Presbyterian theology, so he resigned after one year.[142]

The Rev. Theodore Ledyard Cuyler, a graduate of both Princeton College (1841) and Princeton Theological Seminary (1846), also departed from the Princeton partnership of natural science and theology, though he headed in a different direction than Lesley and Loomis. He served for over fifty years as a faithful Presbyterian minister. Unlike his colleagues, however, he chose never to preach from science in defense of Scripture. "We are no more called upon to defend the Bible than we are to defend the law of gravitation," he observed in his memoir of 1902. "The vast proportion of volumes of 'Apologetics' are a waste of ink and paper." Still, this exception, like the others, was precisely that, exceptional. As Cuyler himself realized, "I have not followed the practice of many of my brethren."[143] Those ministerial brethren seemed never to question Archibald Alexander's advice in the seminary's Didactic Theology course: "theologians should study geology."[144]

Princeton's natural scientists, for their own part, generally kept in step with a particularly Presbyterian understanding of theology, as Henry's veiled reference to the Westminster Confession before the AAAS indicates. One possible exception to this was Arnold Henri Guyot. When the 1848 revolution disrupted campus life at the Neuchâtel Academy in Switzerland, Guyot came to America under the auspices of his fellow countryman Louis Agassiz, who was then working at Harvard. Guyot and Agassiz shared a common devotion to developing an idealist understanding of nature. Agassiz's idealism, which will be discussed

142. Roger S. Kohn, "Elias Loomis (1811–1889): Pioneer American Scientist," (senior essay, Dept. of History, Yale University, 1979), 50. A copy is available in Papers on Yale Collections, Manuscripts and Archives, Yale University Library, Box 1, Folder 17.

143. Cuyler, 69–70.

144. Brown, Lecture Notes on Didactic Theology.

further in the next chapter, regarded all natural phenomena as expressions of God's thoughts, impressed onto nature in acts of creation. In *The Earth and Man* (1849), Guyot treated the earth as an "organism" that developed like an embryo, gradually forming "organs," such as mountains and oceans. Just as Agassiz's program of comparative anatomy identified the divine plan unifying diverse living beings, so also Guyot's "comparative physical geography" detected God's stamp upon the earth's surface.[145] Consistent with this developmental history of the globe, Guyot also was among the leading proponents of the nebular hypothesis in America.[146] At first glance, Guyot would appear to be anything but a Princeton Presbyterian as the "ideal type" of that individual has been presented in this chapter.

But Guyot quickly became a Princetonian. True, his developmental natural history in some ways tended toward the theories presented in *Vestiges*, but his idealism kept him from accepting species transmutation, and his affirmation of God's active involvement in nature enabled him to enter the fortress of God's sovereignty at Princeton with a grand welcome.[147] Guyot was a committed Calvinist, in fact, who actively participated in the Presbyterian church.[148] Princeton botanist John Torrey had regarded Guyot as "a truly pious man" from the moment he meet him.[149] After establishing a reputation as a public lecturer on geology and geography, Guyot joined the faculty at Princeton College in the mid 1850s, and also served occasionally as a lecturer at Princeton Theological Seminary.[150]

Princeton seminarian Henry Clay Cameron found Guyot's idealist philosophy of nature—for example, his view that "the trinities in nature [are] the

145. Arnold Guyot, *The Earth and Man: Lectures on Comparative Physical Geography, in Its Relation to the History of Mankind*, trans. from the French by C. C. Felton (1849; rpt., Boston: Gould and Lincoln, 1858). I thank Philip Wilson for lending me his personal copy of Guyot's *Earth and Man* and for discussing the intricacies of Guyot's philosophy of nature with me on several occasions. On the European influences on Guyot's thinking, and his characteristic pedagogy, see Philip K. Wilson, "Arnold Guyot (1807–1884) and the Pestalozzian Approach to Geology Education," *Eclogae Geologicae Helvitiae* 92 (1999): 321–25.

146. Numbers, *Creation by Natural Law*, 120.

147. Gundlach, 24–28.

148. Gundlach, 21n31; Numbers, *Creationists*, 9.

149. John Torrey to Asa Gray, 5 Oct. 1848 (Torrey Papers, New York Botanical Garden Library).

150. It is unclear whether Guyot joined the faculty in 1854 or 1855. For the conflicting evidence, see Faculty Files (Princeton University Libraries, Arnold Guyot Box, Folder 2).

reflection of God himself"—to be "new, sublime, & true."[151] For several years, Guyot engaged in an attempt to reconcile his old-earth geology with the Mosaic creation account.[152] He assimilated the nebular hypothesis into the six "days" of creation mentioned in Genesis 1 by regarding those as long, developmental eras. Guyot's version of an otherwise commonplace day-age reconciliation treated the six biblical periods as emblematic of the ideal phases of God's creative work, giving special emphasis to three distinct creations: matter, life, and humanity. This approach avoided the problem of discrepancies between the Bible and established views in natural science as to the sequence of creation events. Moreover, Guyot had forged a synthetic natural-theological history that sharply distinguished between God and nature, between organisms and matter, and between humans and the rest of creation.[153] Hodge, not surprisingly, recommended that his seminary students grant favorable consideration to "the theory of the Eminent Christian Philosopher Prof. Guyot, of Princeton College."[154] Guyot's science could be read as a confirmation of God's sovereignty and, therefore, could receive a Presbyterian stamp of approval.[155]

Guyot also could communicate clearly beyond the Presbyterian world. Whereas Hodge's place at the center of Princeton theology distanced him from outsiders who did not share his Old School Presbyterian faith, Guyot managed to develop close ties with Hodge while also participating in a scholarly network with non-Presbyterians. His frequent correspondence with James Dana at Yale and Louis Agassiz at Harvard generally concerned questions about how species may be understood in relation to divinely ordained ideal types.[156] The typology Guyot had in mind unified nature into a common plan, while also maintaining firm

151. Cameron, Diary, 20 Mar. 1859.

152. An elaborate chart, produced after long contemplation of geology in relation to Genesis, may be found in Arnold Guyot to Horace James, 9 Dec. 1854 (Arnold Guyot Papers, American Philosophical Society). See also Arnold Guyot to James Dana, 6 Dec. 1856 (Dana Family Papers).

153. Gundlach, 32–36.

154. Henry A. Harlow, Notes from Lectures on Didactic Theology, by Charles Hodge, Vol. 1, 1856–1857 (Hodge Collection, Box 41, Folder 1).

155. Gundlach, 27.

156. See, for example, Arnold Guyot to James D. Dana, 28 May 1858, 26 Dec. 1856, 17 Mar. 1858, 16 Apr. 1858, 28 May 1858 (Dana Family Papers, Yale University Archives, HM 160).

distinctions where the author of *Vestiges* had suggested smooth, evolutionary transitions:

> Man is only that which belongs to him as a part of Nature. But this man is not the whole man, the true man. This last is spiritual life incarnate in the body of nature; the living representative of the two lives combined in one living unit, the tie between Earth & heaven; the mediator between spirit & matter. Man in this his true nature is a new kingdom of life, contrasted with & superior to both the Vegetable and the Animal Kingdom[s].[157]

Thus, even if Guyot in some ways spoke a new language of idealism that was foreign to the Princeton tongue, he nonetheless voiced a message Princetonians could appreciate: a sovereign God had created humans specially, and *Vestiges* had no standing in science.

Finally, some clarification is in order concerning the widespread Princeton practice, documented above, of de-emphasizing secondary causes in nature in order to promote the sovereignty of God, who alone is the primary Cause of all. This posture, it has been argued, was Princetonians' core strategy against *Vestiges*, a work that they understood to ascribe too much independence and efficacy to secondary causes, that is, to natural laws. Old School Presbyterian theology, however, in fact had an important place for secondary causes in nature. The Westminster Confession declared, in relation to human free will, that God's sovereignty should not be construed in such a way that "the liberty or contingency of second causes [is] taken away." This affirmation of human freedom was qualified, however, with the statement that God predestines individuals to heaven or else to hell "without any foresight of [their] faith or good works . . . or any other thing in the creature[s], as conditions, or causes moving him [i.e., God] thereunto."[158] Taken together, these statements stood on a middle ground between fatalism and Arminianism; they spoke harmoniously of a single truth to Presbyterian ears, though as chapter 3 will demonstrate, Harvard Unitarians found the combination to be absurd.

For the present, it suffices to note that Presbyterians' particular *via media* concerning secondary causes in human nature was applied also to nonhuman

---

157. Arnold Guyot to James D. Dana, 18 May 1858 (Dana Family Papers).

158. Westminster Confession of Faith, chap. 3, arts. 1, 5.

nature: "Although in relation to the foreknowledge and decree of God, the first cause, all things come to pass immutably and infallibly, yet, by the same providence, he ordereth them to fall out according to the nature of second causes, either necessarily, freely, or contingently."[159] The Westminster Confession's seemingly paradoxical articulation is perhaps best understood not as an essentialist solution to an age-old philosophical dilemma, but rather as a strategy for protecting God's sovereignty against a variety of foes. When, for example, Charles Hodge expounded upon Westminster theology to attack deism, he emphasized God's direct action and downplayed secondary causes; but, when he attacked pantheism, his emphasis reversed, affirming secondary causes in order to distinguish nature from God.[160] Since Princetonians read *Vestiges* as a work of deism tending toward atheism, they understandably de-emphasized secondary causes in order to assert the actions of their sovereign God. It was, therefore, not their commitment to a particular understanding of secondary causes that drove their reactions to *Vestiges*, but rather it was their commitment to God's sovereignty that shaped both their articulations concerning secondary causes and their responses to *Vestiges*. God's sovereignty was of ultimate concern.

Concluding Interpretations: Understanding Vestiges at Princeton

During the decade and a half following the first American printing of *Vestiges* in 1845, faculty and students at Princeton College and Princeton Theological Seminary amassed numerous arguments against the development theory. In many respects, their criticisms of the development hypothesis echoed what other Americans were saying beyond Princeton. The author's analogies were sloppy, solid empirical evidence was lacking, reputable men of science did not support the book's conclusions, and the system as whole tended toward atheism. One did not have to be a Princeton Presbyterian to reach these conclusions, as the sundry examples in chapter 1 testified.

The present chapter, however, has sought to understand the particular ways in which a specific version of Protestant theology functioned to regulate a

159. Westminster Confession of Faith, chap. 5, art. 2.

160. Charles Hodge, *Systematic Theology*, 3 vols. (1871–1872; facsimile rpt., Peabody, MA: Hendrickson, 2001), 1:576, 580–636. At p. 580, Hodge admits that questions such as "What is the relation between his agency and the efficiency of second causes?" are "questions which never can be solved."

local academic community's evaluations of *Vestiges*. The goal has not been to claim that only Princeton Presbyterians would regard *Vestiges* as a pretentious, infidel speculation masquerading under the banner of reputable theistic science. Rather, the goal has been to identify with some precision the single-most nagging problem that Princeton Presbyterians had with *Vestiges*. God's sovereignty held central importance in the Old School Presbyterian faith in a way that it did not do so for other denominations. That latter point will become more evident in chapters 3 and 4, which analyze Harvard Unitarians and Yale Congregationalists, two very different theological communities with two very different social situations when compared to one another and to Princeton Presbyterians.

By underscoring the significance of Old School Presbyterian divine sovereignty in shaping Princeton's reception of *Vestiges* during the 1840s and 1850s, this chapter clarifies previous historical interpretations of science and religion at Princeton. Whereas other scholars have identified a broadly Protestant providentialism that objected to the "cold and fatherless universe" of *Vestiges*,[161] the archival evidence discussed in this chapter, much of which has hitherto not been cited in historical studies, suggests that a more specific kind of providentialism was taught at Princeton—God's active and universal sovereignty—and that this doctrine was defended because of its direct bearing upon Old School Presbyterians' understanding of salvation (double predestination). Thus, this chapter offers both a greater specificity in identifying the kind of providentialism operative at Princeton as well as a higher level explanation that suggests why Princetonians regarded their providentialism as non-negotiable.

God's active and universal sovereignty also provides a deeper explanation than an appeal to the design argument, as to why Princetonians objected to *Vestiges*. Although Herbert Hovenkamp has emphasized that Princeton's Albert Dod deployed the design argument against *Vestiges*,[162] he and other historians have not taken seriously the fact that *Vestiges* itself presented readers with an explicitly teleological system for which God's wisdom was given full credit, following the design argument tradition of the Bridgewater Treatises. *Vestiges* also explicitly affirmed the oft-repeated argument of William Paley that apparent design in nature is evidence of a Deity who so designed it:

161. Gundlach, 7; see also McElligott, chap. 1.

162. Hovenkamp, 198–200.

It has been one of the most agreeable tasks of modern science to trace the wonderfully exact adaptations of the organization of animals to the physical circumstances amidst which they are destined to live. From the mandibles of insects to the hand of man, all is seen to be in the most harmonious relation to the things of the outward world, thus clearly proving that *design* presided in the creation of the whole—design again implying a designer, another word for a CREATOR.

It would be tiresome to present in this place even a selection of the proofs which have been adduced on this point. The Natural Theology of Paley, and the Bridgewater Treatises, place the subject in so clear a light, that the general postulate may be taken for granted.[163]

Dod's objection was not that *Vestiges* claimed nature had developed apart from the Deity's design (for that was not what *Vestiges* claimed), but rather that the author of *Vestiges* speculated that nature had developed according to the Deity's design without the continual, sovereign action of the Deity bringing about the required transformations, that is, without God actively superintending the ordinary course of events and miraculously intervening to bring about extraordinary transformations (such as the supposed development of life from nonliving material).[164] The crucial question, then, was not whether *Vestiges* denied divine design, but whether *Vestiges* denied the particular kind of Divine Designer whom Old School Presbyterians identified as God.[165]

Finally, the Old School Presbyterian understanding of God's sovereignty provides a more comprehensive account of Princeton's reaction to *Vestiges* than does the "Baconian" epistemology of Scottish Realism. Scholars have rightly

163. *Vestiges*, 324.

164. [Dod], 524.

165. Historians are beginning to identify different versions of, and uses for, arguments from design during the nineteenth century. For example, Bryan Bademan has distinguished three prominent versions of the design argument among antebellum American intellectuals: the Paleyan mechanism of Princeton Presbyterians and some Harvard Unitarians, the romantic developmentalism of New England Transcendentalists, and a mediate position taken by New Haven Congregationalists. Only the first of these is hostile to organic evolution. Bryan Bademan, "'Let Us Rise Through Nature Up to Nature's God': Nature and Design in Mid-nineteenth-century Protestant Thought," paper presented at the Intellectual History Seminar, University of Notre Dame, Apr. 2001, and, in abbreviated form, to Ecology, Theology, and Judeo-Christian Environmental Ethics, a Lilly Fellows Program National Research Conference, University of Notre Dame, Feb. 2002.

emphasized the pervasive importance of Scottish Realist philosophy in the early-nineteenth-century American colleges, and certainly Princeton can be found near the center of this movement.[166] This chapter has argued, however, that something more than a "Baconian" preference for empirical fact-cataloguing over speculative theorizing, and the attendant "doxological" interpretations of nature's facts as testimonies of the wisdom, power, and goodness of nature's God, held central importance at Princeton. Baconian epistemology on that campus functioned to reveal the empirically discoverable sovereignty of God in nature—invariable natural laws through which God exercised His eternal decrees over creation. The central importance of God's sovereignty at Princeton may help to explain why Scottish Realist epistemology, or *philosophy of knowledge*, received greater emphasis there, whereas Scottish Realist *moral philosophy* (which attributed more to human agency than Old School Presbyterians allowed) received greater emphasis at Yale College, where New Haven theology dominated.[167]

This chapter has emphasized the Old School Presbyterian doctrine of the sovereignty of God as a more precise explanation than Baconianism, the design argument, or providentialism, for understanding Princetonians' reactions to *Vestiges*. Drawn from the Reformed theology of John Calvin, preserved in the Westminster Confession, taught on the Princeton campus, articulated in periodicals, and preached from pulpits, the Princetonians' conviction of God's

---

166. Bozeman, esp. chaps. 1–2; Hovenkamp, esp. chaps. 1–2; Stewart, chap. 2; Vander Stelt, 12–35, 57–64, and *passim*.

167. George Marsden has indicated that New Haven theologians appropriated the Scottish Common Sense basis for morality and moral responsibility more readily than their Princeton counterparts, who remained more cautious about its problematic relation to the Calvinist doctrine of total depravity. See Marsden, 48–49. Mark Noll has helpfully suggested that Scottish Common Sense Realism consisted of a "cluster of convictions" that different evangelical communities could incorporate into their own project by emphasizing one of its elements rather than another. He identified an *epistemological* framework that instills trust in the existence of the material world and the human mind's ability to learn about it directly and reliably through observation; an *ethical* framework that instills trust in the existence of objective morality and its revelation to the human mind through the conscience; and a *methodological* framework that privileges inductions from observed facts and constrains the use of hypotheses. This schematization of Scottish Realism may help historians to distinguish different versions of Scottish Realism according to how different theological communities placed distinctive degrees of importance on the three elements. See Mark A. Noll, "Common Sense Traditions and American Evangelical Thought," *American Quarterly* 37, no. 2 (Summer 1985): 216–38. For a recent comparison of how Scottish Realism was received at Princeton and Dartmouth, see David K. Nartonis, "Locke–Stewart–Mill: Philosophy of Science at Dartmouth College, 1771–1854," *International Studies in the Philosophy of Science* 15, no. 2 (2001): 167–75.

sovereignty meant, with respect to the cosmos, that the planets were passive and their motions around the sun were caused by God's continual action in nature and, with respect to the human soul, that the individual contributes nothing to regeneration, for God alone accomplishes this transformation according to His eternal and sovereign will.[168] The development hypothesis of *Vestiges* threatened this Presbyterian orthodoxy by claiming that transformations in nature were driven by an inner law of development that, though endowed by God at the first moment of creation, operated thereafter without God's continual providential activity. In Old School Presbyterian terms, this would mean that nature can act outside the bounds of God's sovereignty, which in turn required one to admit that God's sovereignty had limits. This admission could not be entertained by Old School Presbyterians, for it called into question whether God was "sovereign" at all. An 1829 article in the *Princeton Review* defined "deism" and "pantheism" as synonymous to "atheism," noting that all three doctrines deny a "free, sovereign, independent being." To ascribe natural events to "no better cause than chance [atheism] or necessity [deism and pantheism]" was to deny that God is the cause of the universe's stability and, therefore, to deny a core tenet of Calvinist theism.[169] In 1847 Wiley and Putnam, the same New York firm that published

168. For a discussion of the resonance between the Reformed tradition of God's "radical sovereignty" in regard to the doctrine of justification and a tradition of voluntarist interpretations of Newtonian natural philosophy (including Newton's own interpretation) during the late seventeenth and early eighteenth centuries, see Gary B. Deason, "Reformation Theology and the Mechanistic Conception of Nature," chap. 6 in *God and Nature: Historical Essays on the Encounter between Christianity and Science*, ed. David C. Lindberg and Ronald L. Numbers (Berkeley: University of California Press, 1986), 167–91. In extending Deason's argument into the nineteenth century, I am asserting that because Old School Presbyterians regarded themselves as stewards of the doctrine of justification as understood in light of Calvin's emphasis on God's sovereignty, their natural philosophy correspondingly emphasized the activity of God *vis-à-vis* the passivity of matter. The "sovereignty" interpretation of natural and theological science at Princeton has also been carried into a later era than that considered here, namely, the late nineteenth century when Princetonians engaged Charles Darwin's *Origin of Species*. By then, scientific consensus had begun to admit empirical evidence suggestive of species transmutation. Those at Princeton who chose to accommodate to evolutionism did so by interpreting species change as God's sovereign activity in nature. See David N. Livingstone and Mark A. Noll, "B. B. Warfield (1851–1921): A Biblical Inerrantist as Evolutionist," *Isis* 91 (2000): 283–304, esp. at 290 and 303.

169. "The Bible, a Key to the Phenomena of the Natural World" (1829), reprinted in *Essays, Theological and Miscellaneous, Reprinted from the Princeton Review, Second Series, Including the Contributions of the Late Rev. Albert B. Dod, D.D.* (New York and London: Wiley and Putnam, 1847), 1–14, at 3.

Vestiges in 1845, issued a reprint of this article together with Dod's recent review of *Vestiges* and other "testimonial[s] to [Dod's] genius and cultivation."[170]

Reprinted in a post-*Vestiges* context and now juxtaposed with Dod's review, the 1829 article reinforced Dod's judgment that the development hypothesis could not be classified by Princetonians otherwise than as a blatant attack against theism. The earlier article warned against the atheistic tendency of the "idea of the production of animals or vegetables, by what was called equivocal generation, that is, without progenitors, or organized seeds and roots." Lamarck's theory that "place[s] man, in his origin on a level with the speechless brutes, from which condition he is supposed to arise by long and assiduous exertion" received identical treatment.[171] When reviewing *Vestiges,* Dod applied the same conclusion to Lamarck's successor, the author of *Vestiges*. Admittedly, *Vestiges* had a kind of creative deity, but one active only at the beginning. "Of what avail is it to give us the idea of a Creator, if He who created us does not govern us?" asked Dod to his Presbyterian readers, whom he could expect knew the answer well.[172] Who at Princeton would question that *Vestiges* promoted atheism, not Presbyterianism?

If the *Vestiges* controversy made one thing clear at Princeton, it was that Presbyterian theological science and natural science were firmly united together in the hands of theologians and naturalists at both the seminary and the college, all for the sovereignty of God. In this regard, Princetonians were successful educators, for in handling *Vestiges* they fulfilled their aims of proclaiming God's sovereignty for the salvation of the elect, and drawing upon all of the intellectual resources available to do so. At Harvard and Yale, however, other systems of theology structured the aims of higher education, creating alternative social settings for interpreting *Vestiges*. The Unitarian encounter of *Vestiges* at Harvard and the Congregationalist encounter at Yale will be explored by the remaining two case studies, in chapters 3 and 4.

170. *Essays, Theological and Miscellaneous*, n.p.

171. "The Bible, a Key to the Phenomena of the Natural World," 6.

172. [Dod], 533.

Study Questions

1. What was the central organizing principle of Presbyterian theology as taught at Princeton? How did Old School Presbyterians emphasize this doctrine differently than New School Presbyterians?

2. Princeton was founded to serve all denominations; in fact, Roman Catholic, Episcopalian, and Methodist students attended the college during the era of *Vestiges*. Why, nevertheless, should Princeton at that time be understood as an Old School Presbyterian academy?

3. What function did the natural sciences have in the training of future pastors at Princeton?

4. Summarize the faults that Prof. Albert Dod identified in *Vestiges* and contextualize his objections within Old School Presbyterian theology.

5. What lessons did Princeton students learn from their encounters with *Vestiges*—both in the classroom and beyond the classroom?

6. When Prof. Joseph Henry became the president of the American Association for the Advancement of Science, did he leave his Presbyterianism behind in order to pursue a career in secular science? Explain how the *Vestiges* controversy shaped the theological expressions of AAAS leaders like Henry.

7. To what degree did Princetonians respond to *Vestiges* as generic American Protestants, and to what degree did they draw specifically from their Old School Presbyterian heritage?

CHAPTER THREE

Harvard: *Vestiges* Meets the Unitarian Character of Science

"If we understand the author rightly, [Vestiges] is an outright denial of the moral nature of man, and the moral signification of life."

Rev. Joseph Henry Allen, reviewing *Vestiges*, 1845[1]

Introduction

"I partake of the feeling,—perhaps a prejudice—against Princeton," wrote Harvard botanist Asa Gray to John Torrey in February of 1847, "and you will feel less at home there now that Henry is gone."[2] Joseph Henry had resigned his full-time teaching position the previous spring to accept a post as the first directing secretary of the Smithsonian Institution in Washington, D.C. "Dod's death is sad news for Princeton," Gray had written Torrey in December of 1845.[3] With Dod, the mathematician, and Henry, the natural philosopher, both absent, Princeton looked in 1847 like a scientific wasteland to Gray, except that Torrey, his long-

---

1. J[oseph] H[enry] A[llen], "Vestiges of Creation and Sequel," *Christian Examiner and Religious Miscellany* 4th ser., 5, no. 3 (May 1846): 333–49, at 345.

2. Asa Gray to John Torrey, 20 Feb. 1847 (Torrey Papers, New York Botanical Garden Library).

3. Asa Gray to John Torrey, Dec. 1847 (Torrey Papers).

time friend and botanical collaborator, continued to teach there. The remaining professors, thought Gray, "are some of the most excellent men, but not likely to sympathize heartily with scientific men." Though he regarded them as "dear friends," and had experienced a religious awakening through the example of Torrey's Presbyterian faith, Gray had little respect for Princeton Presbyterians as men of science.[4] Like them, he spoke and wrote against *Vestiges* repeatedly; unlike theirs, his vision was that of Harvard science, not Princeton science.

As demonstrated in chapter 2, Princeton science amalgamated theology with the natural sciences by ordering each component discipline around the Old School Presbyterian emphasis upon God's sovereignty in nature, including human nature. Harvard science had a different center, and as a consequence members of the Harvard community addressed *Vestiges* in different ways than their counterparts at Princeton. Most strikingly, Harvard science—in both its natural and theological branches—varied from one practitioner to the next. The Unitarian heritage that shaped the campus *milieu* was not so much a heritage of closely shared dogma, but rather one of attitude. The liberal attitude at Harvard was thoroughly Christian, constituents agreed. Indeed, they insisted that Unitarianism was far more Christian than the strict confessionalism that characterized Princeton's Presbyterianism. The liberal attitude encompassed the Congregational empiricist Asa Gray, the non-affiliated theist and German idealist Louis Agassiz, and the Unitarian Scottish Realist Francis Bowen; beyond Harvard's boundaries—although with roots that dug deeply into Harvard's soil— branched the Transcendentalism of Ralph Waldo Emerson, which many Unitarians thought was even more dangerous to Christian education than was Calvinism.

When historians mindful of evolution controversies have turned their attention to mid-nineteenth-century Harvard, they generally have focused upon tensions that developed between Gray and Agassiz from Agassiz's 1846 arrival in America to the 1859 publication of Darwin's *Origin of Species*. In the early days, Gray and Agassiz both opposed the doctrine of species transmutation as popularized in *Vestiges*. Over the years, however, the two men of science recognized more and more that they had different, even opposing, scientific visions.

4. Asa Gray to John Torrey, 20 Feb. 1847 (Torrey Papers).

Agassiz emphasized the Creator's thoughts, which the human mind, being created in the divine image, could discover by studying patterns in nature. To Gray this seemed to be an inappropriate imposition of *a priori* convictions upon the raw data of natural science—"science" for Gray being something that ought to be a more strictly empirical revelation of God's working in nature rather than a supposition about God's thinking behind nature. Gray wanted to follow God's natural facts wherever they might lead, forming theories from facts rather than using theories as lenses through which facts would be observed. From Agassiz's perspective, Gray's raw inductivism rested upon a misunderstanding of how the human mind constructs knowledge, that is, how the mind formulates a science. Schooled in German Idealism, Agassiz had inherited a tradition from Immanuel Kant and Friedrich Wilhelm Schelling that regarded mind as actively structuring perceptions. In Agassiz's theistic idealism, the divine mind structured the world. Hence, people's empirical observations revealed representations of God's thoughts, or divine "ideas," according to which material existence is constituted. Gray's methodology, to the contrary, had roots that were British rather than Continental, and empirical rather than idealist. Though critical of Princeton science, Gray in fact had much in common with the Princetonians' emphasis upon observed facts and their cautious attitude toward hypotheses. By the mid 1850s, Gray and Agassiz found themselves entangled in substantive, as well as methodological, debates. Most notably, they disagreed over the definition of "species." By 1860, it was clear not only to them but also to the American public that Gray was championing two of Darwin's contentions: that species change and, thus, that diverse species may have descended from a common ancestor; meanwhile, Agassiz remained convinced, as Gray long ago had been, that species are fixed.[5]

The diverging relationship between Gray and Agassiz forms an important part of the story of *Vestiges* at Harvard during the mid 1840s through 1850s. Unfortunately, however, previous historical accounts have tended to focus on

5. A. Hunter Dupree, *Asa Gray, 1810–1888* (Cambridge, MA: Harvard University Press, 1959), chaps. 12–15; Edward Lurie, *Louis Agassiz: A Life in Science* (Chicago and London: University of Chicago Press, 1960), chaps. 5–7; David N. Livingstone, *Darwin's Forgotten Defenders: The Encounter between Evangelical Theology and Evolutionary Thought* (Grand Rapids, MI: William B. Eerdmans, 1987), 57–64; C. George Fry and Jon Paul Fry, *Congregationalists and Evolution: Asa Gray and Louis Agassiz* (Lanham, MA: University Press of America, 1989); Paul Jerome Croce, "Probabilistic Darwinism: Louis Agassiz vs. Asa Gray on Science, Religion, and Certainty," *The Journal of Religious History* 22, no. 1 (Feb. 1998): 35–58.

these two figures to the exclusion of other Harvard voices. In their efforts to discover "precursors" to the Darwinism debates of the 1860s, scholars examining the *Vestiges* debates at Harvard have focused so selectively on Gray and Agassiz as to miss the context of those earlier discussions, which involved Harvard moral philosopher Francis Bowen, professorial candidate Henry Darwin Rogers, Harvard-affiliated ministers Joseph Henry Allen, Samuel Gilman, and Andrew Preston Peabody, and the Harvard expatriate Ralph Waldo Emerson, together with his fellow Transcendentalist Henry David Thoreau. To understand the receptions of *Vestiges* at Harvard, and, in particular, the career roles of Gray and Agassiz as Harvard authorities responding to *Vestiges*, one must therefore examine the broader Harvard context in which the voices of Gray and Agassiz were but two among several.

Unlike the situation at Princeton, not every voice at Harvard was expected to sing the same tune. Nevertheless, they all returned at times to some common keynotes, striking a chord, as it were, that resonated with the spirit of Boston Unitarianism. The first section of this chapter lays forth the Unitarian impulse that shaped Harvard and its familial institutions, such as the Lowell Institute, during the era of *Vestiges*. The remainder of this chapter illustrates the manner in which various Harvard spokespersons responded to *Vestiges* within that liberal, at times vague and malleable, Unitarian movement in Boston's intellectual society. Though specific religious doctrines did not relate to *Vestiges* in quite as straightforward a manner as at Princeton, where the Westminster Confession was invoked as a standard for the sciences, Unitarian theology—particularly in its sociological dimensions—nonetheless shaped the practices of Harvard science. Of ultimate concern was not God's sovereignty working for people's salvation, but rather human moral character and the empirical means for identifying its ideal form in order that self-development may have a sure foundation. This was the central concern structuring the theological situation that *Vestiges* entered when it arrived at Harvard, and it was a sufficiently ecumenical concern to find expression also among members of the Harvard community who were not themselves Unitarians, such as Gray and Agassiz.

The Unitarian Setting:
Harvard College and Harvard Divinity School

When, in 1636, the Puritans of Massachusetts Bay Colony established the first institution of higher learning in British America, they initiated an educational tradition that would have broad, and at times mutually contradictory, consequences for American education.[6] If Calvinist confessionalism be regarded as the rightful heritage of New England Puritanism, then Princeton received the full portion of primogeniture, leaving its older brother Harvard to play the role of bastard child. But the Puritans bequeathed more than Calvinism, and Harvard remained rightful heir to much that was Puritan.

New England Puritan theology, in fact, was not purely Calvinist, but "federal," avoiding strict interpretations of predestination and human depravity in order to accommodate enough human freedom to establish a *foedus*, or covenant, between the community and God. A recurring debate within the New England tradition concerned how, exactly, that covenant becomes owned by individuals within the community, that is, how a person becomes saved.[7] By the late eighteenth-century, a liberal wing of Congregationalists (as the descendents of Puritans then called themselves) had adopted an Arminian answer to that question, namely, that a person freely chooses to accept God's offer of salvation. Whereas Old School Presbyterians at Princeton opposed Arminian free choice, stressing God's sovereign predestination in matters of salvation, their liberal counterparts in eastern Massachusetts welcomed Arminianism as a

6. For the most comprehensive account, though dated, see Samuel Eliot Morison, *Three Centuries of Harvard, 1636–1936* (Cambridge, MA: Harvard University Press, 1936).

7. Sidney E. Ahlstrom and Jonathan S. Carey, eds., *An American Reformation: A Documentary History of Unitarian Christianity* (Middletown, CT: Wesleyan University Press, 1985), 16. For the classic study on the federal theology of New England Puritanism, see Perry Miller, *The New England Mind: The Seventeenth Century* (1953; 2d printing, Cambridge, MA: Harvard University Press, 1954), chaps. 10, 13, 14, 15, 16; in that final chapter, at pp. 484–85, Miller underscores his conclusion, which he says came unexpected to him, that in expanding the Covenant of Grace from the individual to the society, and thereby creating the federal theology of New England, Puritans declined from the heartfelt sincerity of their Augustinian faith into a commercialism of forging a new society. In a sequel study, Miller traced the result of the Puritans' expanding of their covenant to encompass the children of the unconverted, concluding ultimately that this left a fragmented, post-Puritan New England society. See *The New England Mind: From Colony to Province* (1953; 9th printing, Cambridge: Harvard University Press, 1998), chaps. 4, 6, 7, 14, 27. In light of these studies, the rise of Unitarianism in Boston may be seen as an attempt to fulfill the long-delayed Puritan vision of an established Christian society in New England.

rationalization of the Christian tradition that could withstand the litmus test of Enlightenment thought: is it reasonable?

The English philosopher John Locke, in his *Reasonableness of Christianity* (1695), had provided a guide for meeting the deistic and atheistic tendencies of the Enlightenment on its own, rational, terms without forfeiting the supernatural revelation, the Bible, which he considered foundational to Christianity. Locke, in effect, reduced salvation to the reasonableness of faith in Jesus as Messiah—reasonable because the New Testament miracles provided compelling evidence concerning Jesus' divine mission.[8] Evangelism thus became the rational persuasion of a person to believe in Jesus as Messiah, or Savior. Locke's tradition of "rational supernaturalism" provided Harvard's eighteenth-century liberal Congregationalists with a legacy that they passed onto their nineteenth-century heirs, the Unitarians. Religion, learned Harvard's Unitarian ministers-in-training, was thoroughly rational—"the science of God."[9] Unlike in France, and far more so than in Britain, Unitarianism in eastern Massachusetts was a moderately paced development within the socially stable faith of the established elite, not a radical dissent from an earlier generations' reigning orthodoxy.[10]

In addition to the "external evidences" of miracles, liberal Congregationalists rested their faith also on the "internal evidence" of the conscience. Here they borrowed from the seventeenth-century Neo-Platonic moral philosophy of Ralph Cudworth and Henry More at Cambridge University. Like Plato, Cudworth and More taught that morality was founded upon eternal ideas of benevolence and justice. They differed in this respect from orthodox Calvinists, who located the source of morality in God's sovereign will. The eighteenth-century Scottish moral philosophers Thomas Reid and Dugald Stewart, developing what Reid dubbed "common sense" philosophy, argued that human moral faculties were capable of perceiving divine standards of benevolence and justice. In New England, Cambridge Platonism and Scottish Common Sense Realism were extended to imply that God deserves worship chiefly because he fulfills the eternal ideals of morality, just as Plato's notion of

---

8. Ahlstrom and Carey, 12.

9. U. B. Tufts, "Why Does an Intelligent Man Reject Christianity?," Divinity School student essay, 1852 (Harvard University Archives, HUC 8851.318).

10. Ahlstrom and Carey, 15.

divinity was a perfect instance of the good, the true, and the beautiful. Liberal Congregationalists realized, however, that the Calvinist God did not live up to their standards of benevolence and justice. It seemed unfair, for example, that God would predestine some people to hell. It also seemed unfair that those who entered heaven arrived entirely apart from any merits or choosing of their own. In short, the Calvinist God seemed irrational and arbitrary.[11]

If rational supernaturalism distinguished New England Congregationalist liberals from deists, then Neo-Platonic moral philosophy separated them from orthodox Calvinists. They were becoming a religious movement *sui generis*, and soon would become known as Unitarians.[12] In 1805, the Harvard Corporation appointed the liberal Henry Ware to replace the recently deceased David Tappan, a moderate Calvinist, as the Hollis Professor of Divinity. The remaining Calvinist minority at Harvard realized that they had little chance of regaining the school, so they left and established Andover Theological Seminary in 1808.[13] During the following decade, the Rev. William Ellery Channing became the chief spokesperson for liberal Christianity in New England. Channing spelled out the Christological implications of a rationalized Christian faith: Christ should not be worshiped as God, but only honored as a man who led an exemplary life of moral character; his miracles do not attest to his Godhead, but merely show the Father's approval of him as the ideal man whom all should seek to emulate. As a consequence of Channing's claim that only the Father is God, and not Christ or the Holy Spirit, Trinitarian Calvinists labeled Channing's circle "Unitarians,"

11. Daniel Walker Howe, "The Cambridge Platonists of Old England and the Cambridge Platonists of New England," *Church History* 57, no. 4 (Dec. 1988): 470–85, esp. 474–77; Ahlstrom and Carey, 14. For an introduction to Cambridge Platonism, see C. A. Patrides, ed., *The Cambridge Platonists* (Cambridge, MA: Harvard University Press, 1970).

12. The name "Unitarian" had earlier been applied in Britain to anti-Trinitarian nonconformists. Theophilus Lindlsey, Thomas Blesham, and Joseph Priestley founded the first explicitly Unitarian church, Essex Street Chapel, in 1774. But the roots of Boston Unitarianism are not to be traced specifically to this or any other British Unitarian movement; rather, Boston Unitarianism arose as a response to New England Calvinism, and it drew inspiration from two centuries of rational Christianity that had been developed both within and without the Anglican Church in Britain. See Ahlstrom and Carey, 11.

13. Leonard Woods, *History of Andover Theological Seminary*, ed. George S. Baker (Boston: James R. Osgood and Company, 1885), 27; Morrison, *Three Centuries of Harvard*, 187–91; Daniel Walker Howe, *The Unitarian Conscience: Harvard Moral Philosophy, 1805–1861*, with a new introduction (Middletown, CT: Wesleyan University Press, 1988), 4.

although their moral philosophy was more central to their theology than their Christology.[14]

To distinguish themselves from orthodox Calvinists during the 1810s through 1830s, the liberal Christians at Harvard did not simply replace the tenets of the Westminster Confession with a Unitarian creed. Rather, they questioned the very notion of confessional subscription and pledged themselves to a liberal attitude that saw strict doctrinal standards as inimical to true Christianity.[15] As a consequence, some Unitarians retained a quasi-Trinitarian understanding of God, others insisted upon God being a Unity, all held that Christ was divine with respect to the character of life he led, and some—the Transcendentalists who emerged in the 1830s—saw this divinity within themselves more clearly than through external evidences, such as the Bible's testimony about the life of Christ.[16] All shared in common a Puritan heritage of piety: Christianity meant, above all, living as Christ had lived.[17]

To live as Christ had lived was a common goal among Christians of varied denominations, but for Harvard Unitarians this goal received emphasis to the exclusion of Christ's vicarious atonement for sinners that had historically been central to the Christian faith. Rational supernaturalism could, and did, allow for Christ to die on the cross and be raised from the dead, but the internal evidences of the conscience could not deem God praiseworthy if He declared sinners to be righteous simply because Christ had died in their place. To the Unitarian mind, Christ was a physical human person and his death on the cross was a physical event, but sin was a moral problem in need of a moral or spiritual, rather than

14. William Ellery Channing, "Unitarian Christianity," facsimile rpt. of 1819 pamphlet, in *The Unitarian Controversy, 1819–1823*, ed. Bruce Kuklick, 2 vols. (New York: Garland Publishers, 1987). For an insider's history, see Joseph Henry Allen, *Our Liberal Movement in Theology: Chiefly as Shown in Recollections of the History of Unitarianism in New England, Being a Closing Course of Lectures Given in the Harvard Divinity School*, 3d ed. (Boston: Roberts Brothers, 1892), esp. 43–54.

15. Francis C. Williams, "What Are the Principal Causes of Infidelity of Modern Times, Especially of the Infidelity of Literary Men?," Divinity School essay, 1 July 1844 (Harvard University Archives, HUC 8844.318). For an historical treatment of Unitarians' disavowal of theological confessionalism, see Howe, *Unitarian Conscience*, 4–5, 7.

16. On the vagaries of Unitarian christology, see Ahlstrom and Carey, 5, 37–38. The distinctions between Unitarianism and Transcendentalism are discussed below.

17. The Harvard emphasis on moral character could be a source of humor as well as obligation. As John Mead recorded in his diary at the start of his junior year, "The Rev. Mr. Pierpont preached at the chapel and made the Freshmen stare and Seniors grin by saying that a positive duty of all young persons was a reverence for seniors." John Mead, Journal, 1848–1850 (Harvard University Archives), 10 Sep. 1849.

physical, solution. It seemed absurd to speak of Christ shedding his *physical* blood to redeem people from their *spiritual* death in sin. When Harvard's ministers preached that Christ was the Savior, they did not mean that he had purchased people's lost souls by giving up his own, but rather that he had lived, and died, as an example for all people, in perfect love to both God and his neighbors—a love so perfect that the Father validated Christ's ministry by raising him from the dead.[18]

Five theological principles, then, characterized the Unitarian movement by the 1830s. First, external evidences served as a foundation for the faith. These included both the findings of the natural sciences and the testimonies of the Old and especially the New Testament. Second, internal evidence held an equally important role in establishing religious truth. Most notably, the "Moral Argument" against orthodox Calvinism insisted that God must fit humanly discernible rational standards of divine benevolence, and human nature must be more capable of pleasing God than the orthodox had claimed. As the Rev. Andrew Preston Peabody emphasized, Christians should offer their worship to "the Creator, not the Sovereign, not the Judge, but the Father."[19] Third, the true spirit of Christianity, and of the Protestant Reformation, could be fulfilled only in the liberation of the human mind to decide for itself, on rational grounds and unfettered by church creeds, who God is and how humans relate to Him. Fourth, human nature is capable of living a moral life, patterned after the example of Christ; indeed, religion was little more than a guide to living a life of Christ-like character. Finally, the salvation Christ provided was not the forgiveness of sins, but rather the perfect example of moral character, in life and in death.

These ideas were not only preached from Unitarian pulpits; they also shaped social practices within a community, the community that educated its sons at Harvard. The twenty-five men who served, for overlapping terms of varying lengths, as Fellows of the Harvard Corporation during the half century spanning Ware's appointment to the Hollis Professorship (1805) through the presidency of the Rev. James Walker (1853–1860) were all Unitarians, except for

18. Channing, "Unitarian Christianity," 31–36; Allen, *Our Liberal Movement*, 65–67; Ahlstrom and Carey, 15–16, 35–37; Howe, *Unitarian Conscience*, 119; Leo P. Hirrel, *Children of Wrath: New School Calvinism and Antebellum Reform* (Lexington, KY: The University Press of Kentucky, 1998), 23.

19. Andrew Preston Peabody, *Sermons Connected with the Re-Opening of the Church of the South Parish, in Portsmouth, New Hampshire, preached Dec. 25 & 26, 1858, and Jan. 30 & Feb. 6, 1859* (Portsmouth, NH: James F. Shores, Jr., & Joseph H. Foster; Boston: Crosby, Nichols, and Co., 1859), 8.

three Episcopalians; politically, they were nearly unanimous as Federalists and Whigs. Beginning in 1840, John Amory Lowell, whose uncle had founded the New England textile factory system, became the most influential member of the Corporation. The Lowell family was one of a handful into which wealth became concentrated during the early nineteenth century. Politics protected this wealth, and first-cousin marriages concentrated it among those families whom Oliver Wendell Holmes later dubbed the Boston "Brahmins." By the close of the 1820s, the Harvard Corporation had shifted hands from professionals to businessmen. With this rising socioeconomic elitism came enough Unitarian philanthropy to fund the Boston Athenaeum (1807), the Massachusetts General Hospital (1821) and its allied Life Insurance Company (1823), the Massachusetts Horticultural Society (1829), the Boston Society of Natural History (1830), the Perkins Institute (for the blind, 1832), and the Lowell Institute (sponsoring public lectures, 1836). Lowells and Cabots, Quincys and Perkinses, Eliots and Higginsons—these and a few other families had created a Unitarian uppercrust in Boston, and their surnames filled the roll calls of Harvard overseers, administrators, faculty, and students for decades.[20]

John Amory Lowell selected distinguished men of science and letters to give lectures each fall and winter for the Lowell Institute. His cousin John Lowell, Jr., the son of textile magnate Francis Cabot Lowell and a trustee of the Boston Athenaeum, had endowed the lectureships in 1836 for "the truth of those moral and religious precepts, by which alone . . . men can be secure of happiness in this world and that to come."[21] Some speakers came from the Harvard faculty. When the Lowell lectures featured an outsider, the opportunity at times served as a means for assimilating new blood into the Harvard establishment.[22] Consistent with the Unitarian emphasis on external and internal evidences, each season of

20. For the most comprehensive account of the social formation of the Harvard elite, see Ronald Story, *The Forging of an Aristocracy: Harvard and the Boston Upper Class, 1800–1870* (Middletown, CT: Wesleyan University Press, 1980). For additional insights into the Havard *milieu* during the 1830s and 1840s, see James Turner, *The Liberal Education of Charles Eliot Norton* (Baltimore: The Johns Hopkins University Press, 1999), esp. 3–18, 22, 41–49, 60.

21. Robert F. Dalzell, Jr., *Enterprising Elite: The Boston Associates and the World They Made* (Cambridge, MA: Harvard University Press, 1987), 145, 146 (quotation of the Lowell Institute bequest).

22. Examples include Gray and Agassiz, who will be discussed below, and Jeffries Wyman, who will be mentioned below, but whose elevation into the Harvard community via the Lowell Institute is discussed in greater detail by Toby A. Appel, "A Scientific Career in the Age of Character: Jeffries Wyman and Natural History at Harvard," in *Science at Harvard University*, ed. Eliot and Rossiter, 96–120.

Lowell lectures included a lecture on "Natural Religion," the "Evidences of Christianity," or some similar title. Other lectures spanned various sciences, from architecture to zoology, including chemistry, electromagnetism, geology, optics, and political economy.[23] Generally these lectures, too, affirmed that God's rational benevolence toward humanity can be discovered in nature.

From a distance, natural theology may look the same the world over, but in Boston it had a Unitarian image, emphasizing human freedom in the moral sciences and God's fulfillment of humans' intuited sense of justice and benevolence in both the physical and moral sciences. Providence would be unjust, for example, to leave the majority of the human race in ignorance of Himself; therefore, He reveals Himself in nature as a loving Father just as surely as He does so through biblical revelation. Far from being a narrow sect established by peculiar doctrinal confessions, Christianity is the universal natural religion, available for all people to accept by free and rational choice.[24] Unitarians conceded that Calvinists could adduce evidence from nature that would support their belief in a predestinarian God whose human subjects are totally depraved and must rely solely on His mercy. Unitarians argued, however, that the same natural evidence could just as well—indeed, could even more reasonably— support a belief in the Unitarian God, who, as a loving Father, guides His children in lives of moral character.[25] So wide was the gulf between Calvinism and Unitarianism, that Unitarian minister James Walker, a future Harvard professor and president, could declare in the early 1830s that Unitarians did not feel the slightest bit threatened by Calvinists' claims that they deny the Gospel by denying that Christ is God, since "Calvinism is not the gospel" anyway.[26]

As the Unitarian establishment solidified in Boston, the remnant of Calvinist Congregationalists sought countermeasures. At the 1820 constitutional convention, the orthodox tried to disestablish the state church, since they saw

23. Hariette Knight Smith, *The History of the Lowell Institute* (Boston: Lamson, Wolffe, and Co., 1898).

24. William H. Furness, *Nature and Christianity: A Dudleian Lecture Delivered in the Chapel of the University at Cambridge [i.e., Harvard], Wednesday, May 12, 1847* (Boston: William Crosby and H. P. Nichols, 1847), esp. 3, 24.

25. James Walker, *Unitarianism Vindicated against the Charge of Skeptical Tendencies* (Boston: James Munroe & Co., 1832).

26. James Walker, *Unitarianism Vindicated against the Charge of Not Going Far Enough*, 3d ed. (Boston: James Munroe & Co., 1832), 15.

that it was quickly falling into Unitarian hands. The liberals forestalled that effort by warning that without public funding of religious institutions, irreligion would sweep the state. By 1833, however, Trinitarian Calvinists had lost control of so many congregations to Unitarians that they moved again for disestablishment, this time successfully.[27] Harvard's Board of Overseers included numerous *ex officio* members from the Massachusetts state government (many of whom were Unitarian). Harvard therefore remained in some sense an "established" church-state school until the legislature removed those *ex officio* members in 1851. During that time, Congregationalists frequently raised complaints that Harvard was a "sectarian" institution, and they managed, on rare occasions, to elect a Congregationalist to the Board of Overseers, resulting in a theological tug-of-war regarding the nature and purpose of Harvard education.[28]

In the midst of this ongoing tension between Congregationalists and Unitarians, Harvard's Unitarians emphasized that their theology was thoroughly biblical. To Calvinists, Unitarians appeared to be sliding down a slope toward deism. Defending the Unitarian faith against this charge, Walker wrote a pamphlet stressing that deists reason upon natural evidences alone, whereas Christians, whether Unitarian or Calvinist, reason upon both natural and scriptural evidences. He conceded that Unitarians and Calvinists reach different conclusions when reasoning on those evidences, but he emphasized that both of them shared the same sources, nature and Scripture, the latter source separating them from deists. Thus, although the Unitarians and deists might appear to be similar in their oppositions to the Congregationalists' doctrine of the Trinity, the two parties in fact rested their cases on different foundations. Not only were Unitarians scriptural (and therefore not deists); they also could claim to be more scriptural than Congregationalists, who would "explain away those passages, which positively, and in so many words, assert Christ's inferiority to the Father."[29]

27. Howe, *Unitarian Conscience,* 216–220.

28. Morison, 257–58, 293.

29. James Walker, *A Discourse on the Deference Paid to the Scriptures by Unitarians* (Boston: American Unitarian Association, 1832), 3, 9 (quotation). The Rev. Andrew Preston Peabody amplified Walker's biblical arguments for Unitarianism several years later. According to Peabody, where Christ says of himself, "I and the Father are one" (John 10:30), the Scriptures do not here teach that Christ and the Father are united into the single Godhead of the Trinity. The Greek term "one" is neuter, and thus must mean "one thing" (i.e., united in one purpose); it cannot mean one Divine Being, which being a personal subject would

Beginning in the 1810s, future Unitarian ministers received training at Harvard Divinity School, a new division of the college.[30] Out of this seminary came a movement that some Unitarians, most notably Andrews Norton, would find even more threatening than Calvinism. The roots of this movement, called Transcendentalism, branch widely in both American and European soil, penetrating deeply enough to reclaim idealist insights from Plato and Plotinus, while also drawing nourishment from the contemporary works of Carlyle and Coleridge, whose own roots were deeply planted in the soil of German idealism.[31] Most proximately, however, Transcendentalism emerged from within the New England context in which Boston Unitarianism itself also was situated.[32] Though Harvard quickly became inhospitable to Transcendentalism, some of that movement's characteristic features had long been present in what philosopher David Nartonis has identified as an "unofficial Harvard curriculum": students and faculty supplemented the Scottish Common Sense, or Realist, philosophy assigned in class with outside readings that represented an idealist philosophical tradition.[33]

require, according to Peabody, a masculine pronoun. See Peabody, *Sermons Connected . . . South Parish,* 26. At the seminary, students learned that Unitarians advanced biblical knowledge by using reason, whereas Calvinists retarded theological progress by opposing the use of reason in general, and of moral philosophy in particular, when interpreting Scripture. See A. Smith, "What in your opinion are some of the most important causes of infidelity in modern times?," Divinity School essay, 1844 (Harvard University Archives, HUC 8841.318.2).

30. The Divinity School was founded in 1811, 1816, 1818, or 1819, depending upon how one defines "founded." Formalized theological instruction emerged piecemeal, but by 1819 the Theological Department had come into existence as a distinct administrative unit within the college. See Conrad Wright, "The Early Period (1811–40)," chap. 1 in *The Harvard Divinity School: Its Place in Harvard University and American Culture,* ed. George Huntston Williams (Boston: The Beacon Press, 1954), 21–77, at 23–27.

31. René Wellek, "Emerson and German Philosophy" (1943), rpt. in *Confrontations: Studies in the Intellectual and Literary Relations between Germany, England, and the United States during the Nineteenth Century* (Princeton, NJ: Princeton University Press, 1965), chap. 6. Wellek argues that much of Emerson's knowledge of German idealism, and even of Plato, was filtered through Coleridge and Carlyle.

32. Perry Miller, "New England's Transcendentalism: Native or Imported?" (1964), rpt. in *Critical Essays on American Transcendentalism,* ed. Philip F. Gura and Joel Myerson (Boston: G. K. Hall and Company, 1982), 387–401; David M. Robinson, "'A Religious Demonstration': The Theological Emergence of New England Transcendentalism," in *Transient and Permanent: The Transcendentalist Movement and Its Contexts,* ed. Charles Capper and Conrad Edick Wright (Boston: Massachusetts Historical Society, 1989), 49–72.

33. David Nartonis, "Idealist Philosophy of Science in Nineteenth-Century Harvard," rev. ms., based on a paper delivered to the Congress of the History of the Philosophy of Science Working Group, Montreal, 21–23 June 2002 (cited with permission).

Scottish Realists argued that truth can be learned from direct observation of the material world and moral duty can be discerned from personal—yet objective—intuitions. Idealists, by contrast, followed Plato in asserting that the material world is but a representation of the essence, or idea, that alone must be the object of true knowledge. During the late eighteenth and early nineteenth centuries, Edward Young's *Night Thoughts* and John Milton's *Paradise Lost* were borrowed frequently at the Harvard library. Young spoke of "Creation's Model in *Thy* [the Creator's] Breast" and Milton described at greater length the ideas in God's mind that served as a model of the material world God created. Influenced by the seventeenth-century Cambridge Platonist Ralph Cudworth, each writer drew from the idealist creation story presented in Plato's *Timaeus*, a dialogue portraying the Deity as a craftsman who molded the cosmos in an effort to substantiate the eternal ideas of goodness, beauty, harmony.[34]

A Neo-Platonic framework for understanding nature found especially clear expression in another work borrowed from Harvard's library, John Norris's *Essay towards the Theory of the Ideal or Intellectual World* (1701–1704). "Tho' the Natural World be the object of sense," wrote Norris, "yet the Ideal World is the proper object of knowledge."[35] During the 1820s, Harvard students and faculty began to borrow the aforementioned volumes with increased frequency, and they also turned to the British romanticist Samuel Taylor Coleridge's recent works, which elevated knowledge of the ideal world above sensual knowledge of the physical world.[36] The earlier Neo-Platonism and Coleridge's Idealist Romanticism were well represented in the books Harvard students and faculty were reading during the decades before *Vestiges*, even if they did not appear in the official curricula.

The influence of idealism at Harvard could be felt especially in the Divinity School, at times to the point of controversy. Harvard Divinity students borrowed Cudworth's idealist writings more often than Harvard undergraduates borrowed Stewart's realist treatises. The 1836 American edition of Cudworth's *True Intellectual System of the Universe* (1678) received a favorable review in the

34. As quoted in Nartonis, 6.

35. As quoted in Nartonis, 8.

36. Nartonis, 8–9.

Christian Examiner, a periodical produced by and for Boston's Unitarian clergy.[37] Understandably, then, the students at Harvard Divinity School invited an alumnus who deeply admired Cudworth's Neo-Platonism to speak for their graduation in 1838.

The invited speaker was Ralph Waldo Emerson, a minister who had resigned from his Unitarian pulpit seven years earlier. Emerson's breach with the Unitarian clergy grew from his fascination with the idealist philosophy of Coleridge, Cudworth, and Plato.[38] Sharing their sharp distinctions between the material and the ideal, or spiritual, and their privileging of the latter over the former as the proper object of understanding, Emerson began to conceive of the Lord's Supper as an ideal supper, a celebration not requiring the earthly elements of bread and wine or any other material sign of Christ's body and blood. Christ was to be a friend who could be known in the soul; his presence no longer needed to be tasted in the mouth. After preaching a sermon to this effect one Sabbath morn in 1832, Emerson shocked his congregation with the announcement that he would no longer serve as their pastor.[39]

During the years that followed, Emerson became a popular lecturer on religion and the philosophy of nature. He emphasized the Unitarian tradition of internal evidences, reshaping this into an introspective exploration of self, God, and nature. His abandonment of external evidences—be they the eucharistic bread and wine, the text of Scripture, or the facts of empirical scholarship—infuriated Andrews Norton, a former Harvard professor who spent the 1830s writing *Genuineness of the Gospels*. Norton intended this multi-volume work of biblical criticism to buttress the supernatural external evidence of the Unitarian faith against attacks from higher critics in Germany and from Transcendentalists closer to home.[40]

37. Nartonis, 9; Review of *The True Intellectual System of the Universe*, by Ralph Cudworth, *Christian Examiner* 27, no. 3 (Jan. 1840): 289–319. The reviewer, at p. 295, took special delight in Cudworth's Neo-Platonic argument, against atheism, that God's existence and nature are as far beyond intelligent doubt as "the Axioms of Geometry, or the Sentiment of Esteem and Love."

38. Ahlstrom and Carey, 29.

39. Ivy Schweitzer, "Transcendental Sacramentals: 'The Lord's Supper' and Emerson's Doctrine of Form," *New England Quarterly* 61, no. 3 (Sep. 1988): 398–418.

40. Sydney E. Ahlstrom, "The Middle Period (1840–80)," chap. 2 in *The Harvard Divinity School*, ed. Williams, 78–147, at 78; Howe, *Unitarian Conscience*, 76–77, 82–83; Jerry Wayne Brown, *The Rise of Biblical Criticism in America, 1800–1870: The New England Scholars* (Middletown, CT: Wesleyan

Despite this rupture, Emerson continued to share in common with Ware, Channing, Walker, and Norton the Unitarian belief that character development is the essence of religion.[41] All of them wanted to know Christ and to live like Christ. The problem was that the parties differed on where to discover Christian morality: Unitarians appealed to external evidences (nature and the Bible) as well as internal evidences (one's own mind and soul), whereas Transcendentalists focused on the latter to the exclusion of the former. Especially in the instances of Emerson and Norton, the differing, even if overlapping, sources of religious insight corresponded to remarkably distinctive epistemological outlooks. Norton continued Locke's tradition of grounding statements about reality ultimately upon sensory encounters with the observable world. Emerson, having imbibed the Romantic Idealism of Coleridge, broke from Lockean sensationalism and pursued the intuitive knowledge acquired through private self-reflection concerning nature.[42] From the perspective of Norton and many Unitarians, Emerson's Transcendentalism was but one afternoon's contemplation away from skepticism, infidelity, and social disintegration. Each man became his own philosopher, dreaming up his own nature and his own God. When speaking to the Divinity School class of 1838, Emerson even dared to suggest that each minister should become his own Christ, abandoning all churchly forms and sacrificing community consensus for individual self-development.[43] He was promptly

University Press, 1969), 82–84. For Norton's review of Emerson's Divinity School Address, see [Andrews Norton], "The New School in Literature and Religion," *(Boston) Daily Advertiser* (27 Aug. 1838): 2, rpt. in *Emerson and Thoreau: The Contemporary Reviews*, ed. Joel Myerson (Cambridge: Cambridge University Press, 1992), 33–35. An analysis of this may be found in Robert D. Habich, "Emerson's Reluctant Foe: Andrews Norton and the Transcendental Controversy," *New England Quarterly* 65, no. 2 (June 1992): 208–37.

41. David Robinson, "The Road Not Taken: From Edwards, through Chauncy, to Emerson," *Arizona Quarterly* 48, no. 1 (Spring 1992): 45–60, at 51. For a fuller exposition of some continuities and transformations that can be traced from Unitarianism through Emerson to Transcendentalism, see David Robinson, *Apostle of Culture: Emerson as Preacher and Lecturer* (Philadelphia: University of Pennsylvania Press, 1982). Robinson's treatment differs from many other studies, and resonates with the present study, by interpreting Emerson as a theologian and preacher more than as a philosopher or poet.

42. James Turner, "Language, Religion, and Knowledge in Nineteenth-Century America: The Curious Case of Andrews Norton," in *Language, Religion, Knowledge: Past and Present* (Notre Dame, IN: University of Notre Dame Press, 2003), 11–30, esp. at 18.

43. Ralph Waldo Emerson, "The Divinity School Address" (15 July 1838), in *The Collected Works of Ralph Waldo Emerson, Vol. 1: Nature, Addresses, and Lectures*, introductions and notes by Robert E. Spiller, text established by Alfred R. Ferguson (Cambrdige, MA: Harvard University Press, 1971), 71–93, esp. at 82, 90, 93.

banished from Harvard, not to be invited back for thirty years.[44] If Unitarianism was to be preserved from Transcendentalism, then external evidences—garnered from empirical science that could gird a Boston community of faith—had to be promoted.[45]

The Lowell Institute, the Boston Athenaeum, and the Boston Society of Natural History kept local intellectuals informed of new developments in the natural sciences. Harvard, however, lagged behind both Princeton and Yale during the 1830s and early 1840s in keeping abreast of scientific developments. Its enrollment declined and its curriculum in the natural sciences remained weak. The opening of Amherst College in 1825 had taken Calvinist students away from Harvard.[46] The remaining Unitarian sons studied volumes of classical literature, in classical languages, but received only a taste of the natural sciences.[47] Botanical science at Harvard may serve as an example.

Boston subscribers had funded the Massachusetts Professorship of Natural History as a joint community-Harvard venture in 1805, but after the death of the first appointee, William Dandridge Peck, in 1822, the President and Fellows decided to leave the chair vacant. By the mid 1830s, even the botanical garden, tended over the years by a series of curators, had suffered from neglect. In 1838, Harvard librarian Thaddeus William Harris maneuvered to have physician Joshua Fisher's 1833 donation of $20,000 used to finance the appointment of a college botanist. Finally, in January of 1842, President Josiah Quincy sought to fill the position. Asa Gray, a New York botanist, who together with his mentor John Torrey had been elected to the Boston-based American Academy of Arts and Sciences during the preceding fall, was invited for a campus visit. There he met both Quincy and George Barrell Emerson, the president of the Boston Society of Natural History who, at the time, was the city's leading man of science. On April 30, 1842, the Harvard Corporation appointed Gray as the first

44. Morison, 244; Benjamin Rand, "Philosophical Instruction in Harvard University from 1636 to 1906," *Harvard Graduates' Magazine* 37 (Dec. 1928): 188–200, at 200.

45. Ahlstrom and Carey, 27.

46. George M. Marsden, *The Soul of the American University: From Protestant Establishment to Established Nonbelief* (New York and Oxford: Oxford University Press, 1994), 184.

47. See Richard Henry Dana, Jr., to James Dwight Dana, 17 Feb. 1856, where Richard reminisces about his student days at Harvard during the early 1830s (Dana Family Papers, Yale University). See also Turner, *Liberal Education*, 18, 42.

Fisher Professor of Natural History.[48] Even this was but a slow beginning to the growth of Harvard natural sciences. Gray could not yet write to Torrey, as he would in 1847, that Harvard science surpassed Princeton science.

Within a few years of his appointment to the Fisher Professorship, Gray, a New School, or liberal, Presbyterian, became formally assimilated in the Unitarian community by his engagement (1847) and marriage (1848) to one of their prominent daughters, Jane Lathrop Loring. Her father was one of Boston's leading lawyers and had served since 1835 on the Harvard Corporation. Gray's Presbyterianism did not include as strict an adherence to the Westminster Confession as the Old School Presbyterianism of Princeton. Indeed, Gray's pastor in New York, the Rev. Erskine Mason, had been one of the leading agitators for a more liberal understanding of the faith during years leading to the 1837 split between the Old and New Schools. At Harvard, Gray chose to attend a Congregationalist Church, where he could maintain his Trinitarian beliefs, but as a liberal Presbyterian-Congregationalist, he shared both the Unitarian emphasis upon God's benevolence and the corollary Unitarian discomfort with Calvinist predestination.[49] There is no record of any theological friction between Gray and the Unitarian community that had become his adopted family.[50] This should not be surprising, since, in terms of theological attitude, liberal Congregationalists and Unitarians had more in common with each other than either one of them did with Old School Presbyterians. Gray, it is true, professed a faith consistent with "the creed commonly known as the Nicene," but so did Andrew Preston Peabody, who insisted—in contradistinction with the Calvinists—that the creed denied the doctrine of the Trinity.[51] Though Gray himself remained Trinitarian, he found Harvard's spirit to be sufficiently ecumenical to tolerate a broader range of

48. Dupree, 104–113.

49. Dupree, 37, 44.

50. Dupree, 113.

51. Asa Gray, *Darwiniana: Essays and Reviews Pertaining to Darwinism* (1876; rpt., New York: D. Appleton and Company, 1884), vi. Some writers have interpreted this statement as an indication of Gray's "conservative" and "evangelical" outlook. See Fry and Fry, 1; Livingstone, *Darwin's Forgotten Defenders*, 61, 66. No one has adequately addressed the question of what Gray's Nicene faith meant to him, or how it related to the similarly stated convictions of Unitarians, such as James Walker and Andrew Preston Peabody, whose Nicene rhetoric is reported by Ahlstrom and Carey, 4. One possible reading of Gray's statement is that he was endorsing the empiricism of the Nicene Creed ("I look for the Resurrection of the dead") as compared to the fideism of the Apostles' Creed ("I believe . . . in the Resurrection of the Body"); see Dupree, 365–66.

theological views than one would find, for example, at Princeton. Gray, too, was tolerant of other faiths, apparently not feeling any injury to his marriage when his wife continued to attend Unitarian services.[52]

Gray met his future wife—could a Boston romance be more picturesque?—after delivering one of his Lowell lectures.[53] Gray had been quickly, though somewhat unwillingly, socialized into the Lowell Institution during his second academic year at Harvard. He neither asked to be a lecturer, nor particularly welcomed it, but instead acquiesced, when Lowell scheduled him for the winter of 1844, "because I thought it was my duty to do so," part of living out "a perfectly consistent Christian course."[54] Gray felt nervous beforehand[55] and was displeased with himself afterwards. He had repeated some points and omitted others, wincing when his eyes met the audience. At times he failed even to utter a coherent sentence.[56] When Lowell assigned him the same role for 1845, Gray was thankful at least that his lectures would come last in the series, thus giving him more time to prepare.[57] Though not exactly a Unitarian himself, Gray was about to become a leading spokesperson on *Vestiges* as he assumed the role of a public intellectual at Unitarian Harvard.

"Cutting up of Vestiges*":*
Asa Gray Slashes Vestiges *at the Lowell Institute*

Asa Gray felt "quite well and hearty" when preparing to begin his second series of Lowell lectures in February 1845,[58] but his memory of the difficult experience of his 1844 series prompted him once again to suppress newspaper coverage of the lectures.[59] Even as late as 1848, his sixth year as Harvard's botany

52. Dupree, 182.

53. Dupree, 177–81.

54. Asa Gray to John Torrey, 25 Mar. [1844], typescript (Asa Gray Papers, Gray Herbarium, Harvard University).

55. Asa Gray to John Torrey, 17 Feb. [1844], typescript (Herbarium).

56. Asa Gray to John Torrey, 1 Mar. [1844], typescript (Herbarium).

57. Asa Gray to John Torrey, 8 Aug. [1845], typescript (Herbarium).

58. Asa Gray to John Torrey, 12 Feb. 1845 (Torrey Papers).

59. Asa Gray to John Torrey, 8 Mar. 1845 (Torrey Papers); Gray to Torrey, 25 Mar. [1844] (Herbarium).

professor, Gray still had difficulty commanding the attention of his students.[60] He never became a great orator, but he did use what skills he could marshal to distinguish for his audience between *Vestiges* and legitimate natural history.

Gray practiced his opening lecture before his Congregationalist minister, the Rev. John A. Albro, and then wrote to Torrey that "half of it is devoted to a serving up of Vestiges of Creation (which [the London-based American physician and botanist Francis] Boott says is written by Sir Richard Vivian), showing that the objectional [*sic*] conclusions rest upon gratuitous and unwarranted inferences from established or probable facts."[61] Unfortunately, the manuscript of that first lecture is not extant and Gray's effort to avoid press coverage apparently succeeded, so for further insights as to Gray's treatment of *Vestiges* one must look to his later statements.[62]

In his second lecture, Gray emphasized two examples of divine purpose in nature, as revealed by the coal deposits in the Carboniferous strata. "The natural theologian," he said, "very properly points to this ancient vegetation and its results, as <u>designed</u> to supply the wants, and advance the civilisation [*sic*] of a predestined intelligent population."[63] Similar claims could be found in William Buckland's widely-read Bridgewater Treatise, *Geology and Mineralogy* (1836).[64] Gray further argued that the plants from which the coal had been formed had originally served the purpose of removing carbon from the atmosphere of the early earth, so that oxygen-breathing animals, including humans, would be able to inhabit the planet at a later time. The author of *Vestiges* had also suggested this,[65] attributing it, like Gray, to divine planning, but the God of Gray was soon to be distinguished from the Deity of *Vestiges*.

60. Jason Martin Gorham, Diary, 1848–1849 (Harvard University Archives), 25 and 30 May 1848; John Mead, Journal, 1848–1850 (Harvard University Archives), 30 Mar., 19 Apr., 30 May, and 29 Sep. 1848.

61. Asa Gray to John Torrey, 12 Feb. 1845 (Torrey Papers).

62. According to Gray, no reports had appeared in the papers as of early March. See Asa Gray to John Torrey, 8 Mar. 1845 (Torrey Papers). The Gray Herbarium at Harvard University lacks the first series (1844) of Gray's Lowell lectures and the first lecture (12 Feb. 1845) of his second series.

63. Gray, Lowell Lecture No. 2, [1845] (Gray Papers, Herbarium, Folder 34), 45.

64. William Buckland, *Geology and Mineralogy Considered with Reference to Natural Theology*, 2 vols. (London: William Pickering, 1836), esp. 1:97–102.

65. *Vestiges*, 54–60. The edition cited here, and of *Explanations*, below, is Secord's facsimile: Robert Chambers, *Vestiges of the Natural History of Creation and Other Evolutionary Writings, Including Facsimile Reproductions of the First Editions of "Vestiges" and Its Sequel "Explanations,"* ed. with an introduction by James A. Secord (Chicago: Chicago University Press, 1994).

By early March, Gray's voice was becoming hoarse, in part from the lectures and in part because he had caught a cold. Against these obstacles, Gray nevertheless could proudly write to Torrey on March 8 that "my cutting up of Vestiges of Creation was a fine blow."[66] Three days earlier, in his seventh lecture, Gray had completed a three-lecture series on fungi. He apologized for going over some of the details too quickly, and for skipping others altogether, but he wanted to reserve this last lecture on fungi for "the question of Spontaneous or Equivocal generation." Alluding to *Vestiges*, Gray noted the importance of this topic, "considering the use which is now-a-days attempted to be made of it." He suggested two reasons why spontaneous generation properly fits in a lecture concerning fungi. First, fungi furnish the chief evidence that advocates of spontaneous generation supply for their case. Second, Gray's own lectures on fungi effectively undermine the arguments for spontaneous generation.[67] He outlined the two rival positions thus:

> The doctrine is, that living beings,—whether plants or animals,—spring up <under ~~certain~~ various circumstances,> without previous seeds or germs.

> The opposite ground is, that all living beings are the offspring of other individuals, precisely similar to themselves,—to antecedent individuals of the same species;—and this ~~but be deemed~~ <law> ~~to~~ appl~~y~~ies to the lower, minuter [*sic*], and simpler, as well as to the ordinary plants and animals.[68]

To settle the issue, Gray presented an example from *Vestiges* in which white clover grew from moss overlaid with lime, even though the nearest patch of white clover was miles away. Gray suggested that the clover did not arise spontaneously, but rather grew from seeds that had been washed in by streams. "The only question is whether the seeds would preserve their vitality for a long time, so as to be ready to grow, whenever the soil <was fit> and <the> temperature are [*sic*] favorable."[69] The Harvard botanist then cited instances in which corn seed had produced a plant after thirty years of dormancy, rye seed had germinated after forty years, and melon seed after forty-one years. These

---

66. Asa Gray to John Torrey, 8 Mar. 1846 (Torrey Papers).
67. Gray, Lowell Lecture No. 7 (Herbarium), 15.
68. Gray, Lowell Lecture No. 7 (Herbarium), 16.
69. Gray, Lowell Lecture No. 7, [1845] (Herbarium), 18–19.

cases were nothing compared to his final examples: kidney beans, after an entire century of dormancy, had germinated into plants; and—could it be more startling?—wheat, mummified for millennia in Thebes, also "has been made to grow!" Gray conceded that in some cases, the Egyptian grain may have been placed in the tombs by Arabs and thus be only 10 or 100 years old, but he argued that other instances are so well documented by recent archaeological endeavors that the antiquity of the grain stores could not be seriously doubted. Moreover, some raspberry bushes had been grown from seed buried with coins of Emperor Hadrian. "This should be kept in mind," suggested Gray, "in explaining the cases which are adduced from these lower tribes, from Fungi particularly, in behalf of the doctrine of a casual or accidental origin."[70]

In late March, with his Lowell lectures now complete, Gray sat back once more to write his friend Torrey, and once more his pen scribbled concerning *Vestiges*. "The 'Vestiges' is full as possible of absurd and wrong statements of facts,—which the author did not clearly understand. Bowen has a long winded article on it for the next North American. This will keep me from touching it there."[71] Francis Bowen, a Harvard tutor, edited the *North American Review*, a venue where Boston's Unitarian elite addressed pressing questions of the day. The April issue spread his review of *Vestiges* across fifty-three pages. "Long-winded," Gray described it once more to Torrey, but one year later Gray himself would write nearly as much. Despite his expectation that Bowen's review "will keep me from touching it there," Gray would tackle both *Vestiges* and *Explanations* across forty-two pages of the April 1846 *North American*. In the meantime, he judged Bowen's review—parts of which had been read to him from the pre-print manuscript—as "very good in many respects,—faulty in others."[72]

"In fact no history at all": Francis Bowen Rejects
Vestiges' *Speculative Natural History*

Francis Bowen had not been born into one of Boston's elite families, but by the 1840s he exemplified the Unitarians' Puritan ethic of hard-working personal

70. Gray, Lowell Lecture No. 7, (Herbarium), 25.

71. Asa Gray to John Torrey, 30 Mar. 1845 (Torrey Papers).

72. Asa Gray to John Torrey, 7 May 1845 (Torrey Papers).

development. He had managed his way through Phillips Exeter Academy and was accepted into Harvard College. When he graduated in 1833, at age 22, he had attained the honor of *summa cum laude*. After graduation he tutored in mathematics at both Phillips Exeter and Harvard, and before long was contributing regularly to the *North American Review*. In 1843 he became editor of that Harvard Unitarian standard-bearer.[73]

As *Vestiges* was becoming known at Harvard in 1845, nine of Bowen's *North American* articles, spanning 1837 to 1842, were reprinted in a volume entitled *Critical Essays*. There Bowen spoke for the Unitarian establishment by warning that Immanuel Kant's opposition to David Hume's skepticism was itself but another path toward skepticism, since Kant rendered all knowledge subjective. In place of Kant's intuitions that mediated perceptions, inescapably insulating the mind from direct knowledge of things-in-themselves, Bowen desired a more straightforward account of observation and favored the Scottish Realist school: what one sees really exists just as it is seen. This inductive philosophy underpinned the Unitarian reliance on external evidences for establishing religious truths. Further promoting Unitarianism, Bowen critically reviewed Francis Wayland's *The Elements of Moral Science* (1836), which he found steeped in Calvinism. Wayland's arguments in favor of original sin, total depravity, and vicarious atonement were, ostensibly at least, drawn from Scripture, but Bowen insisted that those doctrines were too offensive to reason and justice to be accepted as correct interpretations of divine revelation.[74] From Bowen's pen, the internal evidence of the conscience protected Harvard from Calvinism, and the external evidence of inductive science protected Harvard from skepticism. Where would *Vestiges* fit into this framework?

"This is one of the most striking and ingenious scientific romances," Bowen began his review of *Vestiges*, the juxtaposition of "scientific" and "romantic" being a left-handed compliment. He praised the author's style, his "great ingenuity and correct taste," then commented that *Vestiges* presented

---

73. R. Douglass Geivett, "Bowen, Francis," *American National Biography*, ed. John A. Garraty and Mark C. Carnes, 24 vols. (New York: Oxford University Press, 1999), 3:276–77; Howe, *Unitarian Conscience*, 309–10.

74. Francis Bowen, *Critical Essays, on a Few Subjects Connected with the History and Present Condition of Speculative Philosophy*, 2d ed. (Boston: James Munroe and Company, 1845), 33–65 (Kant), 310–30 (Wayland).

"quite a general, but superficial, acquaintance with the sciences." Still, Bowen wanted to seem impartial, so he called for a careful investigation "so as to obtain a just idea" and "a just opinion" of "the theory as a whole."[75] About a third of the way through his article, Bowen felt his fair-minded approach had uncovered enough evidence to reach a firm conclusion. The tone of his article shifted from inquiry to argument:

> We shall endeavour [sic] to show that this hypothetical history of creation is not only faulty in every point, when viewed from the author's own ground, but, when examined in the proper direction, is absolutely unintelligible, or is in fact no history at all.[76]

Bowen's main charge against the author of *Vestiges* was the failure to follow the inductive method. For example, the so-called "higher law" that united *Vestiges*' three developmental phases (nebular, geological, and biological) into a single law of nature could, at best, be supported only by infrequent occurrences, such as the supposed spontaneous generation of life; a genuine law of nature would be based on regular events.[77] Rhetorically conceding the possibility of spontaneous generation (a generous concession in the wake of Gray's Lowell lectures), Bowen still found species transmutation, in the absence of observational evidence, to be vulnerable to ridicule: "No one will pretend that a dog, a horse, or a man can thus be created."[78]

Bowen especially objected to the author's argument from negative evidence, which he characterized thus: "The more perfect organisms have not been discovered in the earlier strata; *therefore*, they do not exist in them." He warned that new discoveries could undermine such an argument, and that geologists constantly are making new discoveries. "In truth, the researches of geologists are every day bringing to light new facts, which compel them to modify or abandon many of the positions they formerly held; so that a considerable portion of the science is a mere quicksand of shifting theories."[79] For example, Edward Hitchcock had recently found fossil footprints of "gigantic birds" in the

75. [Francis Bowen], "A Theory of Creation," *North American Review* 60 (Apr. 1845): 426–78, at 426–27.

76. [Bowen], 441.

77. [Bowen], 436.

78. [Bowen], 434.

Red Sandstone stratum of the Connecticut River Valley, thus indicating the existence of a higher type of avian than previously had been known for that geological era.[80]

Without desiring to discredit geology's reputation for revealing present-day facts about the earth's crust, Bowen cautioned that these facts did not lead to secure conclusions about the earth's past:

> When we come to the formation of theories respecting the past history of the earth, in order to account for the phenomena at present visible on its surface, we are evidently afloat on a sea of conjecture, each hypothesis being valid only till a more plausible one is proposed,—which happens very frequently,—or till it is effectually disproved by some new discovery in the rocky strata.[81]

From Bowen's standpoint, the author of *Vestiges* had committed the double methodological error of employing "hypothetical causes" ("the operation of law," "the energies of lifeless matter," etc.) to explain "hypothetical effects" (nebular development, spontaneous generation, and species transmutation).[82] *Vestiges* amounted to a science without facts, or, in other words, not a science at all. "This is the whole business of the student of nature," affirmed Bowen, "to place together the results which are so similar, that we may attribute them to a common cause, without assuming to know what that cause is. The sole office of science is the theory, not of causation, but of classification. It is all reducible to

---

79. [Bowen], 446. Bowen likely had in mind the intense disagreements among both British and American geologists, during the 1830s and 1840s, concerning issues such as: whether the same strata sequences are found across the globe, or vary by location; whether major fossil kinds (classes, orders) are distributed sequentially within the strata; whether minor fossil kinds (families, genera, species) are distributed sequentially within the strata containing their classes and orders; and, whether any of those sequences indicate progress from simpler to more complex forms. For a general background, see Peter J. Bowler, *Fossils and Progress: Paleontology and the Idea of Progressive Evolution in the Nineteenth Century* (New York: Science History Publications, 1976), esp. chaps. 3–5. Specific controversies are addressed in Martin J. S. Rudwick, *The Great Devonian Controversy: The Shaping of Scientific Knowledge among Gentlemanly Specialists* (Chicago: University of Chicago Press, 1985); and James A. Secord, *Controversy in Victorian Geology: The Cambrian-Silurian Dispute* (Princeton, NJ: Princeton University Press, 1986).

80. [Bowen], 447.

81. [Bowen], 447–48.

82. [Bowen], 466.

natural history, the essence of which consists in arrangement."[83] In a similar vein, T. Wentworth Higginson's lecture notebook for Harvard Divinity School quoted from Lord Brougham that, "'The fundamental rule of inductive evidence is that no hypothesis shall be admitted—that nothing shall be assumed merely because if true it would explain the facts.' It must be the only thing that will."[84]

Bowen represented an extreme inductivist position that few critics of *Vestiges* shared. Princeton's Joseph Henry, for example, wrote to Charleston naturalist Lewis Reeves Gibbes in May 1845 that Bowen's flat inductivism frustrated the progress of science.[85] Whereas Henry taught his students at Princeton that natural science required the careful use of hypotheses, and that evidence could confirm or disconfirm a hypothesis, Bowen was suggesting that good geologists would not form any hypotheses, since the only "confirmation" that favorable evidence can provide is the empty hope that no contradictory evidence will be discovered in the future.

Science for Bowen and others at Harvard consisted of an arrangement of evidences, both external and internal, without any imposition of hypotheses. In addition to the external evidence of geology, Bowen also brought the internal evidence of the conscience to bear against *Vestiges*. Just as the God professed by Calvinists did not live up to humanly discernible rational standards of divine benevolence and justice, so also the Deity presented in *Vestiges* fell short of the Unitarian ideal. Bowen objected to the implication he found in *Vestiges* that the Creator was not powerful enough to create by a means other than a developmental law that allowed him to rest while the universe formed itself and brought forth its own inhabitants.[86]

When Joshua, son of Nun, had reached the end of his life, he addressed the Children of Israel, whom he had led into the Promised Land. "Choose ye this day whom ye will serve . . . but as for me and my house, we will serve the Lord."[87]

83. [Bowen], 467.

84. T. Wentworth Higginson, Notes in Divinity School, Natural Theology, Sep. 1844 (Harvard University Archives, HUC 8847.386), 5.

85. Joseph Henry to Lewis R. Gibbes, 31 May 1845, in *Papers of Joseph Henry*, ed. Nathan Reingold (vols. 1–5) and Marc Rothenberg (vols. 6–) *et al.*, 8 vols. to date (Washington, D.C.: Smithsonian Institution Press, 1972–1998), 6:280–84.

86. [Bowen], 475.

87. Joshua 24:15, King James Version.

Bowen concluded his review of *Vestiges* with a similar appeal: "Choose ye, then, . . . between an intelligent being and a stone, for the parentage and support of this wonderful system. For our own part, we will adopt the conclusion of one of the most eloquent of those old pagan philosophers"—and he closed with a quotation from *De natura deorum,* where Cicero gave a caricature of Greek atomism: "What if a swirling of atoms can construct a world, why not a portico, why not a temple, why not a town, why not a city . . . ?"[88] The author of *Vestiges* had employed such a rhetorical question in order to argue from the probability of the nebular hypothesis toward a development theory that would encompass Victorian man and his civilization; coming at the close of Bowen's article, however, this quotation served an opposite purpose. It was the improbability of species transmutation that rendered the entire scheme of development improbable as well. *Vestiges* relied on "pure speculation and hypothesis." It was not science.[89]

Asa Gray's Dialogue with the Author of Vestiges

As Gray continued to build a career in science at Harvard, he repeatedly felt the need to respond to *Vestiges*. From Gray's standpoint, the work retarded the progress of research in natural history and scandalized the natural sciences in general. In February 1846, James Dwight Dana, a Yale geologist whose views will be treated in chapter 4, suggested to Gray that *Explanations,* the recently published sequel to *Vestiges,* "is a splendid book for a reviewer who might wish to give the author a trimming."[90] Gray soon began the task, receiving further encouragement from both Dana and also the North Carolina botanist Moses Ashley Curtis. Curtis looked to Gray as just the sort of scientific leader who could "slash at the Vestiges" with authority. "I feel that his system is wrong," he wrote to Gray, "though it would not be easy for me to give it a thorough logical refutation, because I have not science enough."[91]

88. [Bowen], 477–78, my translation. Bowen quoted the Latin from Cicero, *De natura deorum,* bk. II, chap. XXXVII, paras. 93–94.

89. [Bowen], 439.

90. James Dana to Asa Gray, 18 Feb. 1846 (Gray Papers, Herbarium).

91. James Dana to Asa Gray, 21 Mar. 1846 (Gray Papers, Herbarium); M[oses] A[shley] Curtis to Asa Gray, 21 Mar. 1846 (Gray Papers, Herbarium). Several other Curtises were known to Gray; this letter has

"Pen, ink, and paper lie in readiness before us," wrote Gray in the spring of 1846, "but no writing yet appears." Writer's block testifies that pen, ink, and paper do not suffice to produce words on a page; an author's genius must arise to direct those implements. Himself the author of a leading textbook on botany, Gray filled forty-one pages in the *North American Review* in response to *Vestiges* and its rebuttal sequel, *Explanations*. In preparation for this public response to the development hypothesis, Gray first privately put pen to paper in the margins of his personal copies of *Vestiges* and *Explanations*. That he filled so many pages with scribbled replies to the author suggests that Gray himself did not suffer from writer's block. Rather, his statement had a rhetorical aim: just as a text requires an author for its production, so also only a creative Author, "*a supreme will, acting through time*," can account for nature's development.[92]

Gray's review may be considered a public reproduction of his private conversation with the author of *Vestiges*, an author-to-author dialogue concerning the Author of nature. Historian James Secord has concluded that Cambridge University geologist Adam Sedgwick's detailed annotations in his copy of *Vestiges* "intervened in the printed text, softening it up for criticism. In this way, the writing of a learned commentary . . . began in the very process of reading."[93] Similarly, Gray's reading of *Vestiges* served as a rehearsal for his preparation of a formal, published review. "Bah!" he penciled into the margin beside the author's claim that "it seems hardly conceivable that rational men should give an adherence to such a doctrine" as that of successive *fiat* creations— a doctrine to which numerous men of science, Gray included, had given their assent.[94] "Must quote fully," he scribbled on the flyleaf under a reference to a passage on "P. 95" in *Explanations*, and the statements marked there appeared

been identified with Gray's botanical collaborator Moses Ashley Curtis both by the initials in the signature and because it bears the place name "Hillsborough[,] NC," which is where Moses Ashley Curtis was based, according to Herbert Hovenkamp, *Science and Religion in America, 1800–1860* (Philadelphia: University of Pennsylvania Press, 1978), 181.

92. [Asa Gray], "Explanations: A Sequel to 'Vestiges of the Natural History of Creation,'" *North American Review* 62 (Apr. 1846): 465–506, at 475.

93. James A. Secord, *Victorian Sensation: The Extraordinary Publication, Reception, and Secret Authorship of "Vestiges of the Natural History of Creation"* (Chicago: University of Chicago Press, 2000), 239.

94. Gray's copy of *Explanations*, 108. (Gray's copies of *Vestiges*, cited below, and *Explanations* are both held at the Gray Herbarium, Harvard University.)

on page 467 of his review.[95] But far more often than writing a note to himself, he wrote a response to the author of *Vestiges*, entering, as it were, into a dialogue with book before him.

Gray had read *Vestiges* when preparing the Lowell lectures in which he debunked spontaneous generation. The following dialogue prepared him for that occasion, and latter was put to use in his review of *Vestiges* and *Explanations*:

> Vestiges: All animated nature may be said to be based on this mode of origin[:] *the fundamental form of organic being is a globule, having a new globule forming within itself,* by which it is in time discharged, and which is again followed by another and another, in endless succession. It is of course obvious that, if these globules could be produced by any process from inorganic elements, we should be entitled to say that the fact of a transit from the inorganic into the organic had been witnessed in that instance; the possibility of the commencement of animated creation by the ordinary laws of nature might be considered established.[96]

> Gray: no, this is too limited a view[;] not the proper expression[97]

> Vestiges: Now the chemist, by the association of two parts oxygen, four hydrogen, two carbon, and two nitrogen, can *make urea*.[98]

> Gray: make something like <u>urea</u> [circling "urea" in *Vestiges*' statement, Gray drew a line to the bottom margin and added:] is a product of decomposition [as if to emphasize that urea cannot generate life][99]

95. Gray's copy of *Explanations*, flyleaf, 94–95; [Gray], 467. The passage dealt with the question of whether a natural law planned by the Deity could adequately substitute for a "special exertion" of the Deity. Gray also debated the author of *Vestiges* regarding this issue on several other pages, some of which are discussed below.

96. *Vestiges*, 170–73 (italics original, and likewise for the subsequent quotations from *Vestiges* and *Explanations* in this section). Though paginated differently, Gray's copies of these works have identical language as the Secord reprint cited here, at least as to the specific passages presently under discussion.

97. Gray's copy of *Vestiges*, 132.

98. *Vestiges,* 168.

99. Gray's copy of *Vestiges*, 128.

VESTIGES: The whole train of animated beings, from the simplest and the oldest up to the highest and the most recent, are, then, to be regarded as a series of *advances of the principle of development*. ... The nucleated vesicle, the fundamental form of all organization, we must regard as the meeting-point between the inorganic and the organic—the end of the mineral and the beginning of the vegetable and animal kingdoms.[100]

GRAY: [pondering where this argument might lead, Gray penciled a question mark in the margin, then crossed it out and rendered his judgment:] denied[101]

VESTIGES: We have already seen that this nucleated vesicle is itself a type of mature and independent being in the infusory animalcules, as well as the starting point of the foetal progress of every higher individual in creation, both animal and vegetable. We have seen that is a form of being which electric agency will produce—[102]

GRAY: [interrupting:] denied[103]

VESTIGES: —though not perhaps usher into full life—in albumen, one of those compound elements of animal bodies, of which [Gray underlined:] another (urea) has been made by artificial means.[104]

GRAY: denied[105]

Gray had equally strong objections to the author's argument for species transmutation:

VESTIGES: Where a special function is required for particular circumstances, nature has provided for it, not by a new organ, but by a modification of a common one, which she has effected in development. Thus, for instance, some plants

100. *Vestiges,* 203–4 (ellipsis indicates additional text present in the original).

101. Gray's copy of *Vestiges,* 154.

102. *Vestiges,* 204.

103. Gray's copy of *Vestiges,* 154.

104. *Vestiges,* 204.

105. Gray's copy of *Vestiges,* 154.

destined to live in arid situations, require to have a store of water which they may slowly absorb. The need is arranged for by a cup-like expansion round the stalk, in which water remains after a shower. Now the *pitcher*, as this is called, is not a new organ, but simply a metamorphose of a leaf.[106]

GRAY: [in the margin:] not so! [to which the Harvard botanist added at the bottom of the page:] none of them grow in dry or arid situations.[107]

Explanations offered a rebuttal to Adam Sedgwick, who had objected to species transmutation when reviewing *Vestiges* for the *Edinburgh Review*.[108] Gray, who sided with Sedgwick, entered into a debate with *Explanations*, making a particular effort to show that his own specialty, botany, in no way lent support to the development theory:

EXPLANATIONS: The objection of the Edinburgh reviewer, to the alleged transmutation of oats into rye, is that he believes it a fable. ... Let us see, on the other hand, what a greater authority on botanical subjects than he—namely, Dr. Lindley —has stated on the same subject.[109]

GRAY: Dr Lindley ought to be more careful,—that is all—[110]

EXPLANATIONS: The learned writer [Lindley] averts to the "extraordinary, but certain fact, that in orchidaceous plants, forms just as different as wheat, barley, rye, and oats, have been proved by the most rigorous evidence, to be accidental variations of one common form, brought about no one knows how, but before our eyes, and rendered permanent by equally mysterious agency. . . ."[111]

GRAY: denied! give this story better—[112]

106. *Vestiges,* 197.

107. Gray's copy of *Vestiges,* 149.

108. [Adam Sedgwick], Review of *Vestiges of the Natural History of Creation, Edinburgh Review* 82 (July 1845): 1–85; facsimile rpt. in *"Vestiges" and the Debate before Darwin,* ed. John M. Lynch, 7 vols. (Bristol: Thoemmes, 2000), vol. 1 (original pagination retained).

109. Gray's copy of *Explanations,* 78.

110. Gray's copy of *Explanations,* 78.

111. *Explanations,* 111–12.

112. Gray's copy of *Explanations,* 79.

The author of *Vestiges* seemed ignorant not only of botany, but also of American race relations:

> VESTIGES: The few African nations which possess any civilization also exhibit forms approaching the European; and when the same people in the United States have enjoyed a within-door life for several generations, they [Gray underlined:] assimilate to the whites among whom they live.[113]

> GRAY: !!![114]

Being an amateur, in Gray's assessment, the author of *Vestiges* was neither correct nor even novel. To Gray, *Vestiges* was merely popularizing the folly of an earlier writer, Jean-Baptiste Lamarck. Where the author of *Vestiges* wrote, "Early in this century, M. Lamarck, a naturalist of the highest character, suggested an hypothesis of organic progress which deservedly incurred much ridicule—," Gray interrupted: "It contained, essentially, all your book!"[115]

Like many naturalists, Gray insisted that species originated by a different means than they were propagated; he resisted the Lamarckian effort by the author of *Vestiges* to conflate these two occurrences into a single process of "development":

> GRAY: The point is that <the> acting cause we recognize, will which account[s] for their propagation, etc. will not account for their origin.—that we look back therefore & infer some other cause, or act, originating.[116]

> EXPLANATIONS: *A time when there was no life* is first seen. We then *see life begin, and go on*; but whole ages elapsed before man came to crown the work of nature. . . . [Gray wrote in the margin: "cite"] . . . The great fact established by it is, that the organic creation, as we now see it, [Gray underlined:] was not placed upon the earth at once [Gray wrote in the margin: "true"]; it observed a PROGRESS.[117]

113. *Vestiges*, 217.

114. Gray's copy of *Vestiges*, 164.

115. *Vestiges*, 230; Gray's copy of *Vestiges*, 174.

116. Gray, marginalia in *Explanations*, 19.

117. Gray's copy of *Explanations*, 21.

GRAY: [now summarizing his thoughts at the bottom of the page:] So the same <u>will</u>, [emphatic about the Creator's repeated intervention:] <u>afterwards</u> created ~~the~~ such & <u>such</u> species & placed them subject to fixed laws[118]

EXPLANATIONS: Now we can *imagine* the Deity calling a young plant or animal into existence instantaneously; but we see that he does not usually do so. . . . Here we have the first hint of organic creation having arisen in the manner of natural order.[119]

GRAY: false . . . all creation—we can conceive <u>must</u> have been instantaneous[120]

For the *North American Review*, Gray reproduced the statement from *Explanations* about organic creation not occurring all at once, and not occurring instantaneously even in its various stages. The reviewer then explained to his readers the metaphysical dilemma that species transmutation would involve:

That different sorts of plants and animals were not *all* placed on the earth at once, we willingly admit. No doubt, they were created at different periods. But the notion, that the Deity does not create a young plant or animal *instantaneously*, is quite new to us. Is, then, the animal in question for a while in a sort of limbo somewhere between *esse* [being] and *non esse* [not being], a *tertium quid* [third thing], neither create nor uncreate . . . ?[121]

Meanwhile, his dialogue in the margins continued:

EXPLANATIONS: There are many other facts that throw a strong light on transmutation, both of plants and animals. So far from there being any decisive proof against this theory, there is no settled conclusion at this moment amongst naturalists, as to what *constitutes a species*.[122]

---

118. Gray's copy of *Explanations*, 21.

119. Gray's copy of *Explanations*, 21–22.

120. Gray's copy of *Explanations*, 22.

121. [Gray], 473.

122. Gray's copy of *Explanations*, 79.

GRAY: But as Mrs. Candle says, "What has that to do with it?"[123]

Within a decade, the definition of "species" would become a central problem for naturalists in both America and Europe. Gray would find himself writing marginal notations on this in an article by James Dwight Dana, "Thoughts on Species," in 1857—to be discussed in chapter 4—and in a book by Charles Darwin, *Origin of Species*, in 1859.[124] His view on species would itself undergo an evolution, and he no longer would insist upon species fixity. Darwin would build a case for species change on the fact that naturalists themselves could not clearly distinguish one species from another, or define a limit of variation that would prevent one species from becoming another. More and more naturalists would begin to wonder whether what they were studying were not fixed kinds at all, but transient species in the process of evolving.

But while reading *Explanations* in 1846, Gray did not see how naturalists' disagreements over species classification could in any way imply that species themselves were not fixed natural kinds.[125] "Spontaneous generation and species transmutation were purely scientific question[s]," he informed his *North American* readers, "to be settled by observation and just inference."[126] Observation and inference—the hallmarks of inductive science—did not support the development theory. To say otherwise at Harvard in the wake of Gray's 1845 Lowell lectures was to speculate contrary to the facts. And now he had reinforced that point in a contribution to the *North American Review*.

123. Gray's copy of *Explanations*, 79.

124. For Dana's article, see James Dwight Dana, "Thoughts on Species," *American Journal of Science* 2d ser., 24 (Nov. 1857): 305–16, at 305. The copy bearing Gray's marginalia has not been located, but Gray did mention his mark-ups, and reproduced many of them, in Asa Gray to James D. Dana, 7 Nov. 1857 (Asa Gray Correspondence, Yale). Their correspondence will be discussed in chapter 4. For Gray's marginalia in Darwin's *Origin of Species*, see the copy available at the Asa Gray Herbarium, Harvard University.

125. Compare this to his later, more Darwinian, views in Gray, *Darwiniana*, 90.

126. [Gray], 484.

The Shibboleth of Speculative Geology:
Blocking a Vestiges Sympathizer from the Faculty

After Gray delivered his fourth Lowell lecture in a new twelve-part series for the winter of 1846, he confided in a note to Torrey, "(Entre nous, from what I learned yesterday from one of our Corporation, I have no doubt but that the [perhaps he was about to write "the snake"[127]] Rogers—thanks to our prompt and decided action—is thrown out of the question for the Rumford Professorship. ...)"[128] The Rumford Professorship and Lectureship on the Application of Science to the Useful Arts was endowed in 1816 by Benjamin Thompson (Count Rumford), a Massachusetts-born experimentalist in physical science who led a successful career of scientific and military pursuits in England and Bavaria.[129] When the Rumford chair was vacated early in 1845 by the resignation of Prof. Daniel Treadwell, it took the university nearly two years to fill the position, in part due to a one candidate's controversial association with Vestiges.

The "Rogers" to whom Gray was referring was Henry Darwin Rogers. He had directed the Pennsylvania Geological Survey in the 1830s and played an active role in the Association of American Geologists and Naturalists, a scientific society that would reorganize itself, in 1848, as the American Association for the Advancement of Science. Rogers was professor of geology at the University of Pennsylvania and also had lectured at the Franklin Institute, both in Philadelphia. Today his surname is inscribed in large Roman letters above a series of neoclassical columns several blocks east of Harvard, at the Massachusetts Institute of Technology, an institution he co-founded with his brother, William Barton Rogers, in 1860. Fifteen years earlier, however, he was

127. The Rogers brothers had a reputation of being "snakes," that is, career opportunists "who go about prowling among the nooks & corners of the land, ever on the look out for a warm spot in which they may locate themselves." See Jeffries Wyman to Morrill Wyman, 21 Dec. 1845 (Wyman Papers, Countway Medical Historical Library, Harvard University).

128. Asa Gray to John Torrey, 26 Jan. 1846 (Torrey Papers).

129. Sanborn C. Brown, "Thompson, Benjamin (Count Rumford)," *Dictionary of Scientific Biography*, ed. Charles Coulston Gillispie (New York: Charles Scribner's Sons, 1976), vol. 13, pp. 350–51; Richard Alan Shapiro, "The Rumford Professorship: An Analysis of the Development of Practical Science in Nineteenth-century America," B.A. thesis, History of Science, Harvard University, Mar. 1985 (Harvard University Archives), 2.

vying for the Rumford Professorship at Harvard; unfortunately for his career, Rogers had expressed sympathy for *Vestiges* while Gray was on the watch.

Exactly what Rogers's views on *Vestiges* were, one cannot easily determine. His recent biographer, Patsy Gerstner, told of "his support of the idea of organic evolution," indicating that Rogers "lectured on the subject of evolution from time to time" and, after delivering his Lowell lectures in January 1845, acquired a reputation in Boston "as a spokesman for the theory." Gerstner speculated that Rogers first learned of the possibility of organic evolution from his father, who so admired Erasmus Darwin that he named his son Henry *Darwin* Rogers.[130] In the absence of any manuscripts of Rogers's unpublished Lowell lectures,[131] the only statement from his hand connecting his views to *Vestiges* is a letter he wrote from Boston to his brother William Barton Rogers on January 24, 1845:

> Have you seen the work just republished by Wiley and Putnam, "Vestiges of the Natural History of Creation"? It contains many of the loftiest speculative views in Astronomy and Geology and Natural History, and singularly accords with views sketched by me at times in my lectures.[132]

Though more open to theory than Bowen, Gray did not consider "speculative views" to be "lofty," at least not in any complimentary sense. After mentioning "my cutting up of Vestiges" to Torrey, referring to his critique of spontaneous generation in the lecture series that immediately followed Rogers's, Gray added, "Peirce, who you know was rather inclined to favor Rogers, a while

130. Patsy Gerstner, *Henry Darwin Rogers, 1808–1866: American Geologist* (Tuscaloosa: University of Alabama Press, 1994), 142–43. Erasmus Darwin, the grandfather of the family's best-known proponent of evolution, Charles Darwin, had suggested organic evolution in his *Zoonomia; or, The Laws of Organic Life*, which appeared in 1794–1796. At p. 143, Gerstner admits that Darwin's *Zoonomia* touched upon evolution only in one chapter. One must remain open to the possibility that Rogers's father admired Darwin for reasons other than evolutionism. For a comparison of the theories presented by Erasmus Darwin, Jean Baptiste de Lamarck, and Charles Darwin, see James Harrison, "Erasmus Darwin's Views of Evolution," *Journal of the History of Ideas* 32 (1971): 247–64.

131. Gerstner notes that no manuscripts from Rogers's Lowell lectures of January 1845 are extant. See Gerstner, 135–36. Likewise, no reports of Rogers's lectures appeared in the *Boston Daily Advertiser* for that month.

132. H. D. Rogers to Wm. B. Rogers, Boston, 24 Jan. 1845, in *Life and Letters of William Barton Rogers*, ed. Emma Rogers, 2 vols. (Boston and New York: Houghton, Mifflin, and Company, 1896), 1:238–39.

ago, is now sound and strong."[133] Exactly what Peirce had previously favored about Rogers, Gray did not specify. Gerstner interprets Gray's statement to mean that "he had persuaded [Harvard mathematics professor] Benjamin Peirce to stop supporting Rogers's position on evolution."[134] This "position on evolution" may have been an outright endorsement of spontaneous generation and species transmutation, but perhaps Rogers (and, with him, Peirce) had a less dogmatic position, suggesting only the *possibility* of these occurrences in nature, or the appropriateness of an open inquiry on the matter.[135] In any case, even the possibility of spontaneous generation or species transmutation was viewed by Gray and Bowen as too speculative, since the available evidence pressed so consistently against it.

Despite his claimed alliance with Peirce, Gray found that the Harvard Corporation continued to consider Rogers a capable candidate for the Rumford post early in 1846.[136] He wrote to Torrey on January 26 that a member of the

133. Asa Gray to John Torrey, 8 Mar. 1845 (Torrey Papers).

134. Gerstner, 145.

135. At the May 1845 meeting of the Association of American Geologists and Naturalists in New Haven, Rogers reportedly admitted that *Vestiges* was laden with "erroneous and speculative views," but also thought the assembly should grant open consideration to "the sublime and glorious views it unfolded of creation." Samuel Haldeman, who had worked under Rogers on the Pennsylvania State Geological Survey, supported him on this point, noting that both *Vestiges* and its critics relied upon "premises that were not entirely and unanimously admitted by scientific men." The consensus, however, was to drop the issue from discussion. See "Association of American Geologists and Naturalists," *New York Herald*, 9 May 1845, 1. See also Gerstner, 145; Kohlstedt, 74. This AAGN meeting will be revisited from the perspective of Yale science in chapter 4.

136. Gray's efforts to block Rogers may have been forestalled by a sudden concern among Harvard's administration and trustees that the college was being black-listed as a dogmatic, sectarian institution. The debate over religious freedom in the academy had been intensifying for a few years, as a Democratic majority took control of the Massachusetts legislature in 1843 and appointed two Calvinist ministers, one Calvinist layman, and the Jacksonian George Bancroft to the Harvard Board of Overseers during 1843–1844. In February of 1845, when Gray was lecturing against *Vestiges* at the Lowell Institute, Bancroft presented a minority report to the Overseers, accusing the Unitarian-dominated college of theological sectarianism and aristocratic exclusivism. See Morison, 258, 387. Responding to Bancroft before the Board of Overseers in the spring of 1845, President Quincy emphasized that Unitarians "insist on freedom from creeds of men's intervention, and independence of all human dictation in the articles of their faith; maintaining the right of every man to search the Scriptures for himself . . . unbiased by party names and technical dogmas." See Josiah Quincy, *Speech of Josiah Quincy, President of Harvard University, Before the Board of Overseers of That Institution, February 25, 1845, on the Minority Report of the Committee of Visitation, Presented to That Board by George Bancroft, Esq., February 6, 1845* (Boston: Charles C. Little and James Brown, 1845), 48. Given this climate, Rogers's openmindedness to the development theory may have been regarded as an asset, or at least not as so grave a liability as Gray wanted to portray it.

Corporation had promised at last that Rogers would be blocked.[137] During February and March, however, the Corporation received at least fifteen letters of recommendation endorsing Rogers.[138] Gray's confidence, therefore, may have been premature. Peirce, too, had approached a member of the Corporation in January, urgently maneuvering to have the vote delayed. He feared that Rogers's "literary friends" in Boston had won the Corporation over. On January 29, he wrote to Alexander Dallas Bache to announce that he had stalled the vote by a week. He asked Bache to suggest the most scientifically capable man for the post, and confided in him his plan to establish a scientific school at Harvard that would be directed by the new Rumford professor.[139]

By January 31, 1846, Jeffries Wyman, of Hampden-Sydney College in Virginia, realized that he was no longer seriously in the running for the Rumford,[140] but what the Corporation would do with Rogers remained to be decided. On February 2, Bache replied to Peirce, who forwarded the letter to the Corporation: he had known Rogers for fifteen years as a distinguished geologist, but the Rumford chair (especially given Peirce's higher visions for it) also required expertise in chemistry, botany, and mechanics.[141] Other letters, however, suggested the opposite. Franklin Bache, a professor of chemistry at Jefferson Medical College in Philadelphia, wrote that Rogers was especially capable in chemistry.[142] J. W. Bailey, West Point Professor of Chemistry, Mineralogy, and Geology, also expressed a "<u>very</u> high opinion" of Rogers, attesting to the

137. Asa Gray to John Torrey, 26 Jan. 1846 (Torrey Papers).

138. These may be found in Letters to the Treasurer, vol. X, 1846–1848 (Harvard University Archives).

139. Benjamin Peirce to A. D. Bache, 29 Jan. 1846 (Peirce Papers, Houghton Library, Harvard University). Peirce had been promoting, ever since he joined the faculty in 1831, a stronger curriculum in applied mathematics and natural science. His own students often felt overwhelmed by the mathematical challenges Peirce placed before them. In 1838, he suggested that the topic be made an elective in order that his energies could be focused upon students who would use mathematics in their careers. By 1843, Peirce was calling for an entire reorganization of the natural sciences faculty, including his botanical colleague Gray, into a separate school of applied science and engineering. Treadwell was suffering from poor health and expected to retire shortly from the Rumford chair; when he did retire in May 1845, Peirce was hopeful that a school of applied science could be established, with the new Rumford chair as the director. See Mary Ann James, "Engineering an Environment for Change: Bigelow, Peirce, and Early Nineteenth-Century Practical Education at Harvard," in *Science at Harvard University: Historical Perspectives*, ed. Clark A. Elliott and Margaret W. Rossiter (Bethlehem: Lehigh University Press, 1992), 55–75, at 63–69.

140. Jeffries Wyman to Elizabeth Wyman, 31 Jan. 1846 (Wyman Papers).

141. A. D. Bache to [Benjamin] Peirce, 2 Feb. 1846 (Letters to the Treasurer).

142. Franklin Bache to William McIlvaine, 10 Feb. 1846 (Letters to the Treasurer).

candidate's ample qualifications in mineralogy, chemistry, natural history, mathematics, and physics.[143] Judging from the remainder of the correspondence received by the Corporation, the decision pivoted not so much on scientific qualifications, but rather on social standing and personal character, with each of these being intertwined with a key watchword during the era of *Vestiges*: "speculation."

Boston attorney George Hillard, whose family had recently merged with Rogers's through a relative's marriage, cautioned the Corporation not to heed the "cliques" in Philadelphia that opposed Rogers. He warned of A. D. Bache in particular: "Any opinion, therefore, directly or indirectly proceeding from him, should be received with caution." In Hillard's view, geology was such a young science that "to generalize is a sort of necessity," so any accusations about Rogers having a "habit of hasty generalization" should not be taken too seriously. Hillard classified Rogers in the "continental school of science" that makes bolder hypotheses than the "English school," which is constrained by the "established Church."[144] Amos Binney, a zoologist who had founded the Boston Society of Natural History in 1830, also supported Rogers's "enlarged and philosophical generalizations" in geology. Rogers's local supporters attracted letters from afar, including one from Philadelphia by William McIlvaine. Having attended Rogers's lectures during the 1830s, McIlvaine could attest that they were "entirely free from loose generalizations and false inductions." He dismissed the charge that Rogers "is apt to be led away, by a fondness of theorizing," as a partisan move by "the selfish and the envious . . . cliques" who opposed Rogers since he refused to join their "dishonest designs."[145]

Emphasizing that Rogers was an outsider when it came to scientific cliques, his supporters defended his moral character. McIlvaine wrote that Rogers's lectures "inculcated the profoundest reverence for Deity, and have most impressively cherished, on every occasion, those exalted sentiment[s] of <u>religion</u> & morality which are characteristic of pure Christianity."[146] At Harvard, of course, "pure Christianity" meant something different than it did elsewhere. George

143. J. W. Bailey to Amos Binney, 15 Feb. 1846 (Letters to the Treasurer).

144. George S. Hillard to R. R. Curtis, [ca. 12 Feb. 1846] (Letters to the Treasurer).

145. William McIlvaine to John A. Lowell, 17 Feb. 1846 (Letters to the Treasurer).

146. William McIlvaine to John A. Lowell, 17 Feb. 1846 (Letters to the Treasurer).

Emerson, president of the Boston Society of Natural History, made clear that Rogers "was easily disgusted with the dogmas and exclusiveness of presbyterianism"—a disgust generally shared among the Harvard Corporation. Like a good Unitarian, Rogers was non-credal, benevolent, and charitable, "and he certainly holds to the morality of the Gospel."[147]

Though Harvard did not require any specific profession of faith by its professors, holding to "the morality of the Gospel" in a nonsectarian way functioned as an unofficial religious test. Beginning in mid-February, a wave of letters reached members of the Harvard Corporation in support of a new candidate: Eben Norton Horsford, director of the New York State Geological Survey. Horsford had been recommended by Harvard anatomy professor John White Webster, who had experienced a falling out with Rogers about two years prior.[148] Those endorsing Horsford certified him as "a highly moral & religious man" and emphasized that his family were Unitarian relatives of Andrews Norton.[149] Terms like "character" and "gentleman" appeared frequently in letters recommending Horsford, at times combined into a phrase such as "a gentleman of great moral worth."[150] As was made apparent also in the letters recommending Rogers, especially the one by George Emerson mentioned above, scientific expertise and reputable moral character were expected to go hand-in-hand at Harvard, where true religion meant little more than gentlemanly character, and gentlemanly character required an aversion to Calvinist dogma on the one hand, and a guarded stance against skepticism on the other.[151]

With so many uncertainties voiced concerning Rogers, yet with several important members of the Harvard community still defending him (including George Emerson and John Amory Lowell), trustee Thomas Eliot recommended that the matter be tabled.[152] Lowell invited Rogers to lecture again at his institute

147. George B. Emerson to J. A. Eliot, 27 Feb. 1846 (Letters to the Treasurer).

148. Gerstner, 146.

149. Webster to Samuel A. Eliot, 18 Feb. 1846 (Letters to the Treasurer).

150. C. Bronson to [unknown], 21 Feb. 1846; Amos Dean to E. Everett, 21 Feb. 1846; A. Crittendon to Corporation of Harvard College, 23 Mar. 1846 (Letters to the Treasurer).

151. For a discussion of the importance of "character" at Harvard and how it related to Jeffries Wyman's career (though he was denied the Rumford, he did eventually receive another post at Harvard), see Appel, "A Scientific Career in the Age of Character."

152. Gerstner, 147.

in October 1846. Meanwhile, Webster continued to campaign against Rogers, throwing his support to Horsford, who was at the time studying chemistry with Justus von Liebig in Europe. What had begun with Gray's concern over speculative geology turned into a prolonged dispute between factions beyond Gray's control, but these ultimately served Gray's needs. When Horsford returned from Europe—he had been an absentee candidate all this time—the Corporation found him personable and appointed him as Rumford Professor in early 1847, finally putting the Rogers matter to rest.[153]

Reporting the news to Torrey, Gray added that "since Agassiz has so unsparingly & thoroughly cut up Lamarckian Vestiges theories, Rogers & friends have found it was high time to disclaim them." Gray suspected that Rogers had done this in hopes of recovering his candidacy for the position, but in any case the maneuvering came too late; Rogers soon was "taken aback by the sudden appointment of another man."[154] Meanwhile, Louis Agassiz, a visiting Swiss naturalist who had pleased Gray by "cutting up" Rogers's "Lamarckian Vestiges theories," was becoming the center of attention at Harvard and beyond.

Louis Agassiz's Lowell Lectures: God's Thoughts against the Development Theory

Historian Robert Bruce has marked Louis Agassiz's arrival to America in 1846 as the beginning of a new era for American science.[155] This easily may be said of Harvard science, where Agassiz would leave his most enduring impression, and yet much of what Agassiz did at Harvard was a reinforcement of particular trends that had begun before he arrived. Like Gray and Bowen, Agassiz opposed *Vestiges*. Like Benjamin Peirce, Agassiz favored the establishment of a school for advanced instruction in the natural sciences. Like Lowell, Agassiz wanted to pursue science in the service of public education. Even the most novel aspect of Agassiz's philosophical views—his idealist metaphysics—had a longstanding presence in the undercurrents of Harvard intellectual life. Indeed,

153. Gerstner, 151.

154. Asa Gray to John Torrey, 20 Feb. 1847 (Torrey Papers).

155. Robert V. Bruce, *The Launching of Modern American Science 1846–1876*, The Impact of the Civil War series, ed. by Harold M. Hyman (New York: Alfred A. Knopf, 1987), 3, 29.

Agassiz had such a profound impact at Harvard precisely because he could so readily assimilate himself with the very community he sought to transform.

Born in 1807 in a small village at the base of the Bernese Alps in Switzerland, Jean Louis Rodolphe Agassiz acquired a deep fascination with nature at an early age. In 1827, Agassiz began studying embryology under Ignatius Döllinger, in whose home he rented a room. The personalized instruction with Döllinger's microscope instilled in Agassiz a keen appreciation for precision in empirical research. During his four-year stay in Munich, Agassiz also attended lectures by Lorenz Oken, a pioneer in the *Naturphilosphie* movement that combined empirical natural history with *a priori* truths from which deductions could be made to construct a unified system of nature. Döllinger's and Oken's influences led Agassiz to regard his observations of the life-histories of embryos as guides for the imagination, enabling him to apprehend the "idea" by which the Deity had constructed nature. In the spring of 1832, Agassiz attended the lectures of Georges Cuvier in Paris, where the latter had been debating Lamarck's disciple Étienne Geoffroy Saint-Hilaire for the past decade. Geoffroy's idealist perspective on nature resonated with Oken's *Naturphilosophie*. It linked species into a "unity of plan" that implied historical transmutation, but Cuvier's rebuttals had exposed empirical inconsistencies in Geoffroy's work. Cuvier, who frequently hosted Agassiz for dinner, thus reinforced the lessons learned with Döllinger's microscope.[156]

Finding some of Geoffroy's views too speculative, Agassiz favored Cuvier's empirical approach to comparative anatomy while still retaining some strong appreciation for idealism. Cuvier emphasized that the animal kingdom included four distinct plans of organization, whereas Geoffroy more readily smoothed these into a unity of material relations, much as the author of *Vestiges* would do later. For Agassiz, nature's unity was ideal, not material, something to be found in the mind of God, who planned each species distinctly; under this ideal unity, nature's material diversity could be detected by comparative empirical investigations, among both living and fossilized species. Some American critics of *Vestiges* cited Agassiz's multi-volume *Poissons fossiles* (1833–1843), in which he established a new classification system for fossil fishes and emphasized, contrary

---

156. See Lurie, esp. 34–37 (Döllinger), 50–52 (Oken), 55–65 (Cuvier and Geoffroy).

to progressionist theories, that some of the more complex fishes could be found in strata contemporaneous with the earliest fossil fishes.[157]

Agassiz's scientific contacts on the Continent, in Britain, and in America together arranged for him to tour the United States. Funding for this venture came from the King of Prussia, who supported a year of research travel, and John Amory Lowell, who invited Agassiz to lecture at his institute during the winter of 1846–1847.[158] While Agassiz was making preparations for his extended travels in America, he received a letter from Cambridge University geology professor Adam Sedgwick, warning that "the opinions of Geoffroy St. Hilaire and his dark school seem to be gaining some ground in England." During the winter of 1845, while Gray was lecturing against *Vestiges* in Boston, English gentlemen of science repeatedly urged Sedgwick, an ordained Anglican and respected authority on geology, to publish a review against *Vestiges*. Few could agree on who had written the book, but many in Sedgwick's circle were concerned about the revival of French transmutation theories, which they associated with anticlericism and civil disruption—bad science of the worst sort.[159] "When [Lamarck, Geoffroy, Oken, and the author of *Vestiges*] talk of spontaneous generation and transmutation of species," Sedgwick objected, "they seem to me to try nature by an hypothesis, and not to try their hypothesis by nature." He urged Agassiz to lend his palaeontological expertise on Old Red Sandstone fishes to the cause against *Vestiges*, asking, in effect: am I not correct that these are of a higher type than the supposedly primitive forms claimed by the development theory?[160] For Agassiz, the evolutionary *Naturphilosophie* of Geoffroy and Oken was not legitimate "natural philosophy," and he "dreaded" it as much as "religious fanaticism." He

157. Lurie, 82–83. For citations of this work in refutation to *Vestiges*, see, for example, [David Brewster], "Vestiges of the Natural History of Creation," *North British Review* 3 (Aug. 1845): 470–515; rpt., *Living Age* 6, no. 71 (20 Sept. 1845): 564–82; facsimile rpt., in *"Vestiges" and the Debate before Darwin*, ed. Lynch, vol. 1 (original pagination retained), at 573–74 (*Living Age* pagination); and, [George Taylor], "Theories of Creation and the Universe," review of *Vestiges* and *Explanations*, *Debow's Review* 4, no. 2 (Apr. 1847): 177–94, at 185–86. Brewster's review, as noted in chapter 1, was appended to multiple American reprints of *Vestiges*; see Appendix B for a listing.

158. Lurie, 116; Edward Weeks, *The Lowells and Their Institutes* (Boston: Little, Brown, and Co., 1966), 52.

159. Secord, *Victorian Sensation*, 233–34.

160. Adam Sedgwick to Louis Agassiz, 10 Apr. 1845, in *Life, Letters, and Works of Louis Agassiz*, ed. Jules Marcou, 2 vols. (New York: Macmillan, 1896), 1:383–87, quoting 383 (dark school) and 384 (hypothesis).

concurred with Sedgwick that the more ancient fish are not always simpler in kind, and that species transmutation has no foundation in the natural sciences.[161]

Agassiz recently had presented his views on the Plan of the Creation for a European audience; soon he would do the same in Boston, thereby strengthening Gray's assaults against *Vestiges*. While touring the eastern seaboard with Agassiz in the fall of 1846, Gray not surprisingly found him to be "an excellent fellow."[162] Before addressing the Lowell audience, Agassiz visited Benjamin Silliman and James Dwight Dana in New Haven, J. W. Bailey in New York, John Torrey in Princeton, and Samuel Haldeman and Samuel George Morton in Philadelphia. That last contact led Agassiz to oppose *Vestiges* in a way that others had not.

While visiting Morton in Philadelphia, Agassiz experienced his first contact with people of African descent. He was horrified, as he later recounted in a private letter:

> Seeing their black faces with their fat lips and their grimacing teeth, the wool on their heads, their bent knees, their elongated hands, their large curved fingernails, and above all the livid color of their palms, I could not turn my eyes from their face in order to tell them to keep their distance, and when they advanced that hideous hand toward my plate to serve me, I wished I could leave in order to eat a piece of bread apart rather than dine with such service.[163]

Though trying to feel "compassion in thinking of them as really men," Agassiz admitted that "it is impossible for me to repress the feeling that they are not of the same blood as us." Morton's presence solidified this judgment, for Morton had a amassed an impressive collection of human skulls and, following a methodology not unlike Agassiz's comparative anatomy, reached the conclusion that humans of different races were separately created by God, as if they are

---

161. Louis Agassiz to Adam Sedgwick, June 1845, in *Life, Letters, and Works of Louis Agassiz*, 1:387–95, quoting 388. Agassiz, though he had distanced himself from the speculative tendencies of *Naturphilosophie*, reportedly expressed support for Richard Owen's appropriation of aspects of that tradition, when the two attended the September 1846 meeting of the British Association for the Advancement of Science, shortly before Agassiz's arrival in America for the Lowell Lectures. See Phillip R. Sloan, "Whewell's Philosophy of Discovery and the Archetype of the Vertebrate Skeleton: The Role of German Philosophy of Science in Richard Owen's Biology," *Annals of Science* 60 (2003): 39–61, at 57n85.

162. Asa Gray to John Torrey, 13 Oct. 1846 (Torrey Papers).

163. Louis Agassiz to Rose Agassiz, 2 Dec. 1846, as translated and quoted by Louis Menand, *The Metaphysical Club* (New York: Farrar, Straus, and Giroux, 2001), 105.

distinct species altogether.[164] If the development theory were true, then a European man of science like Agassiz would be of the same species as the servants of African descent attending him in Philadelphia; but, if they were distinctly created species, then *Vestiges* surely was false. Agassiz soon would deploy the latter logic in an argument against *Vestiges*.

Unfortunately Agassiz had a cold when he delivered his first Lowell lecture on December 8, but Gray reported to Torrey that, despite Agassiz's hoarse voice, "the most intelligent people, were quite delighted and impressed."[165] Agassiz spoke of recent progress in natural history that had revealed "the intention of the Creator in making this world." Similarities among species, whether living or extinct, bear testimony to the "Intellectual Being" whose "idea" of creation can be traced throughout the animal kingdom.[166] Tickets for all twelve of his Lowell lectures were so over-sold that only half of the subscribers gained admittance; a wave of encore lectures had to be scheduled for the remainder.[167] No doubt the very subject of his lectures, "The Plan of Creation in the Animal Kingdom," had attracted public interest precisely because *Vestiges* had become so popular.

As Agassiz continued his lectures during the following weeks, his listeners acclaimed him as a first-rate orator, despite his foreign accent. The audience welcomed Agassiz as a worthy spokesperson for scientific expertise. "What struck me most," wrote Morrill Wyman to his brother Jeffries, who no longer could hope for the Rumford chair, "was his facility of moving at general laws in anatomy and his ability to fix them upon the mind."[168] After completing his twelve-part series, Agassiz treated Boston's intellectuals to a lecture in French, which they loved.[169]

Gray appreciated, above all, Agassiz's "refutation of Lamarckian or Vestiges views." He ranked these "good lectures on natural theology" as being

164. Menand, 105 (translated excerpt from Agassiz's letter), 106–7 (Morton's methodology). See also Stephen Jay Gould, *The Mismeasure of Man* (New York: W. W. Norton, 1981), 42–69.

165. Asa Gray to John Torrey, 9 Dec. 1846 (Torrey Papers).

166. Louis Agassiz, Lowell Lectures, [1846] (Agassiz Correspondence and Manuscripts, Museum of Comparative Zoology, Harvard University), 5, 9. Only a portion of this lecture manuscript is extant, but internal evidence suggests it was Agassiz's opening lecture.

167. *Boston Daily Evening Transcript*, 9 Nov. 1846; Weeks, 52.

168. Morrill Wyman to Jeffries Wyman, 18 Dec. 1846 (Wyman Papers).

169. Asa Gray to John Torrey, 24 Jan. 1847 (Torrey Papers).

"vastly above that of any geological course I have ever heard."[170] Agassiz's affirmation of direct divine planning for the appearance of each new species, as compared to the physical conditions that prompted *Vestiges'* law of development to unfold, satisfied Gray that this Boston newcomer would not be another Rogers. "His references to the Creator," wrote Gray to contrast Agassiz with that Rumford candidate who had lectured in the fall, "were so natural and unconstrained as to show that they were never brought in [merely] for effect." As the *Boston Traveller* excerpted from Agassiz's eighth lecture:

> We see, therefore, that neither the organization of animals
> nor their dwelling place, is the result of accident or of physical
> conditions, but that it is the result of the great plan of the
> Creator, each species of animal being placed upon the earth in
> the spot which it now inhabits.[171]

In relating the matter to Torrey, however, Gray wrote on for some length to introduce and scrutinize one of Agassiz's peculiar views. The visiting naturalist had concluded that animal (including human) species and some of their varieties originated separately in their own particular geographical zones. "He does not believe," explained Gray, "that the Negro and Malay races descended from the sons of Noah, but had a distinct origin."[172] As reported in the local press, Agassiz had even asserted "that there are several different species of man," each created distinctly by God.[173] Gray suggested to Torrey that "we should not receive it (rejecting it on other than scientific grounds, of which he does not feel the force as we do)."[174]

The context does not make clear whether the "it" that Gray did not wish to receive was Agassiz's broad claim about distinct, local creations for each geographically dispersed natural kind, or only the "extension of his general view" as it applied to the supposedly separate origins of the human races. Whichever

170. Asa Gray to John Torrey, 24 Jan. 1847 (Torrey Papers). Gray apparently was using "Lamarckian" as a general reference to theories of species transmutation, rather than a specific reference to any of the aspects of Lamarck's own theory, such as use-inheritance.

171. "Prof. Agassiz's Tenth Lecture: What Is Meant by the Unity of the Human Races?," *Boston Traveller*, n.d. (filed in the Louis Agassiz biographical file, Harvard University Archives, HUG 300).

172. Asa Gray to John Torrey, 24 Jan. 1847 (Torrey Papers).

173. " Prof. Agassiz's Eighth Lecture: On the Tropical Fauna," *Boston Traveller*, n.d. (filed in the Louis Agassiz biographical file, Harvard University Archives, HUG 300).

174. Asa Gray to John Torrey, 24 Jan. 1847 (Torrey Papers).

claim Gray had in mind in 1847, he would oppose both in the coming years. Gray's parenthetical comment also invites further curiosity among today's historians. He referred to "other than scientific grounds"—theological? political? —that were inhibiting him from welcoming Agassiz's novel perspective. Historians generally have portrayed a contrast between Agassiz's "idealist" science, which is said to be laden with preconceptions, and Gray's "empiricist" science, which is said to be open to new insights that additional observations may provide.[175] In this letter, however, Gray seems to have some non-negotiable preconceptions of his own.[176]

Possibly Gray, like his recipient Torrey, felt constrained by his Presbyterian background not to deviate from the Bible's teaching that humans of all races had descended from a single created pair, Adam and Eve.[177] By the 1850s, Agassiz's polygenist theory of separate creations of the human races would become quite controversial for its "unorthodox" tendencies. At the Lowell debut of Agassiz's theory, however, Gray marveled that "so far from bringing this against the Bible, he brings the Bible to sustain his views, thus appealing to its authority, instead of endeavoring to overthrow it." Agassiz had argued —"conclusively," acknowledged Gray—that Noah's family were all Caucasian, unaltered to the present day, unless God by some miracle had changed some of them into other races.[178]

175. Toby Appel has criticized this generalization, indicating that Gray also "shared to a greater or lesser extent an idealist view of nature" and was "an idealist, if by this term we mean someone who sees a plan of creation in the morphology of organisms." See Toby A. Appel, "Jeffries Wyman, Philosophical Anatomy, and the Scientific Reception of Darwin in America," *Journal of the History of Biology* 21, no. 1 (Spring 1988): 69–94, at 71, 80. At 69n1, Appel locates the mistaken idealist/empiricist dichotomy in Dupree, 221–28, Lurie, 82–83, 260–1, 282–83; Edward J. Pfeifer, "United States," in *The Comparative Reception of Darwinism*, ed. Thomas F. Glick (Austin: University of Texas Press, 1974), 168–206; and, Ernst Mayr, "Agassiz, Darwin, and Evolution," *Harvard Library Bulletin* 13 (1959): 165–94.

176. Asa Gray to John Torrey, 24 Jan. 1847 (Torrey Papers).

177. Passages commonly cited against polygenism include Genesis 10; Acts 17:25–26; Romans 5:12; and, 1 Corinthians 15:21–22. The most comprehensive history of American polygenism remains William Stanton, *The Leopard's Spots: Scientific Attitudes toward Race in America, 1815–1859* (Chicago: University of Chicago Press, 1960). For an insightful exploration of the relationship between polygenism, philology, and the receptions of both *Vestiges* and Darwin's *Origin of Species*, see Stephen G. Alter, *Darwinism and the Linguistic Image: Language, Race, and Natural Theology in the Nineteenth Century* (Baltimore and London: The Johns Hopkins University Press, 1999), esp. 40–41.

178. Asa Gray to John Torrey, 24 Jan. 1847 (Torrey Papers).

More likely, then, Gray's "other than scientific" reasons for rejecting Agassiz's polygenism must be sought in politics. "I make the 'Manual' keep clear of slavery," he wrote to Torrey, referring to his *Manual of the Botany of the Northern United States,* which went to press piecemeal during 1847. The work discussed the plants of "New Jersey, Pennsylvania (if little Delaware manumits, perhaps I can find a corner of it), Ohio, Indiana or not as the case may be, leave out Illinois, which has too many Mississippi plants, take in Michigan and Wisconsin. . . ."[179] Not only did Gray oppose slavery, he did not wish to study the plants that grew in slave states. Likewise he felt uncomfortable with polygenic anthropology, which commonly was deployed as a justification for slavery in the American South.

Regardless of Gray's "other than scientific" reasons for objecting to some of Agassiz's views, he still held the newcomer in very high regard; for one thing, they shared a common opposition to *Vestiges.* Gray's correspondence with Torrey excited such enthusiasm for Agassiz that Torrey soon began arranging for Agassiz to lecture in New York.[180] This plan became interrupted, however, when John Augustine Smith, president of the New York Lyceum, began politicking against Agassiz on the grounds that his polygenism was theologically taboo. (Smith, incidentally, opposed *Vestiges* in his presidential address of 1846.[181]) Torrey, hoping to secure enough subscriptions to host the lectures during Lent (when New York's Episcopalians would be avoiding the theater, but would not hesitate to attend a scientific lecture), urged Gray to bolster Agassiz's reputation as best he could.[182] The effort, however, was already stalled.

Meanwhile, a committee in Philadelphia, including Agassiz's polygenist mentor Morton, arranged for the Swiss naturalist to repeat the Lowell lecture series at the University of Pennsylvania in April 1847. A broadside advertisement promised that the visiting scholar would "explain the successive creation of the

179. Asa Gray to John Torrey, 24 Jan. 1847 (Torrey Papers). On the publication of Gray's *Manual,* see Dupree, 180.

180. John Torrey to Asa Gray, 18 Dec. 1846 (Gray Papers, Herbarium).

181. John Augustine Smith, *The Mutations of the Earth; Or An Outline of the More Remarkable Physical Changes, of which, in the Progress of Time, This Earth Has Been the Subject, and the Theatre: Including an Examination into the Scientific Errors of the Author of the Vestiges of Creation,* 1846 Anniversary Discourse, Lyceum of Natural History of New York (New York: Bartlett and Welford, 1846).

182. John Torrey to Asa Gray, 11 Jan. 1847.

several divisions of the Animal Kingdom." The lectures would "show that the whole is in accordance with a plan ordained 'in the beginning,' and gradually developed during the changes that have subsequently affected our globe." In particular, "the order of succession of animals during all former geological periods, agrees with those changes which young animals of the same families undergo at the present time, in the progress of their formation."[183] Development, recapitulation, a grand plan unifying all of nature—these were the insights of natural history that Agassiz would present in Philadelphia, but his lectures would oppose the kind of development, recapitulation, and unity of plan theorized in a book that University of Pennsylvania chemist Robert Hare was reading at about this time.

Hare, who together with Morton was a member of the organizing committee for Agassiz's Philadelphia lectures, had a curious habit of pasting the bottom of one piece of paper to the top of the next, and then rolling the assemblage into a scroll, which served as a depository for his private notes. In one such scroll, Hare expressed his views concerning *Vestiges*. He began with an argument that "general laws," such as the law of development, would give all the planets the same number of satellites and the same orbital properties (velocity, planar inclination, etc.). But since planets varied in these respects, one most posit "special laws" that account for the uniqueness of each planet's orbit and the number of its satellites. Similarly, organic matter requires special laws, both for its creation and for its endurance. These "special laws" were nothing other than the "specific interference on the part of the Deity" that "the author of a recent work entitled 'Vestiges of the Creation'" denied.[184]

Whereas Morton had perhaps welcomed Agassiz for precisely the reason that Smith would not (that is, polygenism), Hare may have appreciated Agassiz

183. "Professor Agassiz's Lectures," [Apr. 1847], (Broadside Collection, American Philosophical Society, Item B:F865.a:17). The date of this document may be inferred as follows. The text identifies the lecture outlines with those "just completed by [Agassiz] before the Lowell Institute" and specifies that he would speak about the animal kingdom. Agassiz lectured on "The Plan of Creation as shown in the Animal Kingdom" at the Lowell Institute during the winter of 1846–1847; his subsequent Lowell lectures dealt with other topics (ichthyology, 1847–1848; comparative embryology, 1848–1849; lower animals, 1850–1851; natural history methodology, 1861–1862; glaciers, 1864–1865; Brazil, 1866–1867; and, sea dredging, 1869–1870). See Smith, *History of the Lowell Institute*, 52, 53, 60, 62, 63, 66. The broadside notes that the first lecture was to be held on "TUESDAY, the 20th inst." Tuesday fell on the 20th of only April and July in 1847. April better fits the phrasing "just completed" in reference to the winter Lowell lectures.

184. Robert Hare, On the necessity of special laws as well as general laws to the origination and endurance of Creation, [ca. 1845–1846] (Hare Papers, American Philosophical Society).

above all as a fellow critic of *Vestiges*. Gray's enthusiasm for Agassiz explicitly centered on this point, and, as noted earlier, American reviews cited Agassiz against the development hypothesis. Even Smith, an opponent of *Vestiges* who nonetheless had tried to block Agassiz from appearing in New York, would soon learn that *Vestiges* was more press-worthy than polygenism, so Agassiz would not be so dangerous to New Yorkers after all. Thus, Agassiz's position against *Vestiges* seems to have played some role, in part planned and in part unforeseen, in establishing his American career.

In October and November, Agassiz delivered a series of lectures at the College of Physicians and Surgeons in New York City, where Torrey's persistent promoting of Agassiz finally came through.[185] "Both sexes, all ages, and all sects of religion and literature," reported the *New York Herald*, "listen in bursting wonder and astonishment to the revelations which he makes in science and natural history." Agassiz was newsworthy because his "new philosophy" seemed in accord with "the general movement of the great mind [*sic*] of Europe and America, which is now transpiring with reference to religion, manners and customs of the age." Just as David Strauss's *Das Leben Jesu* was pushing modern inquiry back in time and suggesting new truths about the ancient past, so also Agassiz's geology took his audience into the depths of time. Fueling this "philosophy which looks back to millions of years," English booksellers had already sold some 50,000 copies of *Vestiges of Creation*, noted the *Herald*. The reporter, in the *Herald*'s typically provocative style, suggested that neither Strauss, Agassiz, nor the author of *Vestiges* posed a threat to "the different sects of Christianity," except perhaps to those "sects which adhere to the creed of the middle ages and to the interpretations of olden times."[186]

The *New York Herald* had conflated Agassiz's geological views with those of *Vestiges* into a single intellectual "movement." In crafty defiance (it doubtless

185. The lectures ran from October 15 through November 25. See John Torrey to Asa Gray, 7 Oct. 1847 and 27 Nov. 1845 (Herbarium).

186. "The New Movement in Philosophy," *New York Herald*, 1 Nov. 1847, 2. The *New York Herald*, founded in 1835 by the Scottish Catholic immigrant James Gordon Bennett, characteristically reported in a sensationalist, loose-cannon style. Bennett claimed that the *Herald* was politically neutral, and indeed editorials did endorse both Democrats and Whigs, but Bennett's press leaned in favor of Southern proslavery Democrats and against religious authorities, including the Pope in Rome, but more particularly the New England Protestant Whigs, whose Puritan legacy reminded him too much of the Presbyterian establishment that had oppressed his family in Scotland. See James L. Crouthamel, *Bennet's New York Herald and the Rise of the Popular Press* (Syracuse, NY: Syracuse University Press, 1989), esp. 4–5, 34–35, 56–70, 94.

was not ignorance) of the numerous mainstream religious presses that had lambasted *Vestiges* and Strauss (whose biography of Jesus nearly reduced the central figure of Christianity to a cultural legend),[187] the *Herald* suggested that only the most backward and sectarian Christians would have a problem with these new works. This report apparently preceded Agassiz's sixth lecture, in which he sought to clarify that in speaking of "organic development" he did not mean to imply the kind of development advocated in *Vestiges*. It is, Agassiz emphasized in this lecture, only in a "metaphoric manner [that] we are accustomed to speak of metamorphoses in the Animal Kingdom through different ages of geological time." When a caterpillar develops into a butterfly, it is "one and the same animal continually living without interruption." But when one class of animals "develops" into another in geological time, this means only that a succession of organic plans has been introduced. New individuals, which are *not* derived from individuals of the preceding plan, are created according to the new plan, and thus constitute a distinct species with no genealogical relation to other species. The kinds of progressive species development revealed by careful research, Agassiz made clear,

> disagree entirely with the views, and have not the slightest alliance with the views of a work which is very much spoken of, but which I consider entirely unworthy of notice by any serious scientific man—because it is made up of old-fashioned views which have been brought before the notice of the public for a half century, by the French school, and are supported only by antiquated assertions, and by no means by facts scientifically ascertained. It must be owing to some particular circumstance that this work has been so much noticed, because really it is not worthy a critical examination by a serious scientific man.[188]

By not mentioning Lamarck, Geoffroy, or *Vestiges* by name, he hoped not to draw more attention than necessary to any of them. The views of Lamarck and

---

187. See, for example, "The Truth of Christianity," *North American Review* 63, no. 133 (Oct. 1846): 382–432, esp. at 413, where Strauss received the left-handed compliment of being one of "the most learned and skillful infidels of the present day."

188. Louis Agassiz, *An Introduction to the Study of Natural History, in a Series of Lectures Delivered in the Hall of the College of Physicians and Surgeons, New York* (New York: Greeley & McElrath, Tribune Building, 1847), 25.

Geoffroy had been, Agassiz was certain, decisively defeated by Cuvier's arguments over a decade earlier. *Vestiges* was but an outdated imitation of those earlier pretensions, and undeserving of a naturalist's time, except that it had prompted some people to question what Agassiz himself meant by "development." At the Lowell Institute's lectures, it was customary for the audience to hold all applause until the very end (a practice later borrowed by the Boston symphony).[189] In New York, however, the audience applauded once during Agassiz's refutation of *Vestiges,* and once again as he concluded this digression and resumed his lectures with the classification of the Gasteropoda.[190]

During the final moments of his last lecture in New York, Agassiz completed the geological saga of vertebrate forms with "Man, [who] is the highest possible development in the progress of Creation." The Creator had planned to capstone creation with man from the beginning, so no further progress should be expected in the geological future, except, of course, moral and intellectual advancements "within the limits of Mankind." The audience responded with "great applause."[191]

The *New York Herald* now recognized a difference between the author of *Vestiges* and Louis Agassiz. "On the whole," reported the *Herald* after Agassiz's final lecture, "the ingenious author of the 'Vestiges of Creation' produces more conclusive proof to us in the idea that we are merely in a progressive state, and that the Divine Power has by no means exhausted all the powers of creation, but will work great changes in the next ten millions of years." Agassiz had disappointed the reporter, who understood him to claim that "the creative power of the Great Maker of all things, had been entirely exhausted in his last great effort to make man in the shape he is in."[192]

Though disappointing the New York newspaper reporter, Agassiz was an immensely successful lecturer, acclaimed in Boston and far beyond. John Amory Lowell arranged for Agassiz to deliver a series on comparative embryology for the 1848–1849 season, and the lecture was published verbatim in Boston, New York,

---

189. Weeks, 46–47.

190. Agassiz, *Introduction*, 25.

191. Agassiz, *Introduction*, 58.

192. "Philosophical Intelligence," *New York Herald*, 5 Dec. 1847, 2.

and Philadelphia.[193] John Torrey's daughter mounted transcripts of these lectures from the *Boston Traveller,* which the family read aloud beside the fireplace in Princeton.[194] Agassiz's debut series in 1846–1847 marked such an impression upon Lowell and other Boston intellectuals that when Benjamin Peirce's vision for a scientific school came to fruition in 1847, through the financial support of Abbot Lawrence, a professorship was created especially for Agassiz. The Swiss visitor had become a Swiss emigré, with Harvard pleased to provide his new home.

Not long previous, Harvard had been something of a scientific backwater. Peirce, Gray, and now Agassiz were changing that. Thus, Gray's 1847 communiqué to Torrey about the low calibre of Princeton science must be understood as a transition marker: Dod and Henry were no longer at Princeton, and Gray and Agassiz were now both at Harvard. Both had made their marks at the Lowell Institute, and Agassiz was heading up the Lawrence Scientific School (effectively the nation's first graduate school for science). As *Vestiges* continued to be reprinted by American publishers and to be discussed by American men of science in both reviews and lectures, species fixity could hardly have asked for a more respected footing in the natural sciences than Gray and Agassiz could provide at Harvard. But the natural sciences were not everything at Harvard. All students at the Lawrence Scientific School were required to attend church,[195] and Harvard theologians, no less than Harvard naturalists, also had something to say about *Vestiges.*

The Rev. Joseph Henry Allen
Probes the Moral Character of Vestiges

By the spring of 1846, "men of high and acknowledged authority in the various branches of science" had evaluated the development hypothesis in both print and lectures, but the Rev. Joseph Henry Allen felt that they had missed "the

193. Louis Agassiz, *Twelve Lectures on Comparative Embryology, Delivered before the Lowell Institute, in Boston, December and January, 1848–9,* stenographic report by James W. Stone, originally for the *Boston Traveller* (Boston: Redding & Co.; Gould, Kendall, & Lincoln; James Munroe & Co.; New York: Dewitt & Davenport; Tribune Buildings; Philadelphia: G. B. Zieber & Co., 1849).

194. John Torrey to Asa Gray, 26 Jan. 1849 (Herbarium).

195. Croce, 40.

importance of the 'Vestiges.'" Its scientific flaws, its literary appeal, its startling conclusions—these all were overrated.[196] *Vestiges* mattered to Allen because it had implications for "the great topics of life and duty," regardless of how earnestly or persuasively reviewers criticized its methodology.[197] "Every statement he [the still anonymous author of *Vestiges*] has made may be denied, and each argument he has used may be successively confuted; but the tendency he represents will still remain." Like other reviewers, Allen identified that tendency as "an excessive generalizing of the scope and agency of natural law, till it seems to absorb and blot out every other power in the universe."[198] For Allen, however, the power beyond natural law that *Vestiges* should not be permitted to strip away was not so much God Himself, but rather the "character, motives, and the freedom of the human will."[199]

"Few could have better claim to represent the Brahmin caste of New England," than Joseph Henry Allen, read a memoir after his death in 1898.[200] His wife's ancestors had immigrated to Massachusetts in 1636, and his own in 1639. Seven of her ancestors were ministers; his living relatives included two ministers and seven teachers. During his senior year at Harvard (1840), he served as correspondence secretary for the Harvard Natural History Society.[201] Graduating from Harvard Divinity School in 1843, he became, like his father before him, a Unitarian minister. His mother's father, Henry Ware, was the Hollis Professor of Divinity whose 1805 appointment solidified the Unitarian hold on Harvard.[202] Opposing the sentimental tendencies of Ralph Waldo Emerson, Allen championed the rational foundations of the Unitarian faith.[203] He edited the

196. A[llen], 333.

197. A[llen], 334.

198. A[llen], 334.

199. A[llen], 345.

200. Charles Carroll Everett, *Memoir of Joseph Henry Allen*, rpt. from *The Publications of the Colonial Society of Massachusetts*, vol. 6 (Cambridge: John Wilson and Son, 1902), 1. See also Ahlstrom, "The Middle Period," 133n134.

201. Harvard Natural History Society, Membership Lists and Catalogues, 1837–1970 (Harvard University Archives, HUD 3599.754).

202. "Joseph Henry Allen," *Boston Transcript*, 21 Mar. 1898.

203. Ann Douglas, *The Feminization of American Culture*, 2d ed., with a new introduction (New York: Knopf, 1977), 20; Williams, *Harvard Divinity School*, 139.

Christian Examiner from 1857 to 1869. In 1878, Harvard Divinity School appointed him to replace Frederick Hedge as professor of ecclesiastical history. While serving the Jamaica Plain parish at Boston in 1846, he identified on behalf of his Unitarian community some of the chief theological problems in *Vestiges*.

The crucial question was not, like at Princeton, whether a sovereign God fit into *Vestiges*' schema, but rather, "is the dignity or the sacredness of human life in any degree lessened by such a result as this?"[204] In emphasizing this question, Allen distinguished readers of the Unitarian *Christian Examiner* from those subscribing to "other leading journals of the day,"[205] as follows:

> To us it makes no difference in the sacred gift of life, whatever channels of transmission it has passed through, from God the giver. . . . The fact, even on the author's hypothesis, remains the same as ever. Man is the summit, the crowning point, of all God's wonderful creation.[206]

Allen, unlike the professors at Princeton, had no reservations about humans and animals being related. "We," he wrote as a spokesperson for Boston Unitarians, "do not understand the reproach cast upon that hypothesis, of reducing man to a level with 'vulgar nature.'" Allen perceived the same supernatural stamp throughout all stages of creation: the "half-reasoning elephant" and the monkey who "will imitate human actions" bear the "'image and superscription' of the Maker's hand." The author of *Vestiges*, explained Allen, had linked diverse species "organically," that is, genealogically, but this was "very far from irreverent" and thus should not startle Unitarians. Since their theology already rested on an idealist foundation, they had readily accepted Agassiz's view that "the *type*, both of organic structure and mental character" indisputably links all organisms "as parts of one system" under the "unity of design."[207] Whereas Agassiz saw those types as fixed, Allen apparently had no trouble allowing for species transmutation—but that did not entail an agreement with other claims made in *Vestiges*.

204. A[llen], 342.

205. A[llen], 342.

206. A[llen], 343.

207. A[llen], 343.

Willing to concede an animal ancestry for humans, Allen parted company with the author of *Vestiges* at a different juncture: he objected to the fatalistic implications for human moral nature of a natural history in which all things were developed by necessity in accordance with the "law of development." Such determinism, such "absolute and entire fatalism"—such Calvinism, he may have thought, given the predestinarian tone of *Vestiges*—challenged "the supremacy of *character* over every other thing."[208]

It is ironic that *Vestiges* at predestinarian Princeton was rejected for being too Arminian, and *Vestiges* at Arminian Harvard was rejected for being too predestinarian. The first accusation focused upon the claim that nature, entirely apart from the direct, controlling action of God, could somehow improve itself; the second accusation focused upon the claim that the Deity had programmed the entire course of development into nature back at the beginning, leaving no room for nature's own novel contributions to the process. Either way, *Vestiges* would have to be challenged, but the Harvard way was thoroughly opposed to the Princeton way, so much so that *Vestiges* could seem suspiciously Presbyterian at Harvard.

Allen did not dispute any of the author's claims of fact, but rather objected from the standpoint of "a primitive, absolute conviction" that a moral character exists which cannot be reduced to necessity. Just as Immanuel Kant's "categorical imperative" stood beyond the testing ground of experience, argued Allen, so the "supremacy of *character*" was "entirely distinct . . . from anything proposed in scientific inquiry."[209] Human moral nature constituted irrefutable internal evidence. Science must presuppose it and should not seek to prove or disprove it. Phrenology, neurology, and physiology can do no more than "tell us of the physical limitations to mental phenomena; and to our mind they rather suggest the bounds which restrain and hamper the operations of spirit, than speak of the essence of human character and life."[210] Allen agreed with Coleridge that it would be "a contradiction in terms" to ascribe the soul to natural law, since "the soul, in however humble degree, is a creative spirit, like its Maker."[211]

208. A[llen], 344, 345.

209. A[llen], 345.

210. A[llen], 346.

211. A[llen], 346.

Calling for a strict demarcation between "science" and "philosophy," Allen sought to protect the latter from the materialist tendencies of the former. By limiting the sciences to the task of identifying regular patterns, or "laws," behind which God was active, Allen, like many American reviewers, wanted to protect science from materialism. Allen further wanted to limit science to an inquiry concerning physical nature, thus preventing "the encroachments of science or of its supporters, when unwarrantably interfering, though in the smallest degree, with the province of spiritual truth." Philosophy, which rightly concerned itself with spiritual truth, rested on the foundation of "the primitive absolute doctrine of the freedom of the human soul," a truth independent of the physical sciences.[212] The author of *Vestiges* had erred not so much in mistaking scientific facts (though Allen agreed that such mistakes may be found in *Vestiges*), but rather in building those asserted facts into a philosophical system—something beyond the proper purview of natural science—that, in contradiction to an *a priori* truth, reduced human will to the result of inevitable natural laws. Allen warned his readers that *Vestiges* made an "outright denial of the moral nature of man, and the moral signification of life."[213]

Allen quoted from *Vestiges* to illustrate the problem: "Mental action passes at once into the category of natural things [and] . . . the distinction usually taken between physical and moral is annulled."[214] Then paraphrasing again, Allen noted that *Vestiges* claims "that *all things* are included and governed by fixed, unalterable, unerring law."[215] He objected that such a "system which reduces man's moral and intellectual life to the foreordained unfolding of a law" would render discretion, nobleness, and penitence meaningless.[216] *Vestiges*, with all its talk of cosmology and geology, amounted to nothing other than assault of fatalism against human character.

Still, Allen wanted the author of *Vestiges* to get a fair hearing. Unitarians trusted human reason equally in matters of theology and of natural science, so they valued open inquiry. He concluded with a criticism not of *Vestiges*, but of

212. A[llen], 347.

213. A[llen], 345.

214. A[llen], 341 (quoting from *Vestiges,* Cheever ed., 232).

215. A[llen], 340.

216. A[llen], 341.

the New York publisher Wiley & Putnam. The firm's second American edition of the work included the "dull absurdity and clumsy ridicule" by Congregationalist minister George Cheever. If the publisher was so afraid of issuing "a work of supposed heretical tendency," then why not choose a respectable man of science to write "an accompanying attack upon its opinions"?[217]

Perhaps Wiley & Putnam heeded Allen's complaint. No subsequent printing of *Vestiges* included the Cheever introduction. Three American reprints did, however, include David Brewster's article from the *North British Review*. Brewster's critique, together with Adam Sedgwick's critical article in the *Edinburgh Review*, had prompted Robert Chambers to defend *Vestiges* with a sequel, *Explanations*, so either piece would be of interest to Americans who wanted to follow the development of British debates over *Vestiges*.[218] But one might wonder whether Allen still felt that Wiley & Putnam had sidestepped the real issue. Brewster, like Sedgewick, had debated *Vestiges* primarily in terms of scientific facts, though each noted also that *Vestiges* had some problematic implications for human nature and morality; for Allen, however, the scientific facts took a quite secondary place to the ultimate significance of *Vestiges*. At Harvard, where human nature was not so depraved as to depend totally upon a sovereign God, but rather could strive toward divinity through moral self-development, the crucial question concerning *Vestiges* became: what does the development hypothesis imply for human moral character? *Vestiges*, with its fatalistic natural law of development, had to be rejected, since it denied Unitarians the ability to reach toward God on their own.

"That book we have wanted so long": Transcendentalists Welcome Vestiges of Creation

Through the late 1840s, a broad consensus had been assembled at Harvard against *Vestiges*. Asa Gray, Francis Bowen, and Louis Agassiz appealed to the

217. A[llen], 349.

218. The *North British Review* article is appended to *Vestiges of the Natural History of Creation*, 4th American ed. from the 3rd London ed. (New York: Wiley & Putnam, 1846) and *Vestiges*, 4th American ed. from the 3rd London edition, bound with *Explanations; a Sequel to "Vestiges of the Natural History of Creeation," by the Author of That Work* (New York: Wiley & Putnam, 1848; New York: Coyler, 1848).

facts of nature, which showed spontaneous generation and species transmutation to be wild speculations at best. Joseph Henry Allen argued that these concerns, while valid, were secondary; what mattered most were the implications of a deterministic law of development for human moral character, which at Harvard was predicated on the freedom of the will and moral responsibility. Allen at least left the door open for a kind of development theory that would not be so deterministic, but the task of embracing the spirit of *Vestiges* itself, was left for those beyond the limits of Harvard science. Ralph Waldo Emerson, Unitarian expatriate and Transcendentalist prophet, was just such a man.

In contrast to Bowen, Emerson envisioned "nature study" (his substitute for the natural sciences) as a study of relations, not of facts. "All the facts in natural history taken by themselves, have no value," Emerson maintained, "but are barren, like a single sex. ... Empirical science is apt to cloud the sight." According to Emerson in his 1836 address entitled "Nature," naturalists must do more than observe nature; they must discover nature through intuition, recognizing "that guess is often more fruitful than an indisputable affirmation, and that a dream may let us deeper into the secret of nature than a hundred concerted experiments."[219] Bowen, of course, objected, retorting in the *Christian Examiner*, "Why not follow the principle of the gambler entirely, by shaking a number of words in a hat, and then throwing them upon a table ... ?"[220]

Drawing from Bowen's nemesis, Kant, Emerson argued that those who merely classify individual specimens according to their species are exercising only their Understanding, while their higher faculty, Reason, remains dormant. Emerson sought to awaken that higher faculty in order that a fuller study of nature might commence:

> It is not so pertinent to know all the individuals of the animal kingdom, as it is to know whence and whereto is this tyrannizing unity in his constitution, which evermore separates and classifies things, endeavoring to reduce the most diverse to one form ... faint copies of an original archetype.[221]

219. Ralph Waldo Emerson, "Nature" (1836), in *Collected Works,* 1:3–45, at 19, 39.

220. F[rancis] B[owen], "Trancendentalism," *Christian Examiner* 21 (Jan. 1837): 371–85, reprinted in *Emerson and Thoreau: The Comtemporary Reviews,* ed. Myerson, 5–13, at 10.

221. Emerson, "Nature," 40–41.

The end of such nature study, concluded Emerson by quoting from a poem by George Herbert, was that "each part [in nature] may call the farthest, brother."[222]

In the 1840s *Vestiges of Creation* united all parts of nature into a universal brotherhood not unlike the one Emerson had envisioned during the preceding decade: each individual was the diverse offspring of nature's grand "law of development." In April 1845, Emerson accordingly wrote to Samuel Gray Ward, a member of his Transcendental Club (founded in 1836), that *Vestiges* makes "a good approximation to that book we have wanted so long & which so many attempts have been made to write."[223] Writing to his cousin Elizabeth six weeks later, Emerson departed from his idiosyncratic penmanship, in which periods often serve as question marks, by inquiring quizzically: "the Vestiges have you read? the Vestiges, the Vestiges?"[224]

Emerson himself had read *Vestiges* as the work, he tentatively supposed, of Richard Vyvyan, who "has outdone all the rest in breadth & boldness & one only want[s] to be assured that his facts are reliable."[225] Emerson's journal records his sentiments concerning both English natural theology, which he thought cowardly for praising the Divine without thoroughly considering the relations among His works, and also *Vestiges*, for which he had mixed praised, since it was not radical enough:

> What is so ungodly as these polite bows to God in English books? He is always mentioned in the most respectful and deprecatory manner, 'that august,' 'that almighty,' 'that adorable providence,' &c &c. But courage only will the Spirit prompt or accept. Everything in this Vestiges of Creation is good, except the theology, which is civil, timid, and dull.[226]

222. Emerson, "Nature," 41.

223. Emerson to Samuel Gray Ward, 30 Apr. 1845, in *The Letters of Ralph Waldo Emerson*, ed. Ralph L. Rusk (vols. 1–6) and Eleanor M. Tilton (vols. 7–10), 10 vols. (New York: Columbia University Press, 1939–1999), 3:283.

224. Emerson to Elizabeth Hoar, 17 June 1845, in *Letters,* 3:290.

225. Emerson to Ward, 30 Apr. 1845, in *Letters,* 3:283.

226. Emerson, Journal W (Mar.-Sep. 1845), entry entitled *"Vestiges of Creation,"* in *The Journals and Miscellaneous Notebooks of Ralph Waldo Emerson*, vol. 9 (1843–1847), ed. Ralph H. Orth and Alfred R. Ferguson, 16 vols. (Cambridge, MA: The Belknap Press of Harvard University Press, 1960–1982), 9:211.

Still, *Vestiges* gave hope for future self-development, a key theme in Emerson's thought. "Well & it seems there is room for a better species of the genus Homo. The Caucasian is an arrested undertype."[227]

In an editorial preface for the premier issue of the *Massachusetts Quarterly Review* (December 1847), Emerson argued that too many journals had dodged the difficult questions and belittled their audiences with mere recapitulations of long-standing platitudes. He likely had in mind the stalwart Unitarian organs, *Christian Examiner* and *North American Review*, which had excluded both Transcendentalism and the development theory of *Vestiges* from favorable consideration. Emerson envisioned a different editorial policy:

> Here is the standing problem of Natural Science, and the merits of her great interpreters to be determined; the encyclopaedical Humboldt, and the intrepid generalizations collected by the author of the "Vestiges of Creation." Here is the balance to be adjusted between the exact French school of Cuvier, and the genial catholic theorists, Geoffroy St. Hilaire, Goethe, Davy, and Agassiz.[228]

While visiting Britain the following year, Emerson continued to contemplate the development hypothesis. He arranged to dine with Robert Chambers, whom he by that time realized was the author of *Vestiges*, and also with Andrew Crosse, whose spontaneous generation experiments he recalled from that book. Then he visited Lady Ada Lovelace, "an excellent mathematician, & . . . one [of the] persons to whom the 'Vestiges' was ascribed."[229]

During winter evenings from 1845 to 1850 Emerson appeared at Boston's Odeon Theatre, the same hall that generally hosted the Lowell Lectures, to deliver a series of lectures on the idea of a general mind that expresses itself with particular clarity at distinct moments in history, such as in the personages of Plato, Shakespeare, Goethe, and Swedenborg. Published in 1850 under the title *Representative Men*, Emerson's lectures suggested that the greatness in these

227. Emerson, Journal W (Mar.-Sep. 1845), entry entitled *"Vestiges of Creation,"* in *Journals and Miscellaneous Notebooks,* 9:212.

228. Emerson, "Editor's Address," *Massachusetts Quarterly Review* 1, no. 1 (Dec. 1847), reprinted in *Miscellanies*, ed. J. E. Cabot (Boston: Houghton, Mifflin, and Company, 1884), 323–34, at 332.

229. R. W. Emerson to Lidian Emerson, 21 (22?) Feb. 1848; R. W. Emerson to William Emerson, 5 May 1848; and, R. W. Emerson to Lidian Emerson, 21 and 23 June 1848; all in *Letters*, 4, no. 17–24, 70–71, 85–88.

heroic men of letters could be found also within the heart of each common man; moreover, it could be traced back to the "trilobite and saurus." The accomplishments of Homer and Columbus, therefore, "attest the virtue of the tree," since nature patiently works through geological history, in trees as in the "ox, crocodile, and fish," until finally the highest result may be found in humanity.[230] Such a view did not require species transmutation, since nature's progressive development could, as Agassiz had shown, be typological rather than genealogical; but in Emerson's mind all of nature flowed back into itself, so species need not have any fundamental distinctiveness. Thus, *Vestiges* could be greeted as a long-awaited friend.

In some respects, Agassiz's emphasis on ideal relations in nature leaned toward Transcendentalism. But for Agassiz, the idea behind nature was an objective reality within the mind of God, whereas for Emerson and his philosophical compatriot Henry David Thoreau (who gathered specimens for Agassiz's lab), ideas were subjective relations—"friendships," Thoreau would say —between nature and the subjective knower who surrounds himself with nature, as Thoreau had done at Walden Pond. Whereas Agassiz had defined fishes as the ideas of God's thought, Thoreau suggested in his journal, "Ideas are the fishes of thought?" Agassiz used sight to understand fishes as ideas, but Thoreau used fishes to understand sight as a form of self-knowledge.[231] It was enough to make a Unitarian dizzy.

If Agassiz was insisting that people of African and European ancestry were distinct special creations, then one might suppose the Walden Pond philosopher deserved his own classification, *Homo thoreauensis*. Thoreau's notebook contains about as idiosyncratic a response to *Vestiges* as one could have. He noted, in September 1851, that Hugh Miller, a well-known Scottish critic of the development hypothesis, had faulted the author of *Vestiges* for building a case out of "exceptions"—instances in the fossil record where simpler forms predate more complex forms. According to *Vestiges*, simpler forms precede the more complex because the latter develop from the former, but according to Miller,

---

230. Ralph Waldo Emerson, *Representative Men: Seven Lectures*, text established by Douglas Emory Wilson, introduced by Andrew Delbanco (Cambridge, MA: Harvard University Press, 1996), 45.

231. Henry David Thoreau, *Journal*, gen. ed., John C. Broderick, 6 vols. (Princeton, NJ: Princeton University Press, 1981–2000), 4, no. 290, entry dated 26 Jan. 1852. For a discussion of this and similar passages in Thoreau's private and public writings, see Laura Dassow Walls, "Textbooks and Texts from the Brooks: Inventing Scientific Authority in America," *American Quarterly* 49, no. 1 (1997): 1–25, at 12–14.

Agassiz, Sedgwick, Brewster, and other critics, complex forms were quite often found in strata as ancient as those in which the simpler forms first appeared, thus undermining the development theory. Thoreau cared little about which fossil could be found in which stratum. He objected instead to Miller's complaint that the author of *Vestiges* had confused the exceptions with the rule, for in Thoreau's view Miller had done the same thing. Specifically, Miller had written, in a different context, that:

> There is a feeling which at times grows upon the painter and the statuary, as if the perception and love of the beautiful had been sublimed into a kind of moral sense. Art comes to be pursued for its own sake; the exquisite conception in the mind, or the elegant and elaborate model, becomes all in all to the worker, and the dread of criticism or the appetite of praise almost nothing.

Thoreau then responded, "he speaks of . . . his rising above the dread of criticism & the appetite of praise as if these were the very rare exceptions in a great artist[']s life—& not the very definition of it." In questioning the individualism of creative genius, Miller had revealed a "latent infidelity more fatal far [*sic*] than that of the Vestiges of Creation," for he had suggested that the natural state of the mind is enslavement to society.[232]

And so one must go to the pond, alone, to see nature through one's own eyes, and therein to discover oneself, breathing one's own spirit, liberated from the civilization that suffocates.[233] A month before recording his criticism of Miller, Thoreau scribbled down a note sympathetic to the development theory: "Singular these genera of plants—plants manifestly related yet distinct[.]—They suggest a history of Nature—a Natural *history* in a new sense."[234] Like the streams of consciousness that characterized Thoreau's writings, each step in the development hypothesis smoothly led into the next, unifying all of nature, however diverse its products, within a single process.

The enthusiasm that Emerson and Thoreau shared for *Vestiges* troubled the Unitarian mind. Transcendentalists had broken from Unitarianism precisely

232. Thoreau, *Journal*, 4, no. 106–7, entry dated 28 Sept. 1851.

233. Henry David Thoreau, *Walden, or Life in the Woods* (1854), reprinted in *Walden and Other Writings* (New York: Barnes and Noble, 1993), 1–275; see esp. 3–14, 267–69, but also frequently *passim*.

234. Thoreau, *Journal*, 4, no. 6, entry dated 21 Aug. 1851.

by favoring personal introspection over external evidences, and now Emerson and Thoreau, caught up in their own dreams about nature, turned a blind eye to the external evidences amassed against the development theory by Bowen, Gray, and Agassiz. From a Unitarian perspective, Transcendentalism denied the efficacy of external evidences and turned the mind inward to its own poetic projections upon nature. Such a practice lent itself to an individualism that fragmented the community. It also welcomed speculations that departed from the historic revelation of moral character in Jesus Christ. As the next section indicates, if Unitarianism was to hold its own against Transcendentalism, then the authority of the Scriptures—the greatest external evidence of all—had to be defended. This, in turn, meant that *Vestiges* had to be opposed.

Defending the Divine Inspiration of Scripture: The Dudleian Lectures against Vestiges

A century before the Harvard community became preoccupied with *Vestiges,* the Honorable Judge Paul Dudley bequeathed 133 pounds, six shillings, and eight pence to Harvard College for an annual lecture series on religion. The topics were to rotate every four years among natural religion, revealed religion, the "Idolatry of the Roman church," and the legitimacy New England ordinations without Anglican bishops.[235] On two occasions in the mid nineteenth century, in 1848 and again in 1856, a Dudleian lecturer, speaking on revealed religion, defended the divine inspiration of the Bible on the basis of external evidences garnered from geological findings that contradicted the author of *Vestiges*. It is important to note that neither speaker voiced any concern about interpreting Genesis literally; rather, both spoke from an Anglo-American liberal tradition dating back to John Locke's rational supernaturalism, the very tradition of external evidences that had been so important in the emergence of Boston Unitarianism.

In May 1848, the Rev. Samuel Gilman, a Harvard-trained pastor who devoted his ministry to the establishment of Unitarianism in Charleston, South Carolina, delivered a Dudleian lecture at his *alma mater* in defense of "Revealed

---

235. Records relating to the Dudleian Lecture in Harvard University, 1830–1860 (Harvard Divinity School Archives, bMS 523), 1–3.

Religion." He focused particularly upon David Hume's argument that supernatural revelation would be an improbable breach of the natural order. Hume and his school of "skeptics," explained Gilman, regarded nature's present course to be free from supernatural interventions; they denied all miracles, including the inspiration of Scripture. Fortunately for Gilman, however, the natural sciences in fact indicated that supernatural interventions in nature were not uncommon. Geologists had "demonstrated that the order of nature is not so uniform, as to preclude distinct and successive interpositions of creative power, whenever the earth was prepared for new species and genera of animals."[236]

Gilman's confidence that God intervenes in nature rested upon an empirical foundation. In theory, nature could have been different, in which case God's miraculous interventions would not be evidenced in nature. For example, "had the theory of the very able author of the 'Vestiges of Creation,' respecting the self-development of the whole animal world out of a single original germ, received support from the general testimony of nature," then Hume, Voltaire, and Thomas Paine might be able to build a compelling argument against miracles. But, fortunately for Gilman and Harvard's Unitarians, men of science had in fact discovered many signs of God's miraculous acts in nature:

> No one can rise from the inspection of those reports which the most eminent geologists have published regarding fossil remains in successive layers of the earth, without an inevitable conviction that new, specific and original impulses of designing and creating power have from time to time interposed to change the pre-existing order of things, and substituted another in its place.[237]

Spontaneous generation served as another test case. "The whole tendency of philosophical observation and discovery at the present day," said Gilman to an audience that no doubt could recall Prof. Gray's Lowell lectures or at least his *North American* article, "is against the doctrine of spontaneous generation." Indeed, even the eighteenth-century skeptic David Hume "could not have appealed to the testimony of experience to show that a single animal or plant was

---

236. Samuel Gilman, Revealed Religion, the Dudleian lecture preached May 10th, 1848 (Harvard University Archives, HUC 5340.148), 12. See Daniel Walker Howe, "Gilman, Samuel Foster," in *American National Biography*, 9, no. 63–64.

237. Gilman, 12–13.

ever produced by a new combination of the elements." To this empirical law there was but one possible exception: Crosse and Weekes had reportedly produced an insect by electrifying nonliving materials. Gilman dismissed this case, since the experiment had not been replicated and no respectable man of science defended it.[238]

The force of Gilman's argument was not directed against the development theory as such, but rather against a broader movement to deny external evidences of the supernatural in the universe, a movement for which *Vestiges* could be categorized as a recent and popular representative. Refuting *Vestiges* became a way of affirming Scripture as divine revelation. If, contrary to *Vestiges*, God intervenes in geological history to produce new species from time to time, then it seemed plausible that God also would intervene in human history to reveal new spiritual insights from time to time. Thus, no one would have any logical grounds to question the likelihood that divine revelation had been delivered in successive installments through the inspiration of the biblical writers.

Unitarians defended Scripture against skepticism not by vindicating the Mosaic record—a project in which no one at Harvard showed any interest—but by defending the very notion that a divine revelation is both possible and probable. Eight years after Gilman's address, the four-year Dudleian cycle again returned to the theme of revealed religion. This time, the Rev. Andrew Preston Peabody, a Harvard alumnus pastoring at Portsmouth, New Hampshire, addressed the audience that gathered annually in College Chapel. Like Gilman before him, Peabody affirmed a progressive, supernatural revelation from God to human spokespersons throughout biblical history. Also like Gilman, Peabody's understanding of this progressive revelation paralleled geologists' understanding of the progressive development in nature. Once again, nature's development had to be understood as the result of God's periodic interpositions, which meant *Vestiges* had to be refuted. For its striking synthesis of a theology founded on progressive revelations and a geology founded on progressive creations—each without any particular hermeneutic commitment to the opening chapters of Genesis—the following passage from Peabody's lecture deserves to be quoted at length:

238. Gilman, 13.

According to the Christian theory, the history of the spiritual creation has been marked by successive forming epochs, at each of which new spiritual agencies have come into action, new trains of spiritual causes have been put in operation, new modes of spiritual life have been brought into being ... Abraham ... Moses ... Christ. ... Now, if we admit the development-theory of the material universe, this hypothesis as to the spiritual creation is utterly untenable. If, without any creative acts or epochs, a mass of nebulous matter, which filled the entire area comprehended within the orbit of the outermost planet, in cooling threw off successive rings that globed [sic] themselves into worlds; if animal life in its lowest forms was the product of fermenting chaos in the infancy of those worlds; if on our own planet the *Acarus Crossii* furnished the parent stock for all animated nature; if man must abandon the genealogy closing with those sublime words [from Luke 3:37], "which was the son of Adam, which was the sun of God," and must trace his ancestry, not upward, but downward, through the ape, the tadpole, the polype, to the microscopic animalcule,—then is man's spiritual history equally a spontaneous development, and the germ of Judaism, the seed of the kingdom of Christ, floated in nebulous vapor, weltered in unformed chaos, was wrapt [sic] in the thin cuticle of the first animalcule whose aspiring *nisus* raised him to a higher grade of being. But the Lucretius *redivivus*, the modern apostle of this theory [i.e., the author of *Vestiges*], found it necessary in his second treatise [i.e., *Explanations*] to appeal from the inhospitality of the scientific world to the larger receptivity of popular ignorance.[239]

Unitarianism was neither popular nor ignorant; it was the religion of an educated elite who sought to affirm the supernatural in the most rational of ways. *Vestiges* served as a useful counter-example, against which Unitarian Christianity could be defended. By defending the authenticity of Holy Scripture as a divine revelation, Gilman and Peabody spoke a non-sectarian language, even if their

239. Andrew P. Peabody, *The Analogy of Nature and the Bible: The Dudleian Lecture Delivered in the Chapel of Harvard University, May 14, 1856* (Cambridge, MA: Metcalf and Co., 1856), 4, 5. See Robert D. Cross, "Peabody, Andrew Preston," in *American National Biography*, 17, no. 181–82.

usage of that revelation went contrary to Calvinism. It was helpful, perhaps, that the author of *Vestiges* still did not have a name. "Lucretius *redivivus*, the modern apostle" of materialism, was all Peabody needed to say to distinguish believers in supernatural revelation from skeptics. A true gentleman, of course, would have had the courage to show his face, but that fact just reinforced the Harvard conviction that the author of *Vestiges* lacked the Unitarians' moral character of science.

Unveiling the Author of Vestiges?

"Who wrote the Vestiges of the Natural History of Creation?" The question remained timely in the summer of 1849, when Francis Bowen thus introduced a *North American Review* commentary on two recent geological works.[240] One of them, *Ancient Sea-Margins*, bore on its title page an authorial attribution to Robert Chambers, announced to be a Fellow of the Royal Society of Edinburgh. "We had read only the first two sentences of the work," wrote Bowen, "when it seemed to us morally certain that the same hand wrote them which wrote the first two sentences of the Vestiges." He then reproduced in parallel columns the following passages[241]:

It is familiar knowledge that the earth which we inhabit is a globe of somewhat less than 8000 miles in diameter, being one of a series of eleven which revolve at different distances around the sun, and some of which have satellites in like manner revolving around them. The sun, planets, and

Taking observed facts for our data, we know that there was a time subsequent to the completion of the rock formations, when this island (not to speak of other parts of the earth,) was submerged to the height of at least 1700 feet. The proofs lie plain and palpable before our eyes, in the soft detrital

240. [Francis Bowen], review of *Ancient Sea-Margins, or Memorials of Changes in the Relative Level of Sea and land*, by Robert Chambers, and *A Memoir upon the Geological Action of the Tidal and Other Currents of the Ocean*, by Charles Henry Davis, *North American Review* 69, no. 144 (July 1849): 246–69, at 256. For a discussion of a similar argument published in Scotland in Nov. 1847, see Secord, *Victorian Sensation*, 291–92.

241. [Bowen], review of *Ancient Sea-Margins*, 257.

satellites, with the less intelligible orbs termed comets, are comprehensively called the solar system; and if we take as the uttermost bounds of this system the orbit of Uranus (though the comets actually have a wider range,) we shall find that it occupies a portion of space not less than three thousand six hundred millions of miles in extent.

Vestiges, p. 1.

masses, mixed in many places with marine shells, which overlie the hardened formations, reaching in some places to that height above the present sea level.

Ancient Sea Margins, p. 1.

Bowen's "moral certainty," ostensibly drawn from the pure induction of comparing two facts—the text from one book with the text from another—had other sources as well. In addition to quoting two more passages for comparison, this time using *Explanations* instead of *Vestiges,* Bowen stated that "the great popularity of the Vestiges must be attributed almost exclusively to the merits of its style." By this he meant that the author of *Vestiges* evidently had much experience in "adapting science to the comprehension of the people." Among science journalists, no one could compete with Robert Chambers, the editor of *Chambers' Journal,* which "diffuse[d] taste and information among the lower classes in Great Britain." Since *Vestiges, Explanations,* and *Ancient-Sea Margins* each presented science in a highly readable form, Bowen felt he could build a case for attributing the first two works to the avowed author of the third.[242] Moreover, Chambers had claimed in *Ancient Sea-Margins* to have originated certain ideas that first were published in *Vestiges,* a work which Chambers slyly neglected to cite in his *Ancient Sea-Margins.*[243] Who but Chambers could have written *Vestiges?*

If not Robert Chambers, then perhaps William Chambers had written it. An undergraduate writing for the students' newly established monthly, *Harvard Magazine,* asserted in 1854 that "*Vestiges of the Natural History of Creation,* a

242. [Bowen], review of *Ancient Sea-Margins,* 258.

243. [Bowen], review of *Ancient Sea-Margins,* 259–61.

strange compound of old and new philosophical systems, has been ascribed to various persons, but is now considered to be from the pen of William Chambers of Edinburgh."[244] Robert and William were well-known publishing partners, so suspicion of one could easily suggest suspicion of the other. For example, an 1848 notice in one of the most popular southern monthlies had suggested an affinity between William Chambers's *Miscellany of Useful and Entertaining Knowledge* and "the brilliant genius of the author of the Vestiges of Creation."[245]

Beyond Harvard, one could not regard the association of Chambers (whether Robert or William) with *Vestiges* as an established fact. The *New York Times* referred to the matter on numerous occasions. An 1852 article that mentioned the British Parliamentarian Richard Vyvyan in passing emphasized that he "did *not* write 'Vestiges of Creation.'"[246] The *Times* reported Robert Chambers's own disavowal of authorship twice in January of 1859, concluding that the Chambers rumor had likely resulted from the fact that someone had forwarded a few proof sheets of the book to him.[247] In April of that year, the *Times* reported that David Page, George Combe, and John Nichol each denied authorship, and claimed Chambers was responsible for *Vestiges*.[248] Not surprisingly, some Americans thought the author of *Vestiges* was "still unknown, and probably ever-to-be-unknown."[249]

It remains an historical mystery whether Bowen, who so confidently (and correctly) pegged Robert Chambers as the work's author, ever read an article entitled "The Author of the Vestiges of the Natural History of Creation," which appeared in the February 1846 *American Whig Review*. Here one may find many more facts gathered than in Bowen's article. Curiously, these facts pointed in a different direction than Bowen's. The *Whig Review* attributed *Vestiges* to the popular British author Isaac Taylor, since it seemed a natural sequel to his

244. [Daniel Webster Wilder], "Anonymous Books," *Harvard Magazine* 1, no. 1 (Dec. 1854): 22–24, at 23.

245. Review of *Chambers' Miscellany of Useful and Entertaining Knowledge*, ed. by William Chambers, *Debow's Review* 5, no. 4 (Apr. 1848): 399.

246. "The British House of Commons," *New York Times*, 9 Dec. 1852, 4.

247. "Personal," *New York Times*, 8 Jan. 1859, 1; and [untitled], *New York Times,* 21 Jan. 1859, 1.

248. "Miscellaneous," *New York Times*, 14 Apr. 1859, 12; "Literary and Other Items," *New York Times,* 14 Apr. 1859, 12; "Personal," *New York Times*, 23 Apr. 1859, 11.

249. "Fish and Fishermen," *Debow's Review* 15, no. 2 (Aug. 1853): 143–60, at 143.

Physical Theory of Another Life (1838). Like Bowen, the *Whig* writer selected statements from the anonymous *Vestiges* for comparison with statements from a work by an avowed author. The *American Whig Review*, however, made a much stronger case for similarity in style of language. *Vestiges* said, "tend strongly"; *Physical Theory* said, "strongly tending." *Vestiges* said, "with as little disturbance as possible to existing beliefs"; *Physical Theory* said, ". . . in disturbing our religious convictions." Not only did the two works employ similar phrases, but they also each favored "Latinized English" over "the simple, intelligible, good old English of the Saxons." *Vestiges* spoke of "retrogression, aberrant, arrestment, persistency, potentiality, tellurian, super-adequacy," and so on, while *Physical Theory* spoke of "incertitude, occult, nascent, potent, vivacious," and the like.[250] Content no less than style confirmed the *Whig* reviewer's suspicions:

> The fundamental idea of the two works is precisely the same; namely, *the development of a lower organization into a higher by law*. In the Theory of a Future Life, the author's aim is to show that the future man, the man beyond the grave, is to be a development by law, or a *natural* development of the present man. In the Vestiges, the present man, with his specific organization, is a development by law, or a natural development of some one of the various lower animals; which, again, was itself a development of some other below; and so down to the simplest forms of existence. The two books are, therefore, but parts of one book; two divisions of the same general proposition.[251]

Finally, the *Whig* writer juxtaposed sample paragraphs from the two "parts" of this "one book," showing that *Vestiges* and *Physical Theory* treated the distance of fixed stars, the constitution of other planets, the immortality of the mind, the galvanic nature of the brain, the merits of phrenological science, and the relationship of the law of development to revealed religion all in the same manner.[252] Indeed, "the real though suppressed title of the 'Physical Theory' is, the Vestiges of the Natural History of the future creation."[253]

---

250. "The Author of the Vestiges of the Natural History of Creation," *American Whig Review* 3 (Feb. 1846): 167–79, at 169.

251. "The Author of the Vestiges of the Natural History of Creation," 173.

252. "The Author of the Vestiges of the Natural History of Creation," 174–77.

Though Bowen did not share in this argument—and possibly was not even aware of it[254]—he and his Harvard colleagues did reach a conclusion similar to the final point emphasized in the *American Whig Review*. The author of *Vestiges* was too incompetent to speak authoritatively concerning the origins of the cosmos, organic life, and human beings. For the *Whig* writer, Isaac Taylor was known to lack specialized knowledge of geology, physiology, astronomy, and Hebrew (the latter being necessary for relating the former to Scripture). The judgment inevitable for a *Vestiges* authored by Taylor is that "it has no authority."[255] Bowen, in his 1845 review, had also brought into doubt the authority with which the author of *Vestiges* attempted to speak, though without attempting to identify the author's identity at that time.

In 1849, however, Bowen treated Chambers, whom he correctly concluded to be the author of *Vestiges*, with respect when reviewing his *Ancient Sea-Margins*. Chambers argued in that work that certain terraces marking the hills at equal altitudes on opposite sides of Glen Roy, Glen Spean, and other valleys in Scotland, had been formed when the sea suddenly dropped in elevation during the geological past. Bowen noted, however, that Louis Agassiz had attributed the formations to glacial activity and William Playfair had concluded that ancient inhabitants carved them to serve as aqueducts. Bowen, never wanting to give an inch to speculation, suspended judgment until geologists could gather more facts "either to confirm or refute his theory" of sea-margin terraces.[256]

Regarding Chambers's other theories—the nebular hypothesis, spontaneous generation, and species transmutation—Bowen believed sufficient evidence already had been gathered. The facts of science, Bowen had asserted in his 1845 review, settled the case against the development theory. Speaking as a Lowell lecturer in 1849, Bowen again criticized the "popular cosmologist," who authored *Vestiges*, for reviving the old Greek doctrine that the cosmos originated

253. "The Author of the Vestiges of the Natural History of Creation," 179.

254. Bowen's *North American Review* reported that Isaac Taylor's "ideas are in the air, being inhaled and exhaled again by others." If Bowen had been aware of the *American Whig Review*'s attribution of *Vestiges* to Taylor, perhaps he would have dismissed it as a conflation of the originator (Taylor) and the re-breather (Chambers) of Taylor's ideas about development. See Review of *Four Lectures on Spiritual Christianity*, by Isaac Taylor, *North American Review* 61, no. 128 (July 1845): 159–81, at 159–60.

255. "The Author of the Vestiges of the Natural History of Creation," 179.

256. [Bowen], review of *Ancient Sea-Margins*, 264.

as a fire mist. Whether applying the law of development to nebulae, animacules, or humans, "a purely speculative notion is here superinduced upon the inductions of experience, though a lingering respect is still manifested for the Baconian method, the theory being defended by a spurious induction from a few monstrosities." This lecture series, which treated "Metaphysical and Ethical Science applied to the Evidences of Religion," was reprinted in 1855 as a textbook for colleges, *Vestiges* being a notable "instance of the corruption of physical science by metaphysical ideas."[257]

"God's thoughts manifested in tangible realities": *Agassiz's Natural Theology of Species Fixity*

Meanwhile, metaphysical ideas were becoming firmly entrenched in Harvard science through the work of Louis Agasssiz. Though unwavering on the question of species fixity, Agassiz's metaphysics pushed his science in some directions that did not please all at Harvard. Not only did he lead his audience away from Bowen's strict inductivism, he also endorsed a polygenist *Festschrift* in honor of Morton, this time with an ultimatum that one must choose between the racist polygenism of Morton or the development theory of *Vestiges*. Despite rising abolitionist sentiment in Boston, Agassiz managed to strengthen his footing in the community, ultimately building a European-style research museum at Harvard.

Agassiz's national fame as a leading scientific authority opposed to *Vestiges* spread also across the Atlantic, as did the news of his somewhat peculiar views on the origin of the human races. A British observer, aware that "*Vestiges of Creation* has been as generally read in the United States as at home, and has made many half-converts," celebrated the impact of Agassiz's lectures on "the true doctrine of development as opposed to that of Lamarck, popularized in the *Vestiges.*" But Agassiz had linked his refutation of *Vestiges* to an endorsement of polygenism, which some found unscientific and many thought irreligious. Southern "negro-haters" welcomed his views, and "a free-thinking community

257. Francis Bowen, *The Principles of Metaphysical and Ethical Science Applied to the Evidences of Religion,* new ed., rev. and annotated, for the use of colleges (Boston: Brewer and Tileston, 1855), 10, 31, 30.

like that of Boston" tolerated them, but otherwise Agassiz had occasioned "new doubts respecting the relations of science to religion."[258]

Agassiz's most enduring statement on polygenism had direct bearing on the *Vestiges* debates. When the Alabama physician Josiah Nott and the English emigré Egyptologist George Gliddon prepared a polygenist anthology to honor the recently deceased Morton in 1854, they solicited a contribution from Agassiz. In an introductory essay for that work, *Types of Mankind*, Agassiz stated quite plainly that investigators of nature had but two options: "Either mankind originated from a common stock," or else "what are called human races, down to their specialization as nations, are distinct primordial forms of the type of man." *Types of Mankind*, which passed through ten editions from 1854 to 1871, presented numerous arguments (from geology, philology, and many other sciences) against the first option, monogenism.[259] Highlighting one of those arguments, Agassiz referred to recent researches in comparative anatomy by Richard Owen and Jeffries Wyman that revealed human races to differ from one another more than the chimpanzee differs from the gorilla; if the latter are distinct species, then why not the former? In addition to this scientific argument against monogenism, Agassiz also provided an ideological one: "The consequences of [monogenism] run inevitably into the Lamarckian development theory, so well known in this country through the work entitled 'Vestiges of Creation.'"[260] It seemed, then, that to reject the development theory, one had to affirm that each human race had been separately created by God.

A separate creation for each race did not logically necessitate that the races be ranked hierarchically, but Agassiz was, after all, writing in nineteenth-century

258. James F. W. Johnston, *Notes on North America: Agricultural, Economical, and Social*, 2 vols. (Boston: Charles C. Little and James Brown; Edinburgh and London: William Blackwood and Sons, 1851), 2:440–41.

259. The scientific arguments in *Types of Mankind* received much greater respect among American readers than did those deployed in *Vestiges*. For one thing, American men of science widely agreed that geology indicated numerous interventions by God for creating distinct animal forms, so extending this to human races at least seemed more plausible than saying that all species had developed from a fire-mist without any interventions by God. See Stanton, esp. 103–4, 162–65; Alter, 40–41.

260. Louis Agassiz, "Sketch of the Natural Provinces of the Animal World and Their Relation to the Different Types of Man," in *Types of Mankind; or, Ethological Researches, based upon the Ancient Monuments, Paintings, Sculptures, and Crania of Races, and upon the Natural, Geographical, Philological, and Biblical History*, ed. J. C. Nott and George R. Gliddon (Philadelphia: Lippincott, Grambo, & Co., 1855), lvii–lxxvi, at lxxv–lxxvi.

America. Natural historians generally viewed the entire animal kingdom in terms of a scale from lower to higher forms and, as Agassiz had noted when visiting Philadelphia in 1846, even white abolitionists did not consider members of the dark-skinned race to be worthy marriage candidates for their daughters.[261] Polygenism thus reinforced racism, giving it a scientific justification at a time when racism was widely tolerated anyway. In August 1854, members of the Boston Society of Natural History discussed some Native American skull samples sent to them by Nott. The proceedings reported "a weak intellect and a strong animal propensity in the American Indian compared with the white races." The members felt that plates 353 and 354 in Nott and Gliddon's *Types of Mankind* had "somewhat exaggerated" the distinction, but, not surprisingly, they concurred with the general conclusion that Indians were racially inferior to Boston's European-descended intellectuals.[262]

At least one Harvard undergraduate was less concerned about polygenism itself than about the foundation of its argument. Thomas William Clarke wrote for the *Harvard Monthly* in 1855 that the Bible was "a compendious body of morals," not "a complete, universal, cyclopedic [*sic*] collection of all knowledge, past, present, and future." Clarke held that the relations of the human races should be sought from an investigation of nature, not Scripture. He criticized Nott and Gliddon for twisting Genesis 10 (which concerns the origin of human tribes and languages) into an argument for polygenism. Clarke considered Scripture to be silent and neutral on the matter. St. Paul's statement in Acts 17:25–26 that God made all nations from one blood could be interpreted as a remark concerning their moral worth, not their biological origin. The origin of the human races should be sought in nature, and from Scripture men should learn to live a moral life.[263]

One finds in Clarke's article an echo from Joseph Henry Allen's review of *Vestiges*, in which theology and the natural sciences were sharply distinguished, the one dealing with moral character and the other with physical facts.

---

261. Stanton, 103.

262. *Proceedings of the Boston Society of Natural History* 5 (1854–1856): 78. For historical and scientific analysis of Morton's data, leading to the conclusion that he committed "conscious fraud," see Gould, 54–69.

263. Thomas William Clarke, "The Unity of the Human Race," *Harvard Monthly* 1, no. 7 (July 1855): 337–40, quoting 338.

Nevertheless, professors of natural science at Harvard still were expected to be theologically sound. Agassiz's research assistant James Elliot Cabot accordingly informed the community that, although the Swiss naturalist formerly had been a pantheist, his studies of "former creations" had convinced him of "the existence of a personal God, the Author and Ruler of the universe."[264] External evidences had more or less made Agassiz into a Unitarian, even if he "never went to church."[265]

His idealism and erstwhile pantheism risked linking him to the Transcendentalist outgrowth of Unitarianism, but Agassiz carefully spoke the language of traditional Unitarianism when appealing to the external evidences of the Creator's successive interventions in natural history. "An acquaintance with more than fifteen hundred species of fossil fishes," Cabot could quote from Agassiz in 1847 to illustrate his empiricism, "has taught me that species do not pass insensibly into each other, but that they appear and disappear unexpectedly, without showing any immediate connection with those preceding them." Thus, any affinities between species must indicate only the common Mind that planned them all, not a genealogy begun by a common ancestor.[266]

The *Christian Examiner*, a Boston Unitarian standard, gave Agassiz the lead article for the January 1851 issue. During the previous spring, he had been united with the Unitarian community by marrying Elizabeth Cabot Cary; this made him the son-in-law of the wealthy Boston attorney Thomas Graves Cary, the brother-in-law of Harvard classicist Cornelius C. Felton, and also a relative to James Elliot Cabot (Agassiz's aforementioned research aide, who served also as the correspondence secretary of the Boston Society of Natural History).[267] Writing on the "character of God," a phrase commonly heard from Unitarian lips, Agassiz affirmed a "successive, gradual, progressive creation, planned by the Almighty in the beginning, and maintained in its present state by his providential

264. J. Elliot Cabot, "The Life and Writings of Agassiz," *Massachusetts Quarterly Review* 1, no. 1 (Dec. 1847): 96–119, at 99.

265. Charles William Eliot, "Address at the Celebration of the One Hundredth Anniversary of the Birth of Louis Agassiz, delivered at Sanders Theatre, 27 May 1907," rpt. from *Proceedings of the Cambridge Historical Society* (pamphlet copy available in Harvard University Archives, HUG 1128.2), 104.

266. Cabot, 106–7.

267. Marcou, 477; Lurie, 99; Valerie A. Thomas, "Inventory of the Records of the Boston Society of Natural History," History 670 project, M.A. student in History–Archival Methods Program at U. Mass.–Boston, 12 May 1984 (filed as the finding aid for the society's records, housed at the Boston Athenaeum), 21.

action."[268] As before, he emphasized that nature's "development" is not a material change occasioned by physical circumstances, but rather an imprint from "the ideal relations in the mind of the Creator" stamped into nature by miraculous interventions.[269]

Just as Lowell had brought Agassiz to Boston, so also Boston's adopted son Agassiz began to bring others. In 1848, his European colleagues Arnold Guyot, Leo Lesquereux, and Jules Marcou arrived to assist with zoological research at the new Lawrence Scientific School.[270] Agassiz could bring in ideas as well as people. In 1850, he arranged for one of the most popular British rejoinders to *Vestiges* to be republished in America: *Foot-prints of the Creator.*

Hugh Miller, a leader in the Scottish Free Church movement who acquired a widespread reputation for both his evangelical and geological writings, had issued *Foot-prints of the Creator: or, The Asterolepis of Stromness* in Edinburgh in 1849. Miller charged the author of *Vestiges* with a lack of first-hand knowledge of geology. For his own part, Miller embarked regularly on geological travels as he took leave of his editorial duties for the evangelical *Witness* each summer. In 1848 he discovered a section of fossilized bone from a fish of the genus *Asterolepis* in the Old Red Sandstone of the Orkney Islands, a geological layer too ancient to contain such a complex kind of fish if the development theory were to be accepted. Consistent with his theological convictions about original sin, Miller argued that geological succession was a history of degeneration rather than of progress: God periodically created new species, but these degenerated unto extinction, until God created new ones in their place.[271]

Though Agassiz did not share Miller's Calvinist evangelicalism, the two researchers had four other things in common: an appreciation for hands-on natural history; a conviction that natural history revealed God's successively performed direct creations of organic species; the conclusion, drawn from the

268. L[ouis] A[gassiz], "Contemplations of God in the Kosmos," *Christian Examiner* 4th ser., 15 (Jan. 1851): 1–17, at 2, 11.

269. Agassiz, "Contemplations of God in the Kosmos," 10.

270. Lurie, 142.

271. For a facsimile reprint of the 1861 posthumous edition of Miller's work, together with an historical introduction and bibliography of secondary literature, see Hugh Miller, *Foot-prints of the Creator: or, The Asterolepis of Stromness,* introduced by Louis Agassiz, rpt. as vol. 2 of *"Vestiges" and the Debates before Darwin,* ed. Lynch.

preceding two points, that the purported science in *Vestiges* was both amateurish and wildly inaccurate; and, a desire to see *Foot-prints* circulated in America. Indeed, Miller's *Asterolepis* specimen added yet more evidence to a claim Agassiz had made in his 1848–1849 Lowell lectures: complex fossil fishes are commonly found alongside simpler fishes in ancient strata, indicating their contemporaneous origin and undermining the development hypothesis.[272]

Agassiz intended, therefore, to write a welcoming introduction for the American reprint of Miller's work, which already was it its third British edition.[273] He praised Miller for "expand[ing] our views of the Plan of Creation" and forging a "successful combination of Christian doctrines with pure scientific truths."[274] For the remainder of his introduction, Agassiz reprinted an edited version of Sir David Brewster's recent article on *Foot-prints* from the *North British Review*. Though striking two of Brewster's paragraphs concerning *Vestiges*' development theory and some references to Brewster's own review of that work (the same review appended to some American reprints of *Vestiges*), Agassiz retained Brewster's judgment concerning the "unscientific parentage of the theories promulgated in the 'Vestiges.'"[275] Specifically, De Maillet "knew nothing of the geology even of his day [the mid 1700s]"; Jean-Baptiste Lamarck, "though a skillful botanist and conchologist, was unacquainted with Geology"; and Lorenz Oken even admitted writing "in *a kind of inspiration!,*" since he, too, lacked geological experience.[276]

The appeal to geology became more important for Agassiz during the 1850s, as his research objectives shifted from refuting the development theory popularized by the author of *Vestiges* to keeping Gray's botany in check with respect to species fixity. The author of *Vestiges* had suggested that environmental conditions could induce a species to evolve into a higher type. Gray, somewhat similarly, was suggesting by the late 1850s that biogeographic diversity could be

272. Agassiz, *Twelve Lectures*, 27.

273. On the arrangements for the American reprint, see Hugh Miller to Louis Agassiz, 25 May 1850, in *Life, Letters, and Works of Louis Agassiz*, ed. Marcou, 2:470–77.

274. Agassiz, "Hugh Miller, Author of 'Old Red Sandstone' and 'Foot-prints of the Creator,'" introduction to Miller, *Foot-prints*, iii–xxxvii, at iii.

275. See Agassiz's annotated copy of Miller, *Foot-prints* (Museum of Comparative Zoology), where his annotations of Brewster's review also are filed; note esp. Agassiz's markings in Brewster, 459–60, 465.

276. Brewster, as reproduced by Agassiz, "Hugh Miller," xxxv–xxxvi.

the result of populations of a common ancestral species that migrated to regions with different environmental conditions and, consequently, became differentiated.[277] Feeling alienated by Gray's botanical projects, Agassiz confided in James Dwight Dana of Yale College that he remained "fully satisfied" that "the location [both geographical and taxonomic] of animals, with all their peculiarities is not the result of physical influences, but lies within the plans of intentions of the Creator." In order to maintain this position, however, Agassiz urged that fossil species be given more weight. A comparative anatomy focused too exclusively on living forms could lead to an over-emphasis on their relations to the environment, whereas an account of fossil forms would be more suggestive of the Creator's organic plans that linked all species, according to their homologies and analogies, into the Plan of Nature.[278]

Agassiz was the sort of naturalist who could be taken seriously when claiming "to discover six and even eight new species of fishes in a single day."[279] He was a careful observer, able to identify the most subtle distinctions between each of the sundry species God had thought in his mind and created on the earth. Assisted by his staff of European immigrants at the Lawrence Scientific School and a network of American contributors, such as Thoreau in rural Massachusetts and James Hall in New York, Agassiz prepared a grand masterpiece, *Contributions to Natural History*. The first volume, which appeared in 1857, began with an "Essay on Classification" that opposed *Vestiges*.[280] The famed Harvard zoologist gathered 2,500 subscribers for the work, an unprecedented number for scientific books.[281] Agassiz's emphasis on discerning the Creator's thoughts in nature, rather than merely cataloguing His actions, pushed Harvard science away from the strict inductivism that Bowen had advocated. As one reviewer, favorable to Agassiz, humorously observed about Bowen's approach:

277. See the discussion of this in the next section.

278. Louis Agassiz to James D. Dana, 16 Feb. 1853 (Agassiz Papers, Houghton).

279. Louis Agassiz, "Directions for Collecting Fishes and Other Objects of Natural History," typeset circular, 1853, filed among the papers in Henry David Thoreau, Extracts mostly upon Natural History, 1853–1858 (Houghton Library, 12.7.10).

280. This has been reprinted, with historical commentary, as Louis Agassiz, *Essay on Classification*, ed. Edward Lurie (Cambridge, MA: Harvard University Press, 1962). Agassiz explicitly referred to *Vestiges* at pp. 117–18, and the essay as a whole served both to refute *Vestiges* and to promote Agassiz's own idealist program of comparative zoology.

281. "Agassiz's Natural History," *Atlantic Monthly* 1, no. 3 (Jan. 1858): 320–33, at 324.

> If a man would watch a thermometer every hour of the day
> and night for ten years, and give a table of his observations,
> the result would be of interest and value. But the bulbous
> extremity of the instrument would probably contain as much
> thought at the end of the ten years as that of the observer.[282]

By identifying God's invisible thoughts behind the visible specimens of natural history, Agassiz earned the reviewer's praise as one who "has tracked the warm foot-prints of Divinity throughout all the vestiges of creation."[283]

In the fall of 1857, Agassiz addressed an audience in Brooklyn with his typical emphasis upon the divine Thought pervading nature. "To show that Nature speaks in favor of the theory of Thought (and of nothing else) is the true object of the Philosophy of Zoölogy." Agassiz acknowledged that not all species were as "strongly marked" as others, but his understanding of species as instantiations of "the A[l]mighty thoughts" allowed for an analogy that preserved his zoology from the transmutation leanings that were creeping into Gray's botany. Agassiz suggested that just as a philosopher's thoughts include both "well[-]defined thoughts" and "metaphysical subtleties that puzzle almost everyone," so also species, which are God's thoughts, could be either well-defined or else more subtly distinguished. A naturalist with Agassiz's expertise could classify organisms by plan of structure (Kingdom), by the way in which the plan is carried out (Class), by complexity of structure (Order), by form (Family), by ultimate structure of parts (Genus), and by relative size and ornament (Species). "Individuals," Agassiz concluded, "have a real, limited, physical existence; the categories of thought they represent have a real, eternal, metaphysical existence."[284]

Agassiz was convinced that he was discovering God's thoughts in nature, not imposing his own thoughts. In March 1859, he addressed an audience in the Representatives' Hall at the Massachusetts State House, where a series of "Educational Lectures" were being held. Agassiz urged that natural history field work and other hands-on experiences with natural specimens be integrated into

---

282. "Agassiz's Natural History," 329.

283. "Agassiz's Natural History," 333.

284. Theodore Lyman, A Digest of Two Courses and a Half of Lectures, on Zoology and of two Courses on Comparative Anatomy, by Profs. Agassiz and Wyman, to which is added some other matter, Brooklyn, Aug.–Oct. 1857 (Harvard University Archives, HUC 8857.398, Box 495), 1, 6, 9–11.

common education. Books can teach only so much; real learning takes place when one holds a grasshopper and contemplates the functioning of its parts. Books are likely to express the opinions of men, but nature herself can teach the very thoughts of God:

> When we thus trace the relations which exist between organized beings, and reach higher and higher generalizations, it is not our thoughts that we put into nature, but the thoughts which are expressed in nature, which we read out of it. It is in fact God's thoughts as manifested in tangible realities which we attempt to decipher.[285]

Gray by this time was losing respect for Agassiz. It was one thing to reject the development theory of *Vestiges*, that absurdly speculative and sloppy work of amateur science. But to appeal to "God's thoughts" in order to insist that God had created Japanese and North American plants, or African and European humans, separately, seemed to presuppose the very facts that naturalists sought to discover when doing empirical research. Not only that, but Agassiz also remained immensely popular, no doubt leaving Gray feeling jealous.

Asa Gray: A Botanist Preferring Natural Development over Agassiz's Theory

Meanwhile, Gray was developing an alliance with a London geologist named Charles Darwin. Darwin, widely known for the scientific travel journal he published after spending nearly five years traveling the world on the *H.M.S. Beagle*, had been privately sketching out a theory of species transmutation. The clamor against *Vestiges*, which was equally strong in Britain as in America, and his unstable health were keeping him silent on the matter, at least publicly. To Gray, whose botanical expertise Darwin had enlisted through their mutual collogue Joseph Hooker, Darwin confided his new views on nature. Writing to the Harvard botanist in July 1857, he informed him for the first time of the reason he had taken interest in Gray's biogeography:

> Either species have been independently created, or they have descended from other species, like varieties from one species.

285. "The Study of Nature: Lecture by Prof. Agassiz," *Boston Weekly Courier* (26 Mar. 1859).

> I think it can be shown to be probable that man gets his most distinct varieties by preserving such as arise best worth keeping, & destroying the others,—but I shd fill a quire if I were to go on. To be brief I assume that species arise like our domestic varieites with much extinction; & then test this hypothesis by comparison with as many general & pretty well established propositions as I can find made out,—in geograph. distribution, geological history—affinities &c. &c. &c.[286]

Darwin's "&c. &c. &c." received fuller expression in his 1859 book, *On the Origin of Species by Natural Selection*. For the present, he boldly—though still only privately—confessed to Gray, "I have come to the heterodox conclusion that there are no such things as independently created species—that species are only strongly defined varieties."[287] Gray did not yet support Darwin, but neither did he oppose him. If anything, Gray was far more concerned about Agassiz. Gray and Darwin shared a common empirical methodology, a common body of evidence (they had been exchanging notes with one another and with Hooker), and a common conviction that species are defined by ancestry.[288] Agassiz defined species by God's thoughts, which were beyond Gray's empirical methodology and at times contradicted Gray's body of evidence. As Gray confided to Torrey early in 1859:

> I am going to hold forth for nearly an hour, upon Japan Botany in its relation to ours and the rest of the northern temp. zone, and knock out the underpinning of Agassiz'[s] theories about species and their origin—show, from the very facts that stumbled DeCandolle the high probability of single and local creation of species,—turning some of Agassiz'[s] own guns against him.[289]

The problem reduced to this: Agassiz regarded each local population as a distinct species, created on site by God, whereas Gray regarded local populations as geographic varieties related ancestrally to a common species. Gray went public on this issue at the January 1859 meeting of the American Academy of Arts and

---

286. Charles Darwin to Asa Gray, 20 July 1855, in Dupree, 244–45, at 244.

287. Charles Darwin to Asa Gray, 20 July 1855, in Dupree, 244–45.

288. Dupree, 247–48.

289. Asa Gray to John Torrey, 7 Jan. 1859, typescript (Herbarium).

Sciences, which convened that year in his father-in-law's home in Boston. He argued that Agassiz's theory of each species being created on the spot "would remove the whole question out of the field of inductive science."[290] Gray's alternative approach integrated geological history with present-day biogeography. He theorized that a land bridge had formerly linked Asia with North America, during which time the same plant species populated the northern portions of each continent. As glaciers advanced southward, new seed took root at increasingly southern latitudes, where the climate was more suited for them. Thus it came about that identical plants inhabit the middle latitudes of each continent. A similar history of nature could also account for the geographic dispersion of similar, but not exactly identical, species: in this case, they would be varieties descended from a common ancestral species.[291]

Darwin differed from Gray in that Darwin was suggesting that not only new varieties could develop from a common ancestor, but also new species—indeed, new genera, families, orders, classes, phyla, and kingdoms. Though Gray certainly saw more room for God's activity in species transmutation than Darwin thought warranted, the two naturalists were beginning to feel much closer to one another than either one of them did to Agassiz. On October 22, 1859, just four weeks before Darwin's *Origin of Species* hit the streets of London, Joseph Henry wrote to Gray, "I see that the forth coming [sic] work of your friend Darwin is alluded . . . at the meeting of the British [A]ssociation [for the Advancement of Science]."[292] Within months, Gray and Agassiz, who had cooperated in a joint opposition to *Vestiges* fifteen years earlier, divided themselves into the most publicly visible scientific opponents in America concerning Darwin's new form of the development hypothesis: the botanist defended Darwin's theory as theologically acceptable and scientifically grounded; the zoologist insisted just the opposite. Each, by this time, had a different notion of theology, the natural sciences, and the proper relation between the two.[293] The partnership Gray previously had shared with Agassiz against species transmutation was now severed. The era of *Vestiges* was drawing to a close, and a new alignment of the

290. Asa Gray, quoted in Dupree, 254.

291. Dupree, 250.

292. Joseph Henry to Asa Gray, 22 Oct. 1859 (Herbarium).

293. Lurie, chap. 7; Dupree, chaps. 14–15; Livingstone, 59–64; Croce, 44–58.

natural sciences at Harvard was emerging. From now on, the fault-line would run a different direction and have a new name: Darwinism.

Concluding Interpretations: Understanding Vestiges *at Harvard*

Whereas Princeton Presbyterians sharply distinguished the physical from the spiritual in order to preserve the agency of a sovereign God (Spirit) over his passive creation (matter), Harvard Unitarians had a different reason for making a similar spirit/matter distinction in their natural sciences. Unitarian Christology separated the divine from the human ontologically, even though Christ the human was still regarded as "divine" in the functional sense of being a faithful prophet of God the Father—a prophet whose ministry was certified by the Father's bestowal of the power to do miracles and rise from the grave. Thus, although the last thing that Unitarians wanted was the sovereign God of Princeton, they found themselves agreeing with Princeton's conclusion that mind and matter, active causes and passive effects, God and nature, are entirely distinct. To say, as the author of *Vestiges* had done, that nature could develop without the Deity's intervention sounded at Harvard like rank materialism, a loathsome implication that matter contained its own causes. On that conclusion Harvard's and Princeton's men of sciences agreed, but remarkably, and self-consciously, they had reached that agreement from fundamentally opposed theological foundations.

At Harvard the Christological separation of the human and the divine may also have contributed to the disciplinary fragmentation between the theological and moral sciences on the one hand and the natural sciences on the other. This separation of disciplines, in turn, led to a different formulation of responses to *Vestiges* at Harvard than at Princeton. At Princeton, all of God's truth was one, but at Harvard the Rev. Joseph Henry Allen sharply separated theology and morality from the physical sciences. For all Allen cared, the physical sciences could demonstrate—as the author of *Vestiges* claimed that they had—that humans were likely the ancestors of animals. The moral and theological sciences, however, would nevertheless separate man from beast by establishing the "supremacy of *character*," a trait the natural sciences could not scrutinize.[294]

294. A[llen], 345.

As for the character of *Vestiges* itself, members of the Princeton and Harvard communities each projected their most deeply felt existential fears onto that work. At Princeton, *Vestiges* was Arminian, denying God's sovereignty by affirming the potential of nature to improve itself. For Harvard's Unitarians, as Allen emphasized, *Vestiges* was "absolute and entire fatalism," which was a common Unitarian invective against Calvinist predestination.[295] Fatalism assaulted both the moral character of God (by making Him seem like an arbitrary judge who sends some sinners to heaven and other sinners to hell) and the moral character of human beings (by denying that they have enough good in themselves to develop lives that will emulate the perfect moral life of Christ).

The Revs. Samuel Gilman and Andrew Preston Peabody, like Allen, grounded their visions for human and divine morality upon external evidences (most notably, Christ's revelation in the Scriptures) as well as internal evidences (such as the conscience, as understood by biblical theology and moral philosophy). Their biblical objections to *Vestiges* were not predicated upon any commitment to a particular interpretation of the creation events described in Genesis 1–2. Nor were their Dudleian lectures generic Protestant natural theology, but rather Unitarian natural theology. From their characteristically Unitarian perspective, the denial in *Vestiges* of God's interventions in natural history denied a *prima facie* argument in favor of scriptural inspiration, since the latter was, analogously, a process that required divine intervention in human history. The defense of Unitarian biblicism was crucial, both in order to refute Calvinists who accused Unitarians of abandoning the Bible, and also to check the progress of someone in the midst of Harvard Unitarians who was willing to fulfill Calvinists' expectations of Unitarians: Ralph Waldo Emerson.

Tossing aside the Bible, Emerson and his Transcendentalist cadre turned from external evidences to explore more deeply the spiritual potential of internal evidences alone. Emerson's Christology consequently regarded the divine as all moral or spiritual, with no physical component. Recall, for example, that Emerson had left the Unitarian establishment precisely because he wanted to stop using the physical elements when celebrating the Eucharist, and that he infuriated Andrews Norton precisely by appealing to the Christ within each man, to the near exclusion of the Christ whose life was testified in the gospel accounts.

295. A[llen], 344.

Emerson was, in a sense, simply following Unitarian Christology to one of its possible logical conclusions: if Christ cannot be both God and man, then neither can truth be found in both spirit and matter.

By so doing, Emerson also moved to a theological position more amenable to the development theory of *Vestiges*. Emerson's resulting theology tended toward pantheism: nature and his own body were but emblems to help his mind contemplate God who flowed through all, God who was all that really is. Whereas Bowen feared that *Vestiges* promoted materialism, Emerson embraced *Vestiges* precisely because it also fit well with pantheism. It would seem, then, that materialism and pantheism are not so much like two extremes on a continuum, but more like two ends of a horse-shoe: much closer to one another than either one of them is to the dualist center. If the two ends be permitted to touch, then one is left only with a word game: should all of reality be called "God," or should it be called "matter"? Either way, *Vestiges* would win and Unitarians would lose.

At the center of Harvard natural science were two non-Unitarians, Asa Gray and Louis Agassiz. Both of their careers, however, may be illuminated by the postulation of an "ideal type" of Harvard Unitarianism. Gray and Agassiz each became a member of the local theological community, not only by marriage to a prominent Unitarian daughter, but also by participation in a public profession of a Unitarian-compatible faith in science at the Lowell Lectures. The creed was by no means formalized; few Unitarians would have wanted it to be. Whether consciously or unconsciously, Gray and Agassiz each supported the local theological community's program of promoting religious truth from the pillars of internal and external evidences in a manner that avoided any taint of sectarianism. This was how Unitarianism structured the local scientific environment, and it involved, of course, refuting *Vestiges of Creation*, which had brought into doubt the moral character of natural history in general and of Gray's and Agassiz's respective specialties of botany and zoology in particular.

By the 1850s Gray's and Agassiz's paths in the natural sciences were radically diverging. Still, each of them continued to speak in a manner that resonated powerfully with Unitarian convictions. Their participation in the natural sciences thus may be interpreted in terms of a common "ideal type" of Unitarian theology that held them within a socially acceptable range of ostensibly non-sectarian Christian opinion.[296] They were, in terms of the methodology employed for this study, anchored to the same socially situated systematic theology. Gray's empiricism guarded the Unitarian gate against speculative encroachments from Transcendentalism: nothing would grow on the lawn of Harvard Yard but actual, empirical species of plants. Agassiz's idealism, though at odds with Gray's methodology, required a strict separation of the physical/material from the spiritual/ideal, just as the Unitarian Christ had to be spiritually divine without being physically God incarnate. Agassiz's science made sense to Unitarians for precisely the reason that Emerson himself had first developed Transcendentalism from within the Unitarian fold: Unitarians were latent idealists, even if it would take Bowen until the 1870s to embrace Kant.[297] Moreover, Agassiz served the public interest—or rather, served the cause of Unitarian public philanthropy—by gracing the Odeon Theatre with one sold-out performance of Lowell Lectures after another.

No matter how well Gray and Agassiz, each in his own way, spoke a language that pleased Unitarian ears, neither of them was reducible to it, and each was trying to take the Harvard community in a different direction. By the late 1850s, the nature of species—a topic on which they had thought themselves agreeing when they jointly began rejecting *Vestiges* in the mid 1840s—became a wedge issue, representing all that was different between their respective approaches to scientific research. As the two naturalists diverged, they each sought to retain close ties with a third American authority on natural history, James Dwight Dana of Yale College. Dana, who will be discussed at length in the next chapter, corresponded frequently with faculty at both Princeton and Harvard. Neither an Old School Presbyterian nor a Boston Unitarian, Dana's existential concern looked neither to the sovereignty of God nor to the moral character of humanity, but rather in a third direction, which his fellow Congregationalists called the "moral government of God." Dana and his companions at Yale would, accordingly, bring yet another theological perspective to the dilemma of the day: "What must I do with *Vestiges*?"

296. Cf. Croce, 57.

297. On Bowen's eventual acceptance of idealism, see Ahlstrom and Carey, 32; Rand, 197–200.

Study Questions

1. Unitarians at Harvard were a more diverse group than the Presbyterians at Princeton. Nevertheless, the Harvard community was very cohesive. Identify the social and intellectual priorities that anchored Harvard to the Unitarian establishment.

2. How did Unitarians distinguish themselves from deists and skeptics, on the one hand, and from Calvinists, on the other hand?

3. Explain the tension between Scottish Realism and Neo-Platonic Idealism as these two movements shaped Unitarianism and Transcendentalism at Harvard.

4. Contrast the empirical approach to botany pursued by Prof. Asa Gray with the idealist approach to zoology pursued by Prof. Louis Agassiz. Explain why each of these approaches reached a common verdict against the development theory in *Vestiges*.

5. What did Transcendentalists, such as Ralph Waldo Emerson, appreciate about *Vestiges*?

6. Explain how the debate between Agassiz and Gray over polygenism repositioned Gray, who had staunchly opposed *Vestiges*' development theory, as a willing supporter of Darwin's theory of evolution.

7. Although Gray and Agassiz disagreed sharply with one another, and neither one of them was a Unitarian, they each in some sense supported "the Unitarian character of science" at Harvard. Explain how this was so.

CHAPTER FOUR

Yale: *Vestiges* Meets
the Moral Government of God

"I have inferred that the theory of the Vestiges, *if proved,*
would not affect the truths of Christianity.—Am I rig[ht?]"
Prof. James Dwight Dana, confiding to Asa Gray, 1846[1]

Introduction

In January 1845, Benjamin Silliman, Yale College's Professor of Natural History, left his home on Hillhouse Avenue in New Haven, Connecticut, for a lecture tour in the American South—a course of action involving "anxious & religious thought."[2] Consistent with his long-standing interest in using geology to improve theology, he addressed an audience at First Presbyterian Church in Mobile, Alabama, with the hope that his geological "proofs . . . are in harmony with the higher use of this Temple of God."[3] Meanwhile, Gideon Mantell, an

---

1. James Dana to Asa Gray, 21 Apr. 1846 (Gray Papers, Gray Herbarium, Harvard University). The final letters of "right" have been inferred, since a tear in the original document has removed them.

2. Benjamin Silliman, Diary, 16 Jan. 1845 (Silliman Family Papers, Yale University, Reel 3).

3. Benjamin Silliman to the Rev. W. T. Hamilton, 28 Mar. 1845, copied into Silliman's diary entry for that date (Silliman Family Papers, Reel 3).

English geologist, sent a letter to Silliman concerning "a little volume of 380 pages, <u>anonymous</u>, called 'Vestiges of the Natural History of the [*sic*] Creation.'" The book had "made a great sensation," in England. Its "most extraordinary speculations" appealed especially to readers who were not sufficiently trained in geology to recognize its "insufficient data, or mistaken facts."[4] It was not until May that Silliman replied. "I read the work in the steam boats going out," he wrote, "& my impression corresponded well to yours."[5] Silliman, in fact, published Mantell's letter in the June edition of the *American Journal of Science*, of which he was the founding editor. Mantell's critical assessment thus became the second notice of *Vestiges* to appear in America's leading scientific journal.[6]

Some mystery surrounds the first notice of *Vestiges* that appeared in Silliman's *Journal*. In the March issue Silliman's assistant editor, James Dwight Dana, had published a single paragraph on *Vestiges*, "strongly recommend[ing] the work to our readers." Dana, who taught geology and mineralogy at Yale, provided no specific warning of scientific errors or theological heresies. He did, however, allude to some minor reservations. "We cannot subscribe to all of the author's views," he conceded vaguely, characterizing *Vestiges* as "novel and interesting," its claims at times "startling."[7] The discrepancy between Dana's generally positive evaluation and the hostile stance shared by Mantell and Silliman invites the historian's curiosity, and the issue becomes more puzzling as new evidence is uncovered.

Dana admitted privately to Asa Gray, the Harvard botanist who during this time was obsessed with defeating both *Vestiges* and the author's supposed American ally, Henry Darwin Rogers, that he had prepared his brief notice for the *American Journal of Science* without having ever seen the book. In his letter to

---

4. Gideon Algernon Mantell to Benjamin Silliman, 28 Feb. 1845 (Silliman Family Papers, Reel 13).

5. Benjamin Silliman to Gideon Algernon Mantell, 23 May 1845 (Silliman Family Papers, Reel 13).

6. [Gideon Algernon Mantell], "Vestiges of the Natural History of Creation," *American Journal of Science* 49, no. 1 (June 1845): 191. Silliman's *Journal* was among the few American periodicals collected by the library at Oxford University during the 1850s. It was highly regarded by scientific men both in the United States and Europe. See "The Present State of Oxford University," *Princeton Review* 26, no. 3 (July 1854): 409–36, at 431; "Journalism in the United States," *Southern Quarterly Review* 3, no. 6 (Apr. 1851): 500–18, at 516; "The American Journal of Science," *Southern Literary Messenger* 9, no. 2 (Feb. 1853): 126. For a study of Silliman's editorial career, see Simon Baatz, "'Squinting at Silliman': Scientific Periodicals in the Early American Republic, 1810–1833," *Isis* 82 (1991): 223–44.

7. [James Dwight Dana], Review of *Vestiges of the Natural History of Creation*, *American Journal of Science* 48, no. 2 (Jan.–Mar. 1845): 395.

Gray, Dana claimed to have based the paragraph on a "<long> notice by the Professor," meaning Silliman. Thus, it was not Dana, but "Prof. S.," who "criticized some few geological facts, but not the fundamental error of the work." Dana himself had "no faith in the <author's ~~work~~> ~~his~~ views."[8] Or did he? If he had to edit his statement with two inserted words and two deletions, perhaps one should not be too surprised to find his mood shifting in the weeks that followed. Dana's subsequent letters to Gray expressed a tolerance, almost an admiration, for the theological foundations of *Vestiges*. Meanwhile, Silliman, who reportedly had written "Dana's" generally favorable notice of *Vestiges*, would later travel the nation delivering lectures against *Vestiges*.

The historical artifacts that connect present-day readers to the reception of *Vestiges* at Yale during the 1840s and 1850s thus do not lend themselves to a simple portrait of who thought what, or when and why they did so. Silliman's summary of *Vestiges*, upon which Dana claimed to have written the first *American Journal of Science* notice, cannot be located in the collected papers of either Silliman or Dana, nor are there editorial records of the *Journal* itself for this time. Perhaps Silliman's draft was among the "hundreds of useless and obsolete letters" that he reduced to ashes on January 6, 1858, when "revising" his correspondence, as he termed the process in a diary entry for that date.[9] It may even be possible that Silliman, if he really did draft the review that Dana condensed and published, had, like Dana, not yet read *Vestiges*. Louis Agassiz and Andrews Norton at Harvard each, on occasion, published notices of books they had not yet read,[10] and Silliman, who was busy lecturing in the South, may have desired that his *Journal* print an early notice of *Vestiges* in order to sustain subscriptions in a tight market. Perhaps actually reading the book was not top priority.[11]

If one clear lesson can be learned from the *Vestiges* controversies, it is that good history—whether of nature or of people who study nature—should not be

8. James Dana to Asa Gray, 19 June 1845 (Herbarium).

9. Chandos Michael Brown, *Benjamin Silliman: A Life in the Young Republic* (Princeton, NJ: Princeton University Press, 1989), 359.

10. Andrews Norton to James Walker, 7 Dec. 1835, as cited in Robert D. Habich, "Emerson's Reluctant Foe: Andrews Norton and the Transcendental Controversy," *New England Quarterly* 65, no. 2 (June 1992): 208–37.

11. On Silliman's longstanding struggle to secure sufficient subscriptions, see Baatz, 223–44.

too speculative. Thus, one must cease entertaining endless possibilities about who read and wrote what, and return to what can be more reliably documented concerning the receptions of *Vestiges* at Yale. This chapter will identify two dominant themes in the careers of Silliman, Dana, and others at Yale who responded to *Vestiges*. First, as at Princeton and Harvard, a local theological context shaped the ways in which members of the Yale community read and replied to *Vestiges*. Second, both Silliman and Dana—who was Silliman's favored pupil, his co-editor, his long-term house-guest, and his son-in-law—desired that their careers in geology would not be hampered by the expertise of those who specialized in the sciences of theology and philology. This did not mean that the two geologists felt any conflict between their own science and theology, for indeed both were deeply pious and shared a common religious framework with their academic adversaries, such as the philologist Moses Stuart and the theologian Nathaniel William Taylor.

In fact, Dana, like his mentor Silliman, took great care to maintain a unity of truth between the natural sciences and theology. In Dana, especially, one may see the influence of what his contemporaries called the "New Haven theology," which was developed during the 1820s and 1830s by Yale theologians Eleazar T. Fitch and Nathaniel William Taylor. Also known as "Taylor's system," the New Haven theology emphasized God's moral government in both natural history and human affairs. In some ways it paralleled Princeton's emphasis on God's sovereignty, but the kind of moral philosophy embedded in Taylor's system allowed a greater a role for human agency than could be accepted at predestinarian Princeton. In fact, the influence of New Haven theology upon the New School wing of the Presbyterian Church had been the chief doctrinal issue leading to the 1837 split between the Old and New Schools. Yale's theological *milieu* thus fit somewhere between Princeton's and Harvard's: the Trinitarian faith and Calvinist heritage of Yale's Taylorite Congregationalists drew them toward Old School Presbyterian Princeton, but their more liberal attitude, particularly with respect to human free will and humanly discernable standards for God's moral character, led them also to appreciate aspects of Harvard Unitarianism.[12]

---

12. Leo P. Hirrel, *Children of Wrath: New School Calvinism and Antebellum Reform* (Lexington, KY: The University Press of Kentucky, 1998), 29, 41–42, 52, 182.

Chapters 2 and 3 revealed that specialists in various sciences—natural, moral, and theological—at both Princeton and Harvard repudiated the development theory that was popularized by *Vestiges*. Those chapters also demonstrated that *Vestiges* took on a different meaning in each of those two academic communities, and was accordingly rejected for different—though at times overlapping—reasons. In this chapter, a third local case study will be conducted, this time exploring the New Haven theological context at Yale College. To discover how Taylor's system shaped responses by Yale spokespersons to the development theory, this chapter begins with an account of Taylor's distinctive theological tradition, followed by its implications for the design argument. The chapter next proceeds into discussions of how Silliman and Dana built their careers as religious geologists in the Taylorite tradition, and how their responses to *Vestiges* emerged at the meeting point between their theological convictions and their career concerns. Special attention also will be devoted to Amherst College's geologist-theologian Edward Hitchcock, who studied under both Fitch and Silliman at Yale, and whose reactions to *Vestiges* closely paralleled those of Dana. Of particular significance, neither Dana nor Hitchcock experienced much theological discomfort with *Vestiges*. Yet, although Dana and Hitchcock each felt their pious Congregationalist faith could accommodate *Vestiges*, they maintained an opposition to the development theory based upon their commitment to natural science, a commitment inspired by Silliman and developed within the context of New Haven theology.

The "New Haven Theology" Setting: Yale College and Theological Department

In 1701, nine Harvard alumni formed a trust to establish Yale College in New Haven, Connecticut. They desired the institution to train men in the arts and sciences for service to both church and state; this included the training of ministers who would serve in church pulpits and also of other professionals who would serve in civic and political stations. A century later, during Timothy Dwight's presidency, the Yale Corporation was having trouble filling the Professorship of Divinity, so Dwight assumed a dual role as professor and president. By this time, ministerial students generally took private instruction from Dwight for two years following their graduation. Dwight's theological protégés included the philologist Moses Stuart, the evangelist Lyman Beecher, and the systematic theologian Nathaniel William Taylor.[13] In 1802 Dwight's revivalist preaching in the college chapel led to the conversion of one-third of the student body and sparked the Second Great Awakening in New England.[14] Among the stirred spirits was Benjamin Silliman, a recent graduate whom Dwight appointed as Professor of Chemistry and Natural History that same year.[15]

In the years following Dwight's death, in 1817, Yale undergraduates petitioned the Faculty to organize a theological department. In 1822, the Corporation resolved to appoint four professors in order to establish such a department. They justified this move based upon the 1701 charter (which stated that Yale would train ministers), the need for pastors in the West (meaning western New York and the upper Midwest of today), and the threat of a rising Unitarian movement in Massachusetts (which made "true doctrine" more necessary than ever). The Corporation unanimously selected Taylor, who had been Dwight's favored pupil and subsequently served as Moses Stuart's successor in New Haven's First Church, to become the divinity school's Professor of Didactic Theology.[16]

Beyond question, Taylor's presence dominated Yale's Theological Department until his death in 1858. Students recalled him as an innovative and systematic thinker. His lectures rolled from his mouth like sermons.[17] Eleazar T. Fitch, by contrast, who replaced Dwight as Professor of Divinity for the undergraduates and taught Greek for the Theological Department, had a nervous

13. John T. Wayland, *The Theological Department in Yale College, 1822–1858*, American Religious Thought of the 18th and 19th Centuries series, ed. Bruce Kuklick (rpt., New York and London: Garland Publishing, 1987), 7–9, 46–50.

14. George M. Marsden, *The Soul of the American University: From Protestant Establishment to Established Nonbelief* (New York and Oxford: Oxford University Press, 1994), 81–82.

15. Anson Phelps Stokes, *Memorials of Eminent Yale Men: A Biographical Study of Student Life and University Influences during the Eighteenth and Nineteenth Centuries*, 2 vols. (New Haven: Yale University Press, 1914), 1:19 (conversion); Leonard G. Wilson, "Benjamin Silliman: A Biographical Sketch," in *Benjamin Silliman and His Circle: Studies on the Influence of Benjamin Silliman on Science in America*, ed. Leonard G. Wilson (New York: Science History Publications, 1979), 1–10, at 1 (professorship).

16. Wayland, 52–61, 75.

17. Wayland, 84–85.

style about his speech. Students joked that if ever he were to forget his manuscript, he would not be able to speak a word. Behind the scenes, however, Fitch and Taylor worked as a team, developing what soon became known as "New Haven theology." It was Fitch's 1826 sermon on the nature of sin that first drew suspicion that something heretical was happening at Yale.[18] Taylor's *Concio ad Clerum* address to New Haven ministers in 1828 confirmed these suspicions for Old Calvinists, as the able speaker set forth his "moral government theory" that denied the received understanding of the Calvinist doctrine of original sin.[19]

The controversy over how to interpret and apply Calvinism divided along a deep fault-line in Yale's mixed theological constituency.[20] Ministers who studied at Yale often moved back and forth between Congregationalist and Presbyterian churches, as had been arranged by the 1801 Plan of Union. Their pulpit-sharing was reinforced during the 1820s and 1830s by the growth of cross-denominational evangelical reform societies.[21] Congregationalists tended to place greater emphasis on human agency in joining the covenant of grace that God had established through Christ Jesus, whereas Presbyterians, anchored to their Westminster Confession, worried that such talk of human agency denied the total depravity of humans and diminished the absolute sovereignty of God. The most extreme academically established instances of these two positions could be seen at Harvard, where liberal Congregationalism had spawned Unitarianism, and Princeton, where Old School Presbyterianism was separating itself from the Congregationalist-influenced New School. Yale's theologians thought of themselves as neither die-hard Old School Presbyterians, nor Unitarian Arminians. They sought a *via media* that would soften aspects of Calvinism that were susceptible to Unitarians' rationalist criticisms, while avoiding the Arminian

18. Wayland, 90, 92–93.

19. Wayland, 325–26.

20. For example, see [David N. Lord], "A Letter to the Corporation of Yale College, on the Doctrines of the Theological Professors in that Institution," *Views in Theology* 4, no. 16 (May 1835): 291–341; Lord, though he had studied theology under Fitch in the years immediately following Dwight's death, sided with the Old School Presbyterians during the split that emerged in the late 1820s and early 1830s. At p. 295, he objected, much as a Princeton man would, specifically to Taylor's apparent denial of God's sovereignty. As noted in chapter 2, his brother Eleazar had served as a trustee of Princeton Theological Seminary.

21. George M. Marsden, *The Evangelical Mind and the New School Presbyterian Experience: A Case Study of Thought and Theology in Nineteenth-Century America* (New Haven: Yale University Press, 1970), 14–19.

implication that individuals somehow save themselves by choosing to live for God, rather than receive salvation entirely by God's own decree.

When Taylor commenced his duties as Professor of Didactic Theology in 1822, he desired to teach Calvinism, but without making God seem arbitrary and inhumane, as critics of predestination claimed. Simultaneously, he sought to promote a conception of human nature that conformed to intuitions about free will and moral responsibility, but without succumbing to Arminianism. He was not the first to pursue these goals. In the late 1700s, Samuel Hopkins and Joseph Bellamy, two Congregationalist ministers in New England, had suggested that God allowed sin to happen as a necessary prerequisite for salvation. This, they hoped, would answer the Enlightenment skeptics who asked: "How can an omnipotent and benevolent God allow evil to exist?" Bellamy further addressed an apparent inconsistency in the Calvinist claim that salvation was both won by Christ's suffering on the cross and also was a free gift. If God gave it freely, then why should Christ (or anyone) have to pay for it? Bellamy suggested that Christ's death was not propitiatory, but was merely a dramatic demonstration of God's freely given love. President Dwight had not taken sufficient interest in dogmatic theology to state a definite opinion in these matters, but in practice his revivalism departed from the "Old Sides" Calvinism of his predecessor, President Ezra Stiles, and started Yale on the trend of adapting Jonathan Edwards's understanding of human freedom and moral virtue into a more anthropocentric language.[22]

22. Hirrel, 27. Historians disagree substantially in their characterizations of Taylor's theology in relation to "Old Calvinism" (an eighteenth-century New England theological tradition that found strong expression in the nineteenth-century Old School Presbyterianism detailed in chapter 2) and the rival "New Divinity" school (a movement inspired by Jonathan Edwards, and established during the late 1700s by his protégés Samuel Hopkins and Joseph Bellamy).

Historian Allen Guelzo regards Taylor as "the embodiment of Old Calvinism carried into the new generation" and reads his *Concio ad Clerum* (1828) as a deathblow to Edwards's *Freedom of the Will* (1754). He argues that Taylor was even more opposed to Edwardsians than to Unitarians, and that insofar as his theology at times sounded Edwardsian, this was due to his attempt to adapt aspects of that tradition for the benefit of promoting Old Calvinism in a new historical context.

Douglas Sweeney, by contrast, insists that Taylor was "a symbol of the vitality of Edwardsian Calvinism throughout the first half of the nineteenth century." In particular, he emphasizes that Taylor's moral government theory was derived from the Edwardsians' New Divnity theology.

This deep disagreement between the historians results in part from different characterizations of the two movements. Rather than emphasizing Taylor's moral government ties to the Edwardsians, Guelzo claims Taylor for Old Calvinism by emphasizing his promotion of the "church-in-society," the use of means (preaching, Bible reading, prayer, baptism, and communion), and God's wisdom in allowing for sin—three things generally not promoted by New Divinity theologians. Clearly Taylor was somewhere between these

Taylor energetically worked out a precise systematic theology to make Calvinism compelling, both to people who accepted the rationalist standards of the Enlightenment and to those concerned with the practical exigencies of revivalist evangelism. Since it would, for Taylor and his constituents, be absurd for a revivalist to tell his listeners that they are totally depraved and can do nothing to approach God, and then to urge them to step forward and believe in God, Taylor departed from Old Calvinism's doctrine of total depravity. For Taylor, Christ had died in order to urge people to choose freely to repent of sin and to live selflessly—and people had enough free will to make that choice.[23]

Taylor's theology, defined on its own terms, was far more profound than merely balancing Old Calvinism and Unitarianism (though that, in itself, was no easy feat). He contributed to American religious thought a distinctive understanding of God's moral government. According to this schema, freely acting human agents discern and pursue their greatest good within God's moral order, but, recognizing their shortcomings, they also can be expected to conclude that they need redemption. From both natural intuitions in the human mind and supernatural testimony in Holy Scripture, any reasonable person can conclude that God, being a perfect moral governor, would not have created the existing moral order, which allows for sin, unless He also was going to provide a means for redemption after people used their free will to commit sin. Christ's crucifixion thus demonstrates the perfection of God's moral government and inspires the penitent to renew their lives with moral living. Human sin and Christ's death, though each undesirable in isolation, fit a necessary part of God's moral system,

two movements, with closer apparent sympathies for one rather than the other, depending, in part, upon how the historian defines each of these schools of theology.

For the purposes of the present investigation, it suffices to note that Taylor's language—even if intended to serve Old Calvinist objectives—sounded to many Old Calvinists like a turn toward Arminianism, and that his theology centered not on the sovereignty of God as Princetonians understood God's sovereignty, but rather on the moral government of God. Indeed, Guelzo and Sweeney, together with many other historians, agree that Taylor's theology for these reasons led to the 1837 split between the Old and New Schools of the Presbyterian Church. See Allen C. Guelzo, *Edwards on the Will: A Century of Theological Debate* (Middletown, CT: Wesleyan University Press, 1989), esp. 240–44, 277 (quoting 240); Douglas A. Sweeney, *Nathaniel Taylor, New Haven Theology, and the Legacy of Jonathan Edwards* (Oxford: Oxford University Press, 2003), quoting 4–5, but see esp. his introduction (for a literature review of other historians' interpretations of Taylor), chap. 5 (on moral government theory), and p. 149 (Presbyterian split).

23. Marsden, *The Evangelical Mind*, 34–39, 46–51; Paul K. Conkin, *The Uneasy Center: Reformed Christianity in Antebellum America* (Chapel Hill: University of North Carolina Press, 1995), 90–91, 101–13, 210–12; Wayland, *The Theological Department at Yale*, 314–22.

which, by allowing for human free will, serves a higher moral good than would be possible under a predestinarian system.[24]

The proceedings of the Rhetorical Society, a weekly gathering of students and faculty from Yale's Theological Department, reveal that Taylor's system structured the questions that were raised and the answers that were adopted. Often, Taylor himself presided over these Wednesday meetings. Like many debate clubs of the era, the Rhetorical Society addressed contemporary concerns about slavery, polygamy, and common schooling. This society, however, devoted special attention to questions that were characteristic of New Haven theology. In November of 1845, the members debated: "Could God prevent sin in a moral system?" Other topics for debate included: "Does the tendency to universal disobedience to God on the part of men, furnish any evidence that he does not prefer their obedience to disobedience?" In each case, the answer was decidedly Taylorite: God had to allow sin if his system was to be moral, yet that by no means indicates that God prefers disobedience over obedience.[25] Like Taylor, the members of the Rhetorical Society felt that "those who adopt the New Haven views in theology" can "conscientiously assent to the standards of the Presbyterian church."[26] Although Old School Presbyterians accused Taylorite Congregationalists and New School Presbyterians of deviating from the Westminster Confession, the latter groups thought they had done nothing more than make orthodoxy presentable to a new generation.

Fitch's sermons at Yale College brought the "system" that Taylor taught in the Theological Department to the students of the Academic Department, or college proper. In 1839, Fitch expounded, in typical Taylorite fashion, upon Hebrews 9:27 ("It is appointed unto men once to die"). Arguing from both reason and Scripture, Fitch developed three propositions. First, Providence planned from the beginning that humans would not be immortal upon the earth. Second, that since humans must be at some time removed from the earth, this could occur either by supernatural means or by a natural, temporal death. Third, temporal death is the better of the two possible means for humans' removal from this

---

24. Nathaniel W. Taylor, *Essays, Lectures, Etc. upon Select Topics in Revealed Theology*, ed. Noah Porter (New York: Clark, Austin, & Smith, 1859), esp. chaps. 12–14.

25. Secretary's Book (Rhetorical Society of the Theological Department, Yale University, Box 1, Folder 3), 12 Nov. 1845, 16 Dec. 1846.

26. Secretary's Book, 10 Feb. 1847.

earth, and therefore it is the means God in his wisdom has chosen. Taken together, these propositions supported Fitch's general thesis that God, as the moral governor of the world, has established an economy in which those who "concur with the designs of [God's] wisdom" can feel "forever safe, forever happy." In other words, those who regard death as God's chosen pathway for each person to pass into blessed immortality need not fear it.[27] In an 1846 sermon, Fitch focused upon the death of John the Baptist, who was beheaded. Continuing in Taylor's tradition, he concluded that the martyrdom of John the Baptist was the best way for God's moral government to serve the needs of all. It removed John from this world to the next. It ennobled John, who willingly became a martyr before Herod just as Elijah had been persecuted by Ahab. It provided closure to John's ministry exactly when Jesus, for whom he was to prepare the way, was beginning his own ministry. In all these ways and more, the course guided by God's moral government was the best one possible.[28]

The New Haven Design Argument: Taylor's Systematic Theodicy

In their presentation of God's moral government, Taylor and Fitch were promoting a kind of design argument. Taylor's notion of divine planning differed, however, from what historians of this period generally mean when they refer to the "design argument." Scholarly literature on the nineteenth-century British and American intellectual life has emphasized the influence in both countries of the design argument presented in William Paley's *Natural Theology* (1801). In 1958, historian Ralph Gabriel concluded that Paley's works were Yale students' chief sources on religion for the sixty years preceding the Civil War.[29] Writing on science and religion in American for the same period, Herbert Hovenkamp concluded, in 1978, that Paley was the number one force shaping American natural theology—with second place going to the Bridgewater Treatises, which

27. Eleazar T. Fitch, "The Wisdom of God in the Appointment of Temporal Death," sermon on Hebrews 9:27 (1839), in his *Sermons, Practical and Descriptive, Preached in the Pulpit of Yale College* (New Haven: Judd and White, 1871), 335–52.

28. Eleazar T. Fitch, "The Death of John the Baptist," sermon on Matthew 14:3–11 and Mark 6:17–29 (1846), in *Sermons*, 217–34.

29. Ralph Henry Gabriel, *Religion and Learning at Yale: The Church of Christ in the College and University, 1757–1957* (New Haven, CT: Yale University Press, 1958), 109–13.

were, moreover, written by people schooled in Paley's thought.[30] Paley's *Natural Theology*, as well as his work on moral philosophy, was frequently reprinted, assigned in college courses, and mimicked by many writers and lecturers during the first half of the nineteenth century. Historians after Gabriel and Hovenkamp have understandably repeated the suggestion that Paley's arguments characterize the mindset of that era.[31] Their account of history rings true in today's ears in part because Paley's opening example in *Natural Theology* remains a popular argument among present-day opponents of naturalistic evolution: just as the carefully coordinated parts of a watch must have been designed and built by a watchmaker, so also the universe is full of well-designed parts—"contrivances," as Paley called them—that bear evidence of the Deity who fashioned them.[32] Curiously, however, the archival records at Yale do not support the common historical perception view that Paleyan design reigned during the era of *Vestiges*.[33]

Although students at Yale College, like their counterparts at Harvard, and Princeton, were required to read Paley's *Natural Theology*, local professors' evaluations of Paley's writings may have held more influence than Paley's text itself. Taylor, for example, told his theological students that "Paley gives a <u>wretchedly</u> <u>loose</u> definition of Omnipotence and his <u>Moral</u> <u>Philosophy</u> is a <u>Second hand</u> book."[34] As for Paley's *Natural Theology*, Taylor found the design argument —"Design proves a Designer," as his students phrased it—to be tautological. In 1843, he instructed his class that, "This needs explanation. It is [an] identical

30. Herbert Hovenkamp, *Science and Religion in America, 1800–1860* (Philadelphia: University of Pennsylvania Press, 1978), 219.

31. See, for example, Jon H. Roberts, *Darwinism and the Divine in America: Protestant Intellectuals and Organic Evolution, 1859–1900*, paperback ed., with a new foreword (Notre Dame, IN: University of Notre Dame Press, 2001), 4, 8–10.

32. William Paley, *Natural Theology; or, Evidences of the Existence and Attributes of the Deity, Collected from the Appearances of Nature*, 12th ed. (1st ed., 1801; Weybridge, Eng.: S. Hamilton, 1809), chap. 1. Paley's "watchmaker" argument has carried sufficient conviction during recent decades that one of the chief proponents of naturalistic evolution felt the need to attack it head-on. See Richard Dawkins, *The Blind Watchmaker: Why the Evidence of Evolution Reveals a Universe without Design* (New York: Norton, 1986), 4, 37–41, and *passim*.

33. For a similar critique, focusing on Britain, see Jonathan Topham, "The Bridgewater Treatises and British Natural Theology in the 1830s," Ph.D. diss., Lancaster University, 1993.

34. William D. Ely and David T. Stoddard, Lectures on Mental Philosophy, Pastoral Theology, Etc., by Eleazar T. Fitch, Chauncey A. Goodrich, Nathaniel W. Taylor, 1837, 1842 (Course Lectures, Yale University, Box 31, Folder 148), 442.

proposition, as much so as 'production ~~as~~ needs a producer[.]' Unless then Dr Paley means to beg the question. . . ."[35] President Theodore Dwight Woolsey voiced similar criticism of Paley. In his 1857–1858 political philosophy course, he noted that Paley confuses a right to a thing with a right to have a thing.[36] Taylor, likewise, had told his moral philosophy students during the 1843–1844 academic year that Paley had a "very feeble" definition of "moral obligation."[37]

True, Yale's students had Paley's works readily available at the several libraries of student organizations, such as those of the Brothers in Unity and the Linonian Society,[38] but members of those societies tended not to discuss contrivances in nature. Records of their weekly debate meetings indicate a greater interest in contemporary politics, such as the Wilmot Proviso.[39] Student writings from the period suggest that Paley was studied with a low degree of enthusiasm, if studied at all. George Sherman, Yale Class of 1839, recorded in his diary during his senior year: "At 4 went for recitation in Paley without having looked at the lesson & flunked."[40] In 1852, the undergraduates published a "revised edition" of the academic catalogue in which they listed "Theology Naturally Pale" as a required course for seniors.[41] The following year's spoof catalogue continued the joke: "Pale Ale discussed" was listed for seniors in the

35. Alexander MacWhorter, Notebook of Nathaniel W. Taylor, Lectures on Moral Philosophy, 1842–1843, with Lectures on Natural Theology, 1843 (Course Lectures, Yale University, Box 30, Folder 147), 103. Cf. Samuel W. Barnum, Lectures on Theology by Taylor, Fitch, Bacon, and Goodrich, 1843 (Box 29, Folder 144), 47.

36. E. Greenough Scott, Notebook, Theodore D. Woolsey, Lectures on History and Political Philosophy, 1857–1858 (Course Lectures, Yale University, Box 23, Folder 130), 3.

37. Samuel J. M. Merwin, Lectures on Expository Preaching and Missions, Astronomy, Moral Philosophy, by Chauncey A. Goodrich, Denison Olmsted, and Nathaniel W. Taylor, 1838, 1842, 1843–1844 (Course Lectures, Yale University, Box 31, Folder 150), 3.

38. *Catalogue of the Library of the Society of Brothers in Unity, Yale College, April, 1846* (New Haven: B. L. Hamden, 1846); *Catalogue of Books Belonging to the Linonian Society, Yale College, November 1846* (New Haven: J. H. Benham, 1846).

39. Brothers in Unity, Catalogue of Members and Questions for Debate, 1840–1856 (Records of Clubs, Societies, and Organizations, Yale University, Box 10, Folder 49); Calliopean Society, Questions Presented for Debate during Meetings, Nos. 815–1082 (Records of the Calliopean Soceity, Box 3, Folder 14).

40. George Sherman, Diary (Yale Student Diaries, Yale University, Box 1, Folder 4), 26 Feb. 1839.

41. *Catalogue of the Officers and Students in Yale College, 1852–53* , rev. ed. (Springfield, MA: B. L. Thunderclap, 1852), 22;

second trimester, and "Evidences of Christianity—pale and indistinct" for seniors in the third trimester.[42]

Quite in opposition to Paley's design argument, Horace Bushnell, schooled at Yale Divinity, concluded in 1832 that "All attempts to prove the existence of a God from the necessity of a first cause and from those marks of design which we observe in nature seem to me to prove either too much or nothing at all."[43] Bushnell later wrote that even if all the Bridgewater Treatises in the world could be piled up until they reached the moon, their natural theology still would amount to nothing worthwhile.[44] Admittedly, Bushnell's romanticist preference for a culture of "Christian nurture" (in place of the ordinance of baptism), and his dismissal of dogmatic details made him, one might say, the New Haven equivalent of Harvard's expatriated Emerson.[45] Nevertheless, Bushnell's dissatisfaction with the very Paleyan design argument that historians have suggested most Americans cherished, was not uncommon among those schooled at Yale. Whether one looks at educators such as Taylor and Woolsey, at their students, or at renegade alumni such as Bushnell, one must conclude that Paley's

42. *Catalogue of the Officers and Students in Yale College, 1852–58*, rev. ed. (New Haven: R. H. Sawbones, 1857), 27.

43. Horace Bushnell, "There Is a Moral Governor," 1832 (Bushnell Papers, Yale Divinity School, Box 3, Folder 22).

44. Horace Bushnell, "Preliminary Disquisition on Language," in *God in Christ: Three Discourses Delivered at New Haven, Cambridge, and Andover* (Hartford, CT: Brown and Parsons, 1849), 9–17, at 30. Bushnell's point was that the human language, because it was emblematic of divine mysteries, bears greater testimony to God's existence and character than the physiological contrivances catalogued *ad nauseum* in so much of Anglo-American natural theology. The larger thesis of this work was that language itself, though the most revelatory instrument for knowledge concerning God, does not provide sufficient specificity to settle the fine-grained doctrinal debates between various Calvinist and quasi-Calvinist factions in New England.

45. Bushnell promoted in his most popular treatise, *Discourse on Christian Nurture* (1849), a romanticist, organic understanding of the Christian life, influenced by his reading of Coleridge and intended to oppose the more logically rigid, mechanical mode of Taylor's New Haven theology. According to Bushnell, for example, conversion was not a singular event, but rather one was raised as a Christian, growing into that role gradually through the nurture of parents and church communities. See Horace Bushnell, *Christian Nurture*, introduced by Luther A. Weigle (1849; rpt., New Haven: Yale University Press, 1967). Bushnell served as one of the editors for the New Haven-based *New Englander* during the 1840s, but at Yale's Theological Department he was always watched with suspicion. In 1854, Nathaniel Taylor summoned a formal inquiry of Bushnell's doctrines, though in the end the Christian nurture advocate was exonerated from any heresy charges. See Stokes, 1:68. For a discussion of Bushnell's theology in comparison with those of Charles Hodge at Princeton and Nathaniel William Taylor at Yale, see Conkin, chap. 7. No evaluation by Bushnell of *Vestiges* is known. For a discussion of his difficult to discern response to Darwinism, see Thomas Paul Thigpen, "Bushnell's Rejection of Darwinism," *Church History* 57 (1988): 499–513.

oft-cited design argument had far less influence on New Haven natural theology than one would infer from previous historical studies.

The center of New Haven theology—both natural and revealed—was God's moral government. "Dr. Taylor's theory," not Paley's, gave structure to student's class notes.[46] Taylor did not so much agree with the "contrivance" design argument presented in chapters 1–22 of Paley's *Natural Theology*, as he did with the utilitarian theodicy of chapters 23–27, which are seldom mentioned by historians of science. There Paley suggested that violence in nature's prey-predator relations, even if it at times resulted in extinction of a species, and economic inequalities in Britain's industrial society, even if it at times left the poor suffering hopelessly, nevertheless brought about the greatest good for the greatest number of individuals in God's government of both nature and society.[47] Taylor, who felt squeezed both geographically and theologically between Harvard and Princeton, expanded this theodical portion of Paley's writings into a system of God's moral government that he hoped would make Calvinism more presentable in a climate tempered with objections that Calvin's God was arbitrary and unjust, and that Calvin's man was too depraved to commit a moral act.

Benjamin Silliman: The Precedence of Geology over Theology and Philology

In addition to Taylor's New Haven theology, a second theme distinguished Yale from Princeton and Harvard during the era of *Vestiges*: the various sciences tended to compete, rather than cooperate, in their efforts to reveal God's ways in nature.[48] On paper, at least, Yale promoted a harmonious integration of the disciplines. The faculty's curriculum report of 1828—widely known as the "Yale Report"—formalized the standards not only for Yale, but for many other colleges as well. The report, occasioned by student protests and Connecticut legislators' concerns during the late 1820s, contained two sections. In the first part,

46. Samuel J. M. Merwin, Lectures on Expository Preaching and Missions, Astronomy, Moral Philosophy, by Chauncey A. Goodrich, Denison Olmsted, and Nathaniel W. Taylor, 1838, 1842, 1843–1844 (Course Lectures, Box 31, Folder 150), 42.

47. Paley, chaps. 26, 27.

48. This distinction should be understood only in the limited sense of whether or not interdisciplinary conflict had a bearing upon the reception of *Vestiges*.

President Jeremiah Day identified two tasks for college education in the liberal arts: to "discipline" (or expand) the mind, and to "furnish" (or fill) it with facts. These tasks were to serve the larger goal of balancing the various mental faculties (reason, memory, moral judgment, and so on) in order to form young men of balanced character. The second part of the Yale Report, written by Yale's professor of ancient languages, James Kingsley, addressed the students' and legislators' concern that the "dead languages" were a waste of time. Kingsley insisted that the study of classical language disciplined and furnished the mind better than any modern subjects, and that the study of classical cultures instilled American political values, such as the principle of liberty and the duty of patriotism.[49]

Silliman's *American Journal of Science* reprinted the Yale Report in 1829, a year after it had appeared in pamphlet form. Excerpts from that report also appeared in Yale curriculum catalogues for decades. As Yale's leading professor of natural science, Silliman was, no doubt, pleased that the report advocated the study of chemistry for lawyers, of geometry for pastors, and of astronomy for physicians—all with the expectation that a well-rounded training of the mind would make for a better man no matter what his chosen profession.[50] But this consensus report of the Yale faculty, which was concluded by an affirmation from Connecticut Governor Gideon Tomlinson on behalf of the Yale Corporation, concealed something else about a Yale education that would have a far-reaching impact upon the reception of *Vestiges* twenty years later. Professors who specialized in one discipline, providing their students with its pertinent furniture for the mind, did not always regard the different disciplines and mental furnishings of other professors to be of equal value as their own in acquiring a correct understanding of God's world.

Unfortunately, the conflict between theologians and geologists at Yale has been misunderstood in previous historical studies as a conflict between religion and science, rather than between two disciplines that each were religious and scientific in their own way. In 1986, Louise Stevenson's close study of the "New

49. Frederick Rudolph, *Curriculum: A History of the American Undergraduate Course of Study since 1636* (1977; rpt., San Francisco: Jossey–Bass Publishers, 1992); Marsden, *The Soul of the American University*, 81–82.

50. Excerpt from *Reports on the Course of Instruction in Yale College, by a Committee of the Corporation, and the Academical Faculty* (1828), in *The Educating of Americans: A Documentary History*, ed. Daniel Calhoun (Boston: Houghton Mifflin Company, 1969), 229.

Haven scholars" portrayed a deep animosity between theologians like Taylor and natural scientists like Silliman and Dana. She suggested that Taylorites were conservative biblical literalists who opposed geologists' denial that the earth was created six thousand years ago in six 24-hour days.[51] This account perpetuated the view of Daniel Coit Gilman, whose 1899 biography of Dana depicted the mid nineteenth century as a time when narrow-minded philologists like Moses Stuart and old-fashioned theologians like Eleazar Fitch feared scientific progress because it challenged their traditional interpretations of the Mosaic record.[52] A more careful examination of Silliman's career in geology suggests that the friction between his personal combination of theology and natural science, and the ways that others at Yale related theology and natural science, had less to do with biblical literalism than with competing specializations in the diverging sciences of theology, geology, and philology.

James Hadley's diary entry concerning the 1850 meeting of the American Association for the Advancement of Science in New Haven, which Stevenson cites to illustrate a straightforward conflict between ignorant theologians and progressive scientists, in fact illustrates a more complex relationship at Yale between theology and other sciences. Hadley, a Yale alumnus, member of the AAAS, and Assistant Professor of Greek Language and Literature, remarked about "Dr. [Leonard] Bacon's forenoon sermon—extemporaneous—intended specially for the Scientifics, Agassiz, etc., who were there." The sermon launched a "side blow as usual against Taylor—liberal in tone—held that, although the Sacred Scriptures are infallible, our interpretation is fallible and must be modified to answer the demands of science." From these words it would seem that Taylor, a conservative theologian, had been insisting upon interpreting Scripture apart from the conclusions natural science, and that Bacon, a liberal minister, was supporting the desires of AAAS members to revise their interpretations of Scripture in view of scientific insights. But Hadley's own judgment on the matter reveals a third position. "True," he said in regard to Bacon's suggestion that science could rectify a mistaken interpretation of

51. Louise L. Stevenson, *Scholarly Means to Evangelical Ends: The New Haven Scholars and the Transformation of Higher Learning in America, 1830–1890* (Baltimore and London: Johns Hopkins University Press, 1986), 20.

52. Daniel C. Gilman, *The Life of James Dwight Dana: Scientific Explorer, Mineralogist, Geologist, Zoologist, Professor in Yale University* (New York: Harper and Brothers, 1899), 181.

Scripture, "if you mean exegetical science, but if you mean astronomy, geology, physiology, etc., most false. . . . If Moses calls a thing black, the fact of its being white does not prove that he meant white when he said black."[53] Hadley, a philologist, would allow for new advancements in the study of Greek or Hebrew to suggest new interpretations of Scripture, but he would not permit the same license among geologists, whose science he thought less pertinent to theology than his own.

Yale's founding professor of geology, Benjamin Silliman, had for some decades grappled with the cross-disciplinary conflict embodied in Hadley's diary entry. When conducting geological fieldwork in Scotland during 1805–1806, he became convinced that the earth's history stretched back eons prior to the 4004 B.C. creation date printed in the margins of many Bibles and regarded by many American Protestants as the true meaning of the text of Genesis.[54] By the 1820s he was hopeful, however, that the "new views" of geology would become acceptable to a greater number of theologians once they learned what Silliman had learned about geology. "No mere divine—no mere critic in language," he wrote to Edward Hitchcock, a Yale-trained minister who now was under Silliman's geological tutelage, "can possibly be an adequate judge of the subject or deserve unqualified deference, however able in other respects." Geology, for Silliman, was not just about mapping the earth's strata; it was about mapping disciplinary turf so that he could study God's earth without regard to what theologians and philologists might think.[55]

In 1829 Silliman arranged for William Blakewell's popular English textbook on geology to be reprinted in America. He appended an article of his own on the relations between geology and theology. Blakewell's work, with Silliman's commentary, went through several American printings in the years that followed. Moreover, Silliman's discussion also was reprinted as a stand-alone pamphlet. "None but geologists study it [the earth] with diligence," wrote Silliman in order to carve out an institutional space for his career, "and none who

53. James Hadley, *Diary (1843–1852) of James Hadley*, edited with a foreword by Laura Hadley Moseley (New Haven: Yale University Press, 1951), 97–8. See also Stevenson, 20.

54. John H. Giltner, "Genesis and Geology: The Stuart–Silliman–Hitchcock Debate," *Journal of Religious Thought* 23, no. 1 (1996–1967): 3–13, at 4–6.

55. Benjamin Silliman to Edward Hitchcock, 18 Aug. 1820 (Edward and Orra Hitchcock Papers, Box 3, Folder 37).

have not made themselves masters of the facts, are qualified to judge of their importance, and their bearing."[56] Not only did Silliman criticize philologists for intruding ignorantly into his own specialty of geology, he also attempted to outdo philologists at their own game. He argued that the "days" mentioned in Genesis cannot be literal, since the sun was not created until the fourth day and since "day" in many other biblical passages clearly refers to a longer epoch. Moreover, even a solar day can last up to six months for people living near the poles.[57] Silliman concluded that theologians and philologists should either become geologists, or else keep silent on the relationship between geology and Genesis, receiving whatever demonstrations trained geologists may deliver to them.[58]

Silliman's outspokenness drew criticism from the Yale-trained theologian Moses Stuart, who had pastored First Church in New Haven prior to accepting a professorship at Andover Theological Seminary in 1810.[59] Founded just after Harvard had taken a decided turn toward Unitarianism, Andover was the stronghold of New England Congregationalism. Stuart has been portrayed as a "conservative literalist," but that should not suggest that he agreed with every person who in a later generation received similar labels.[60] Stuart, who had studied German criticism and prepared his own Hebrew textbook, urged his hermeneutics students at Andover to interpret the Scriptures *historically*, not *literally*. "Let every writer be placed in his own age, and if possible, transfer yourself back there, with him." Stuart's philology was not a quest for timeless essences encapsulated in words that have unchanging significance, but for historically situated meanings that the language of that earlier time had sought to record. When Moses wrote "sun," it meant what it meant to Moses, which was not necessarily what it meant to Newton. In order to fulfill his task of recovering from an early epoch the way that God had once revealed himself, the philologist had to

56. Benjamin Silliman, *Suggestions Relative to the Philosophy of Geology, As Deduced from the Facts and to the Consistency of Both the Facts and Theory of This Science with Sacred History* (New Haven: B. L. Hamden, 1839), 40.

57. Silliman, *Suggestions,* 94, 111–15. Silliman cites Gen. 2:4, Job 18:20, Isa. 30:8, Luke 27:24, John 8:56, 2 Pt. 3:8, Job 14:6, Eze. 21:25, and Prov. 6:34.

58. Silliman, *Suggestions,* 117.

59. For the most comprehensive account of Stuart's life, see John H. Giltner, *Moses Stuart: The Father of Biblical Science in America* (Atlanta: Scholars Press, 1988).

60. Conrad Wright, "The Religion of Geology," *New England Quarterly* 14 (June 1941): 335–58, at 344.

avoid commingling his own science of the past with the present theories of other sciences. Stuart did not expect Moses to know anything about Newton, and likewise did not appreciate Silliman's efforts to accommodate the Mosaic writings to modern geology. "Divine Revelation was not designed to teach geography or physics, or astronomy, or chemistry."[61]

Stuart's *Hebrew Chrestomathy,* which appeared in 1829, the same year that Silliman issued an American edition of Blakewell's geology, included an objection to "recent critics and geologists!" who had meddled with the first two chapters of Genesis.[62] Stuart already had written a private criticism to Silliman in 1824, but only part of their exchange is extant. Based upon this fragmentary record and later discussions during the 1830s between Stuart and the Silliman-trained geologist Edward Hitchcock, Stuart's biographer has concluded that he regarded the laws of biblical exegesis as more certain than the theories of geologists, and that he did not feel either Silliman or Hitchcock was competent to comment upon such things as the meaning of "day" or the Hebrew conjunction *vav.* Hitchcock had followed the suggestion of an earlier generation of German philologists who argued—incorrectly, according to Stuart—that the Hebrew particle *vav* (translated "And" in the King James Version) could mean "Afterwards." Hitchcock's preferred translation would introduce a gap between the first two verses of Genesis 1, thus allowing for a long span of geological time to intervene between the initial creation of the cosmos and the subsequent creations of present-day living kinds.[63] The Stuart–Silliman–Hitchcock debate never polarized in terms of science versus religion, but remained instead a debate over disciplinary expertise. All three men were pious Congregationalists; they differed only in their decisions of which science—geology or philology—ought to take precedence when interpreting Genesis.

Silliman also faced conflict from another front, and this one was closer to home. Eleazar Fitch, who pastored the College Church where Silliman kept regular attendance, did not feel comfortable with all of the liberties Silliman was taking in his geological interpretations of Genesis. According to Gilman's 1899 report, Fitch preached before Yale's student body, during the mid nineteenth

---

61. Quoted in Giltner, *Moses Stuart*, 54.

62. Quoted Giltner, "Genesis and Geology," 3.

63. Giltner, "Genesis and Geology," 6–12.

century, that "days" in Genesis were 24-hour, solar days.[64] Silliman recorded at the close of his notes for the final lecture in his 1851 geological course that he was hesitant to emphasize his own day-age reconciliation of Genesis and geology, "because I was immediately upon the track of Prof Fitch who the sabbath before had endeavored to retain the common understanding of time."[65] It would seem, then, that the tension between Silliman and Fitch fits the classic model of a conflict between natural science and theology. A closer look at Fitch's sermon manuscripts indicates, however, that something more complex was happening at Yale.

When preaching on Genesis 1, Fitch did defend the reading of the creation "days" as 24-hour periods, and he did insist that humans had lived on the earth for only about 6,000 years. But he quite willingly allowed for the earth itself to be much more ancient. He advocated something similar to Hitchcock's gap theory, suggesting that the material creation preceded the organic by innumerable ages during which successive geological formations, and their ancient populations, now fossilized, came into existence. Fitch was not, therefore, a theologian restlessly attacking modern geology. He was a theologian progressively accommodating to geology, but in a manner that a nearby geologist, Silliman, did not approve.

Ironically, the target Fitch had in mind was not geologists like Silliman, but rather those who would claim matter to be eternal. Creation *ex nihilo*, not creation in 4004 B.C., was his thesis. His gap theory, though it differed from Silliman's day-age theory, still was respected by Silliman, at least when held by Hitchcock, a fellow geologist.[66] A note attached to Fitch's manuscript refers to the 2 Peter passage that Silliman had listed in his Blakewell commentary, in which a day is said to be like a thousand years. It seems, therefore, that Fitch may have been contemplating the day-age interpretation, but it is unclear whether he read that note when delivering his sermon. Another inserted page suggests that

64. Gilman, 181–82.

65. Silliman, Yale Lecture on Geology, 1851 (Silliman Family Papers, Box 38, Folder 13).

66. Edward Hitchcock to Benjamin Silliman, 26 Oct. 1852 (Edward and Orra Hitchcock Papers); for a third-party analysis, see Review of *On the Relation between the Holy Scriptures and Some Parts of Geological Science*, by John Pye Smith, *Princeton Review* 13, no. 3 (July 1841): 368–94, at 384.

geologists could have all the time they wanted, so long as they put it in the "gap" passage, rather than in the "days" passages.[67]

Quite possibly these emendations represent different versions of a Genesis sermon that Fitch delivered multiple times over the span of several years.[68] Perhaps Fitch modified his presentation over the years in response to some concerns voiced by Silliman. Though the available evidence is insufficient for establishing exactly what Fitch said, and exactly what Silliman heard, one may at least conclude that Fitch was not persistently committed to a specific age of the earth. Furthermore, his sermons on death, referred to earlier, indicate that he regarded death as necessary for God's moral government; it was not, as more conservative theologians were saying, the consequence of sin, and therefore it could precede Adam and Eve's first sin. Thus, he could agree with Silliman that the extinction of pre-Adamite species would cause no theological alarm.[69] They also both rejected polygenism, or the theory that God had created each human race by a distinct act; in other words, a theory that not all humans could be traced back to Adam and Eve.[70] The disagreements between Fitch and Silliman seem to have resulted, more than anything, from a rivalry between two departments at the college, theology and geology, rather than between any ideas that were peculiar to either department.

In fact, Silliman generally felt at home where Taylor's system was preached. Genesis 2:7 recounts that "the LORD God formed man of the dust of the ground, and breathed into his nostrils the breath of life; and man became a living soul." Fitch, emphasizing Taylor's views of God's moral government, exclaimed in an 1845 sermon on that verse: "How is the benevolence of God manifest in instituting his government over man, directing man to pursue that course which will conduce to his greatest happiness. Nothing else does he

67. Eleazar T. Fitch, Sermon No. 158, on Gen. 1:1, n.d. (Fitch Papers, Yale Divinity School, Box 4, Folder 16).

68. Sermon manuscripts from this era frequently list numerous delivery dates and locations. Unfortunately, the date of Fitch's sermon cannot be judged except by paper type, which places it in the 1840s or 1850s.

69. Fitch, "The Wisdom of God in the Appointment of Temporal Death," sermon on Hebrews 9:27, [3 Aug.] 1839, in *Sermons*, 335–52; Silliman, Yale Lectures on Geology, Lecture XXIX, 4 July 1842 (Silliman Family Papers, Box 38, Folder 11), vol. 3, p. 91.

70. Eleazer T. Fitch, Sermon No. 259, on Acts 17:25–26, n.d. (Fitch Papers, Box 6, Folder 28).

require. He only requires man to act according to the constitution of his nature."[71] Silliman agreed that God had established a system of intertwined physical and moral laws, such that man could find his greatest happiness if only he lived in accordance to this divine government. "The remedy" for the health problems brought about by the intemperate use of alcohol, he told Yale alumni in 1842, "is chiefly in the hands of the sufferers themselves, for the physical laws of God must be obeyed as well as the moral, and he who faithfully obeys *both*, will, ordinarily, live long, obtain and do much good, and die at last happy in Christian hope and joy."[72] And more pertinent to the reception of *Vestiges* at Yale, Fitch and Silliman both agreed with Taylor that "those who deny the providential government of God," are "atheists."[73]

Silliman's Crusade for Geology and against Vestiges

By May of 1845, Silliman had returned from his lecture tour in the South. Together with Yale's men of science—including the astronomer Denison Olmsted, the geologist-turned-physician James Davenport Whelpley, and of course Silliman's close colleague James Dwight Dana—he welcomed visiting practitioners for the annual meeting of the Association of American Geologists and Naturalists. Silliman's son, Benjamin, Jr., who specialized in geology and assisted Dana with the editorial work for the *American Journal of Science*, served as the society's secretary. On May 6, "Little Ben," as he was affectionately called, read a paper entitled "Views on Attraction." This had been submitted by Prof. Samuel Webber, a Harvard-trained physician who resided in Charlestown, New Hampshire, where he devoted much of his schedule to the study of nature.[74] Webber's topic may have seemed innocent enough, but one phrase prompted a

71. Fitch, Sermon on Gen. 2:7, 12 Oct. 1845, in Notes of Chapel Sermons by Eleazar T. Fitch and Nathaniel William Taylor, 1845–1848 (Yale University Archives, Item Ynr 17 848f).

72. Benjamin Silliman, *An Address Delivered before the Association of the Alumni of Yale College, in New Haven, August 17, 1842* (New Haven: B. L. Hamlen, 1842), 39.

73. Nathaniel W. Taylor, Sermon on Psalm 14:1, Afternoon Chapel, [2 Nov. 1845], in Notes of Chapel Sermons by Eleazar T. Fitch and Nathaniel William Taylor, 1845–1848 (Yale University, Item Ynr 17 848f).

74. "Webber, Samuel," *Appleton's Cyclopaedia of American Biography*, ed. James Grant Wilson and John Fiske, 7 vols. (New York: D. Appleton and Company, 1889), 6:405.

lively debate: "In this view I am sustained by a new work called 'Vestiges of Creation.'"[75]

The society's official proceedings, stating simply that the secretary read a paper by Webber, omitted any mention of Webber's reference to *Vestiges* or the discussion that this prompted. These proceedings were prepared for publication by Silliman, Jr., Denison Olmsted, Jr. (the astronomer's son), and Dana.[76] An unofficial report appeared in the *New York Herald*, relating in some detail the exchange that took place among members of the AAGN when the phrase "*Vestiges of Creation*" was uttered. Webber had suggested that chalk may be formed from the decayed remains of animalcules. In passing he referred to *Vestiges*, suggesting that if chalk is composed of decayed animalcules, then perhaps it could somehow fertilize the ova of living animalcules, thus accounting for what appears, in the experiments by Crosse and Weekes that *Vestiges* cited, to be a spontaneous generation.

Charles Jackson, a professor of medical physiology from Harvard, "doubted the orthodoxy of the work," meaning the scientific orthodoxy of *Vestiges*. Prof. Olmsted, for his part, "rose more particularly for the purpose of enquiring [*sic*] if any member present was cognizant of any case of gelationous matter being discharged from meteors." No one replied to this, for Henry Darwin Rogers soon "observed that he could not dissent from the remarks made as to the production of animaculae." Noting that spontaneous generation was both an obscure and an important topic, he urged the society to devote attention to it. He insisted that any errors in *Vestiges* "were far overbalanced by the sublime and glorious views it unfolded of creation." Samuel Haldeman, a naturalist from Philadelphia and colleague of Rogers, defended the author of *Vestiges* by suggesting that he had merely "adopted the course followed by all orthodox writers; that is, of reasoning on premises that were not entirely and unanimously admitted by scientific men." But Silliman, Sr., objected heartily, "cit[ing] the authority of Dr. Mantell . . . against the book, which . . . was of a dangerous and irreligious tendency."[77]

75. " Association of American Geologists and Naturalists," *New York Herald*, 9 May 1845, 1.

76. *Abstract of the Proceedings of the Sixth Annual Meeting of the Association of American Geologists and Naturalists, Held in New Haven, Conn., April, 1845* (New Haven: B. L. Hamlen, 1845), 83 (Webber's paper), 84 (publication committee).

77. "Association of American Geologists and Naturalists," *New York Herald*, 9 May 1845, 1.

This would not be the last time that the senior Silliman would voice his opposition against *Vestiges*, or attempt to counteract the support voiced by others. Within weeks, he wrote to Mantell that "Prof. H. D. Rogers and Mr. Haldeman the conchologist both expressed great admiration of the <u>Vestiges of the Creation</u>." Someone in Baltimore suggested to Silliman that Richard Vyvyan had written the work, but he expected that Mantell, who was working in England, would have a better sense for the authorship than anyone in the States.[78] In reply, Silliman's London contact could not identify who had written *Vestiges*, that "tissue of errors," but he did hope it would "soon fall into oblivion."[79] In fact, Mantell was troubled that Silliman's own *Journal* had commended the work in March.[80] (The March notice, published by Dana, will be discussed again later.)

In a follow-up letter, Mantell mentioned John Herschel, Britain's premier man of science and the president of the British Association for the Advancement of Science, which was everything that the Association of American Geologists and Naturalists hoped to become.[81] Herschel—*Sir* John Herschel, ever since he was knighted in 1831—was apparently too much of a gentleman to debase the BAAS meeting by mentioning *Vestiges* in name.[82] Alluding to it carefully, he "condemn[ed] the work in as strong terms as I have done," reported Mantell. "His remarks would be well worthy of insertion in your journal."[83]

Although Silliman did not reprint Herschel's address, his *Journal* did discuss *Vestiges* in future issues. In 1851, for example, the *Journal* favorably reviewed Adam Sedgwick's *Discourse on the Studies of the University of Cambridge*, of which "but 90 pages of the 760 . . . constitute the volume," the remainder being some supplementary notes, plus an oversized preface in which "the learned author" explores the interrelated truths of theology and the natural

---

78. Benjamin Silliman to Gideon Algernon Mantell, 23 May 1845 (Silliman Family Papers, reel 13).

79. Gideon Algernon Mantell to Benjamin Silliman, 10 June 1845 (Silliman Family Papers, reel 13).

80. Gideon Algernon Mantell to Benjamin Silliman, 10 June 1845 (Silliman Family Papers, reel 13).

81. Indeed, the latter would reorganize itself in 1848 as the American Association for the Advancement of Science, explicitly borrowing from the structure and scope of its British sister society. See Sally Gregory Kohlstedt, *The Formation of the American Scientific Community: The American Association for the Advancement of Science 1848–60* (Urbana: University of Illinois Press, 1976).

82. James A. Secord, *Victorian Sensation: The Extraordinary Publication, Reception, and Secret Authorship of "Vestiges of the Natural History of Creation"* (Chicago: University of Chicago Press, 2000), 406–7.

83. Gideon Algernon Mantell to Benjamin Silliman, 21 June 1845 (Silliman Family Papers, reel 13).

sciences and puts "the argument of the Vestiges of Creation . . . in its true light." The *American Journal of Science* then reproduced two full pages of ridicule, in which Sedgwick exposed the "weak logic" in *Vestiges*. Most notably, the author of *Vestiges* had borrowed some absurd ideas from the Lamarckian school, about short-necked elephants desiring longer necks so much that somehow their progeny ended up looking like giraffes.[84]

Silliman also proclaimed his message beyond the printed page. During the evenings of December 22 through 27, 1845, he drafted out a lecture entitled "The Influence of Science and the Arts upon the Physical, Intellectual, and Moral Condition of Man." He rehearsed the presentation before his family in preparation for the February 3, 1846, meeting of the Philadelphia Mercantile Library Association.[85] Meanwhile, the New Haven Young Men's Institute also invited him to speak on January 29, so he resolved to use the same manuscript for that occasion. He also accepted, at this time, a long-standing invitation to return to Baltimore for another series of geological lectures. "I have asked divine direction," he recorded privately, "and trust that I go in the fear & favor of the God." He regretted having to leave his family, and wanted to avoid being absent from Yale when President Jeremiah Day and Chaplain Eleazar Fitch were both ill, but he also felt strongly that his work in geology was work in God's service.[86] It also was a work against *Vestiges*.

"All Science, whether <u>intellectual</u>, <u>physical</u>, or moral is founded upon truth," he began when addressing the Young Men's Institute.[87] In physical nature, man could observe the ocean, earthquakes, birds, insects, fish, quadrapeds, reptiles, and—by the aid of a microscope—animalcules. The economy of these natural kinds serve as "proofs <u>inumerable</u>, and <u>decisive</u> <of> <u>purpose</u> and <u>design</u>, and <u>wisdom</u>, and <u>power</u>."[88] Silliman was not arguing from physiological contrivances, as Paley had done in the early chapters of *Natural Theology*, but

---

84. Review of *A Discourse on the Studies of the University of Cambridge,* by Adam Sedgwick, *American Journal of Science* 2d ser., 11 (May 1851): 144–46, at 144.

85. Silliman, Diary, 11 Jan. 1846.

86. Silliman, Diary, 25 Jan. 1846.

87. Silliman, "The Influence of Science and the Arts on the Moral, Intellectual and Physical Condition of Man Especially in Relation to Our Own Country," with delivery dates spanning 29 Jan. 1845 through 31 Jan. 1853 (Silliman Family Papers, Box 45A, Folder 59), 1.

88. Silliman, "Influence," 7.

rather from the broader pattern of design in nature, in which earthquakes and animalcules each had a place, with everything working toward the highest good. This latter form of the design argument, presented toward the end of Paley's work, held central importance to the New Haven theology taught by Taylor and preached by Fitch. "We feel justified," continued Silliman, "in declaring our mutual conviction, the result of no brief course or observation, that all our science is lame and even presumptuous, without the admission and habitual recognition of a first cause." Moreover, the responsibility falls to men of science to "check speculations, that <may> seem to be presumptuous."[89]

The presumptuous speculation Silliman had in mind was spontaneous generation. "If no possible congeries and arrangements of atoms can result in the production of an elephant <or a whale> no more could it cause a mouse <a butterfly> or an infusorial animalcule to exist." Silliman thought that species transmutation was equally presumptuous:

> The man of Eden was, in his faculties, the same as the man of Mount Vernon; the centaurs of ancient fable and the Callibans of modern fiction with <all other> creatures of a distempered imagination, partly brute & partly human, are only <the phantasms of> night mare [sic] dreams; and notwithstanding the transmutations, so zealous–contended for by a recent popular <the> author, we <of the Vestiges of the [sic] Creation, we> must still believe, that no possible metamorphoses, continued through uncounted cycles of <years> could produce <change> a Newton from <into> an Ape, or convert an Ape into a Newton.[90]

The corrections made to this manuscript likely were spread over a period of years. Silliman delivered this lecture to the New Haven Young Men's Institute (January 1846), the Philadelphia Mercantile Library Association (February 1846), an audience for geological lectures in Baltimore (February 1846), a public meeting in New Haven "by special invitation of the citizens" (March 1846), and the "large and attentive" audience of the New York Mechanics Society (January 1853).[91] Quite possibly Silliman did not feel the need to mention the much-

89. Silliman, "Influence," 48.

90. Silliman, "Influence," 52.

91. A final, unnumbered, page in Silliman, "Influence," contains a note that revisions were made in preparation for the 1853 presentation.

discussed *Vestiges* by name in 1846, though in 1853 the allusion may have been less obvious, thus requiring a specific mention.

In January of 1847, Silliman went to Boston to hear Louis Agassiz's Lowell lectures. "Agassiz wins the hearts in Boston," wrote Silliman in his diary. Silliman's own heart also was won over. "He finds in nature all the types of the different orders and classes of animals introduced almost as early as there was any animal life . . . and thus he removes . . . the foundations of the 'Vestiges of Creation.'"[92]

Later in this chapter it will become clear that Silliman's own theological community did not find the New Haven theology to be incompatible with the docctrines of the *Vestiges*. This observation raises the perplexing question of what *Vestiges* must have meant to Silliman if, unlike Dana and others connected to New Haven theology, he did not welcome the theory. A closer look at his religious life may suggest some answers. As indicated earlier, Silliman was a deeply pious member of Fitch's Christ Church on the Yale campus. He prayed about whether to travel for a series of lectures, and he desired that those lectures be a means for using geology to enhance theology. In March of 1845, just as the *Vestiges* controversy was entering his life, he wrote a letter to Sidney Johnson, thanking him for arranging for his recent lectures in New Orleans. Silliman was not writing merely out of politeness, however; his faith prompted also a deep concern, for he had received several reports "that you have relinquished your religious views & have substituted natural religion." Natural religion, he went on to explain, is merely "a basis," upon which the Bible must be placed, the Bible which alone bears "infinite importance" for the afterlife. Having known Johnson and his "Christian training" from childhood, Silliman must have been close to tears. "I cannot well understand how it has been brought about, unless it has been by the gradual and perhaps unperceived influence of the world."[93]

Did Silliman fear that *Vestiges* was one such "influence of the world" that could lead people away from the Bible, and from Christ, toward a merely natural religion? Possibly, but if so he would have been going against the flow of his Yale community. Dana's favorable theological assessment of *Vestiges* will be treated

92. Silliman, Diary, Jan. 1847 and 13 [*sic*: 20?] Jan. 1847.

93. Benjamin Silliman to Sidney L. Johnson, 9 Mar. 1845, ms. copy in Silliman, Diary, 9 Mar. 1845.

below. Meanwhile, it is worth noting that when Silliman attended the inauguration of President Theodore Dwight Woolsey in October 1846, he heard Pastor Leonard Bacon emphasize the same "moral government of God" of which Fitch and Taylor so frequently spoke, and he heard Woolsey himself suggest that *Vestiges* need not be feared. "Is it any the less a proof of a divine intelligence at work, that what we have called a contrivance is the evolution of a law?" Answering his own question, Woolsey concluded, "We need not fear, then, that any new form of science will take away from the teacher his privilege of conducting his pupils up to God."[94]

Woolsey apparently was attempting, without mentioning *Vestiges* by name, to embrace the development hypothesis. That he could plausibly do such a thing is profoundly indicative of the theological mood at Yale, since he had been ordained into the Congregationalist ministry just four hours before giving this, his first public address as a minister of the gospel and the President of Yale College.[95] Although Woolsey did not address the issues of spontaneous generation or species transmutation explicitly, he did make direct reference to the first plank in *Vestiges'* development theory, the nebular hypothesis. Here Woolsey alluded to LaPlace's godless theory, proposing in its place a theistic development hypothesis: "No blind and unconscious dynamics—no 'mechanique celeste'—but celestial law, emanating from the highest intellect, controls the world; and being understood, awes the mind into reverence and harmony."[96] It seemed, then, that at New Haven all the development theory required was a clear emphasis upon the Divine Intellect whose plans were manifest in nature's laws. Taylor's whole system was, in fact, a lawful manifestation of God's wisdom and benevolence, in both the physical and the moral realms. The God of Paley's early chapters worked through physiological contrivances, but the God of Yale, and of

94. Leonard Bacon, "Sermon at the Ordination," in *Discourses and Addresses at the Ordination of the Rev. Theodore Dwight Woolsey, LL.D., to the Ministry of the Gospel, and His Inauguration as President of Yale College, October 21, 1846* (New Haven: Yale College, 1846), 5–40, at 17–18; Theodore Dwight Woolsey, "Inaugural Address," in *Discourses and Addresses*, 73–100, at 97–98.

95. According to the standards of Congregational polity, in order for Woolsey to serve as president of a Christian college, which would include the pastoral duty of shepherding students and faculty, he first had to be ordained as minister of the gospel. This was done at a 10 a.m. service on October 21, 1846. At 2 p.m. that afternoon, he was inaugurated as Yale's president. See "Preface," *Discourses and Addresses*, 4–5.

96. Woolsey, 98.

Paley's later chapters, worked through a system of laws that allowed evil but also kept evil in check, all for the highest good.

When Silliman lectured against *Vestiges* in New Haven, he was not, therefore, speaking from Taylorite theology. Silliman's clearest reason for opposing *Vestiges*, and celebrating Agassiz's opposition to the same, stemmed, then, not from theological reservations (like those harbored at Princeton, or, in a very different way, at Harvard), but rather another source. The most likely source was his concern that *Vestiges* threatened his career in geology. In a sense, however, that concern also was theological. Silliman viewed himself as a geologist for God. *Vestiges* risked drawing all of natural history under suspicion, both by its speculative methodology and by the negative press that it tended to attract nationwide. In order to promote the authority of geology over philology and theology in matters of natural history, Silliman had to ensure that *Vestiges* would not be regarded as legitimate natural science, even if President Woolsey could dance around the issue in such a way as to leave the door open for a careful accommodation to *Vestiges*.

When Silliman spoke for geology and against *Vestiges*, he spoke with a prophetic voice. He traveled the nation to deliver lectures, and the nation responded by filling sanctuaries to learn about geology. Silliman, it may be suggested, contributed to the Second Great Awakening that historians generally have associated with evangelists but not with evangelical geologists.[97] Just as Elijah spoke for the Lord who revealed himself through prophetic inspiration, so also Silliman was a holy man who testified of God's self-revelation in the works of nature. When Elijah was carried up into heaven by fiery, angel-driven chariots, his mantle dropped to his disciple Elisha, who thereby received a double portion of his spirit (2 Kings 2:9–15). So also, after Silliman's retirement, when James Dwight Dana assumed the newly endowed Silliman Professorship at Yale in 1856, it was said by one correspondent that "the mantle of Elijah hath worthily fallen" upon Dana.[98] In the end, Dana would succeed Silliman as a leading national opponent of *Vestiges*, but this role for Dana by no means came inevitably; for a

97. On the Second Great Awakening, see Charles Roy Keller, *The Second Great Awakening in Connecticut* (New Haven: Yale University Press, 1942); Nathan O. Hatch, *The Democratization of American Christianity* (New Haven: Yale University Press, 1989). Keller identified Silliman as "a persistent supporter of evangelical Protestantism" (8), but did not develop this point, and made no reference to Silliman's evangelical geology presentations. Hatch likewise focused on preachers and itinerant revivalists.

98. A. W. Kinner to Benjamin Silliman, Sr., 25 Feb. 1856 (Silliman Family Papers, reel 3).

time, at least, he acknowledged the possibility, observed also by Woolsey, that *Vestiges* was not in itself a threat to New Haven theology.

James Dwight Dana: The Formation of a Geological Prophet

Personal religious convictions, shaped by Yale's theological climate, profoundly informed the career of James Dwight Dana, who was perhaps America's premier homegrown earth scientist during the era of *Vestiges*. Dana's impact in the allied fields of geology and mineralogy can hardly be overstated. A placard still labeling a display of one of Dana's books at Amherst College's Pratt Museum of Natural History reports: "Dana's *System of Mineralogy*, first published in 1837 and now in its seventh edition, is still the standard reference in the United States."[99] Since that placard was written, an eight edition (1997), numbering 1872 pages, has become available for purchase.[100] The targeted consumers are not antique collectors or historians of science, but serious students of the earth sciences.[101] Dana also produced a *Manual of Mineralogy* (1848) and *Manual of Geology* (1862), which served as standard American textbooks during the middle through late nineteenth century and received widespread European acclaim as well. Moreover, it was Dana who succeeded Silliman, both in his capacity as a Yale professor and as the *American Journal of Science* editor. To be sure, Dana assumed only half of each post, the remaining duties being fulfilled by Benjamin Silliman, Jr. This fact testifies, however, not to any weakness on Dana's part, but rather to the increasingly specialized nature of scientific work during the middle part of the century.

By any standards—including the private judgments of Harvard luminaries Asa Gray and Louis Agassiz and Princeton's Arnold Guyot—Dana was a man of great capacity, equaled or surpassed by few. As his pupil and biographer Daniel Coit Gilman observed, "he gave the impression of incessant energy, forced

99. Exhibit of James D. Dana, *System of Mineralogy* (1837 ed.), Room D: Minerals, Crystals, Meteorites, at the Pratt Museum of Natural History, Amherst College, Amherst, MA, visited by the author on 14 March 2002.

100. *Dana's New Mineralogy: The System of Mineralogy of James Dwight Dana and Edward Salisbury Dana*, ed. Richard V. Gaines, et al., 8th ed. (New York: Wiley and Sons, 1997). As advertised online, *www.amazon.com*, 8 Jan. 2003.

101. Julian C. Gray, "Finding the Right Mineralogy Text: Dana's System of Mineralogy," *Tips and Trips*, 28, no. 4 (Apr. 1999): 7. Also available online: *www.gamineral.org/Dana-system.htm*.

sometimes to rest, but bounding back to his work as a ball rebounds from a wall which has interrupted its progress."[102] This personality trait would, ironically, make Dana delay for some years before even reading the copy of *Origin of Species* that Charles Darwin mailed to him in 1859, but that is a story for a later time. Prior to 1859, Dana applied much energy to the controversies in which *Vestiges* had become entangled, controversies that like a Venus fly trap had seized and long would continue to hold the attention of many an American man of science.

Dana, not unlike Darwin, jump-started his career with a scientific seafaring exploration. The Congressionally funded United States Exploring Survey and Expedition to the South Seas (1838–1842), led by Naval Lieutenant Charles Wilkes, retraced much of the course of Darwin's earlier voyage, during 1831–1837, on board the *H.M.S. Beagle*. Dana took particular interest in the coral reefs the Wilkes crew visited while exploring the South Pacific. At Australia Dana received news of Darwin's theory of reef development and also was working out one of his own. Like Darwin, he took to writing and publishing upon his return, completing the *Geology of the United States Expedition* in 1848.[103]

But the Wilkes Expedition marked a transformation far more profound than any intellectual discovery could bring to the mind of young Dana. The voyage was the occasion of Dana's religious awakening.[104] Filling ten pages with closely written words, Dana described the fierce tempest that the crew encountered in the south seas near Cape Horn in the winter of 1839. This letter reveals what it meant for a Yale alumnus, who on the eve of his departure had met for prayer and felt himself converted with a close cadre of New Haven Congregationalists, to become a man of science on the eve of *Vestiges*. Dana

102. Gilman, 280.

103. Gilman, 209. For a detailed treatment of the expedition , see William Stanton, *The Great United States Exploring Expedition of 1838–1842* (Berkeley: University of California Press, 1975).

104. Dana's 1838 conversion experience and the 1839 shipwreck deliverance that reinforced his faith have been mentioned, but not analyzed in any detail, by Gilman, 69; William Stanton, "Dana, James Dwight," in *Dictionary of Scientific Biography*, ed. Charles Coulston Gillispie, vol. 3, pp. 549–54 (New York: Charles Scribner's Sons, 1971), 550; David N. Livingstone, *Darwin's Forgotten Defenders: The Encounter between Evangelical Theology and Evolutionary Thought* (Grand Rapids, MI: William B. Eerdmans, 1987), 71; Margaret W. Rossiter, "A Portrait of James Dwight Dana," in *Benjamin Silliman and His Circle*, ed. Wilson,, 105–128, at 110–11. Dana's religious experience of the 1839 storm was passed over by Stanton, *The Great United States Exploring Expedition*, although some references to Dana's 1838 conversion and his religious sympathies during the voyage may be found at pp. 129–30, 144, 223, and 320.

received a strong signal from heaven that the Wilkes Expedition was journeying in God's world, studying God's handiwork, and doing so under God's protection. "I have constant cause for thankfulness," wrote Dana to Robert Bakewell, who promptly passed the letter to the Dana's partners in prayer: the Silliman family, President Day, Mr. Whelpley, and Mr. Fisk.[105] "It is a tear of gratitude—after a flood of tears—for the loving-kindness of God in rescuing us, when but a few days since danger thickened around our vessel and appeared to threaten certain destruction." Dana lamented "the privations of a long separation from a Christian community," but was thankful at least to "sail under a Christian Captain" who "conducts divine services Sunday morning." Dana himself retired to his private room for additional devotions each day. He had clear reason to do so. "At that moment, when destruction of our vessel appeared the most certain, when all appeared to await it with breathless silence, an Almighty hand interposed and guided us safely on our course. . . . We were left with our bow anchor . . . our only dependence—No, not our only dependence; for a God that heard prayer still ruled." Dana concluded the letter by confessing his sins, thanking God for bringing him to a saving faith in preparation for this difficult voyage, and exhorting Bakewell and company to praise God together back in New Haven.[106] Still feeling himself transformed by the experience, he wrote to Asa Gray in May that "a merciful God answered our prayers," and he announced his intention to support a mission society.[107]

Dana could never thereafter doubt that by dedicating his life to the natural sciences he was devoting himself to the study of God's moral government in nature. This conviction would structure his arguments in the *Vestiges* debates during the late 1840s through 1850s and shape his reactions to years of secondhand reports before he finally read Darwin's *Origin* for himself in 1864. For Dana, no new development in science could deny what he had learned so decisively in the south seas: "a God that heard prayer still ruled." This conviction did not, of course, preclude the possibility of evolution. It did, however, insist

105. James Dana to Robert Bakewell, 24 Mar. 1839; Robert Bakewell to James Dana, 8 Aug. 1839 (Dana Family Papers).

106. James Dana to Robert Bakewell, 24 Mar. 1839. See also James D. Dana to Edward C. Herrick, 15 Apr. 1839 (Herrick Papers, Yale University, Box 1, Folder 10).

107. James Dana to Asa Gray, 6 May 1839 (Natural Science Manuscripts, Yale University, Box 1, Folder 61).

that whatever means the Creator had employed in developing the earth and its life forms, God remained within hearing distance for human prayer, and God responded graciously to rescue those who were pleading for His aid.

With this particular life event and Taylor's theology both in mind, Dana's initially favorable review of *Vestiges* may be more perceptively examined. In the March 1845 issue of the *American Journal of Science*, Dana published a one-paragraph notice of Wiley and Putnam's *Vestiges*, "strongly recommend[ing] the work to our readers."[108] As noted earlier, the writer (whether Silliman, Dana, or the two jointly) provided no specific warning of scientific errors or theological heresies and for that reason attracted criticism, directed to editor-in-chief Silliman, from Gideon Mantell. Dana's own correspondence with Harvard botanist Asa Gray also was critical of *Vestiges*—but not always. As will be explored next, Dana at times could sympathize with that work, particularly as to its theological implications.

Dana and Gray regarded one another as scientific partners: Dana was a geologist and mineralogist who assisted Silliman in producing the *American Journal of Science*, and Gray was a botanist with connections to the Harvard community's *North American Review*. Dana thanked Gray in January of 1845 for offering to review his recent book on mineralogy in the *North American*.[109] In June Dana wrote again, expressing to Gray, "I am much indebted to you for your kindness[.] I am going to beg another favor of you—In the first place, however"—Dana suddenly changed topics—"a word about the notice of the Vestiges, which Silliman put on my shoulder in a letter to you." Here Dana began to distance himself from the brief notice that had appeared in the March 1845 *American Journal of Science*:

> It is true that I penned it, but I had before me a <long> notice by the Professor—which was the basis of it—for I had never seen the work itself—Prof. S. criticized some few geological facts, but not the fundamental error of the work—about which he only expressed ~~about~~ the amount of doubt in the notice. I have myself no faith in the <author's ~~work~~> ~~his~~ views—(derived from the German School).[110]

108. [James Dwight Dana], Review of *Vestiges of the Natural History of Creation*, *American Journal of Science* 48, no. 2 (Jan.–Mar. 1845): 395.

109. James Dana to Asa Gray, 16 Jan. 1845 (Herbarium).

In February 1846, as Gray was finishing up another series of Lowell lectures, Dana wrote again to support his partner in science against the "groundless—utterly so" development theory of *Vestiges*. Dana announced that another review of the work would soon appear in the *American Journal of Science*, and he urged Gray to write a review against the sequel, *Explanations*. "He [the anonymous author] deserves a lashing for it," suggested Dana; nevertheless, "I view the subject as one for calm scientific discussion—which I think is the only means of treating it proper for a scientific journal."[111] Faithful to Dana's promise, the March 1846 *American Journal of Science* carried a review of *Vestiges* and *Explanations*. There the complaints of Gray were echoed: *Vestiges* contradicted the findings of geology and speculated wildly about spontaneous generation and species transmutation.

A footnote on the final page of Dana's review referred to his own book, *Zoophytes*, which was just then in press.[112] As he just explained to Gray while polishing the manuscript, he had chosen to analyze "corals as <u>animals</u>," first discussing their characters as "<u>living</u> zoophyte[s]," and next considering the "peculiarities of the skeleton."[113] In studying zoophytes in terms of comparative physiology as well as comparative anatomy, Dana found that other researchers had failed to notice specific distinctions and confounded numerous species under the same name. "I have consequently undertaken to reconstruct the science," he informed Gray when sending some draft sheets for his review.[114] Ten days later—on March 2—Dana forwarded additional sheets, and remarked that "I shall be glad to see your review of the Vestiges" in the *North American Review*.[115] In a follow-up letter one week later, Dana requested that Gray would write "a statement that the work [*Zoophytes*] is what it ought to be—that in a Science in such an unsettled state as Zoophytes, patching in new species was impractical."[116]

110. James Dana to Asa Gray, 19 June 1845 (Herbarium).

111. James Dana to Asa Gray, 18 Feb. 1846 (Herbarium).

112. [James Dana], "Sequel to the Vestiges of Creation," review of *Explanations*, *American Journal of Science* 2d ser., 1, no. 2 (Mar. 1846): 250–254, at 254n.

113. James Dana to Asa Gray, 20 Feb. 1846 (Herbarium).

114. James Dana to Asa Gray, 21 Feb. 1846 (Herbarium).

115. James Dana to Asa Gray, 2 Mar. 1846 (Herbarium).

116. James Dana to Asa Gray, 8 Mar. 1846 (Herbarium).

Dana was proposing a revised system of classification, not just the insertion of a few new species into the old system.

Part of the problem stemmed from the intellectual ancestor of *Vestiges'* author, Jean-Baptiste Lamarck, who had built a theory of species transmutation from the gradation of characters linking zoophytes to both plants and animals. "By way of Lamarck," complained Dana to Gray, "descriptions cover several of my species. . . . Genera were also in a most uncertain state, & so on." It was no wonder that Lamarck, and the author of *Vestiges* in more recent days, had promoted species transmutation: their classification schemes tended to merge separate species into the same category.[117] And zoophytes, which resemble both animals (Greek: *zoe*) and plants (Greek: *phyta*), could, if sloppily classified, easily implicate both Asa Gray's plant kingdom and Louis Agassiz's animal kingdom into a singular, sweeping scheme of development. "I shall look at your slash at the Vestiges with much interest," wrote Dana to Gray on March 21. Dana's *Zoophytes* had gone to press and, within a few weeks, so would Gray's article in the *North American.*[118]

Dana, not surprisingly, "was very much pleased" with Gray's review of *Vestiges* and *Explanations*, and he sent a long letter to tell him so. In this letter, however, Dana began to address the development theory from a different perspective than he had previously adopted. His *Zoophyte* work sought to establish a firmer footing for a science in which practitioners had been inconsistent in classifying their specimens. *Vestiges*, by contrast, had capitalized upon the vagueness of zoophyte taxonomy in order to build a case for development. These were the stakes for the progress of natural science, but what about theology? "I have never been afraid of the book," wrote Dana in reference to *Vestiges*, "although I have not doubted that it would make some infidels."

Dana even went so far as to write, "I believe that the publication of the Vestiges will be the means of much good to Religion & to Science in general." He gave examples of how other scientific advances, such as in archaeology, had shed light on how to interpret the Bible. Dana then explained why in the March 1845 *American Journal of Science* he had stated that *Vestiges* did not threaten his religion:

117. James Dana to Asa Gray, 8 Mar. 1846 (Herbarium).

118. James Dana to Asa Gray, 21 Mar. 1846 (Herbarium).

> I remarked that <u>should</u> the <theory> prove true, it would not affect the truths of Religion.—My reasoning is this.—That the condition & character of man requires for his improvement and sanctification the principles we, as Christians, believe and this fact will not be modified by any view of our Creation.— We are sinful beings, and no other plan could restore us. The system should be one which will make us feel our dependence on God; perfect resignation to his will, and filial love & fear, while pride & selfishness are subdued.[119]

Dana's words could have come straight from the pen of Nathaniel Taylor. Five times in his religious vindication of *Vestiges*—which continued for more than has been quoted above—he used the term "system," and the context revealed that Dana had in mind what his contemporaries knew as "Taylor's system." The Christian message that Christ was God come to earth in order to rescue man was a necessary conclusion given Taylor's premise that God, being a perfect moral governor, would provide for the salvation of those who become lost through disobedience to Him. As Dana explained to Gray, "The coming of Christ, as we believe, is an essential part of the system," and how humankind originated cannot interfere with this gospel essence. Perhaps recalling the days in the south seas when God answered his prayers for deliverance, Dana emphasized that God's moral system "makes us children of the heavenly father, to whom we may appeal under all circumstances." So long as *Vestiges'* development theory could be understood as a system in which God morally governs his universe and responds to the prayers of His people, the exact method of human origins was theologically irrelevant. "On this ground," concluded Dana in his letter to Gray, "I have inferred that the theory of the Vestiges, if proved, would not affect the truths of Christianity—Am I rig[ht?]"[120]

No reply by Gray has been found in the archives of their correspondence. Gray later did adopt a theory of organic evolution while still maintaining a religious faith. As will be suggested in the conclusion of this book, Gray seems to have done so on different lines than those sketched here by Dana. The Taylorite approach that Dana himself adopted likely had two origins. The first, of course,

---

119. James Dana to Asa Gray, 21 Apr. 1846 (Herbarium).

120. James Dana to Asa Gray, 21 Apr. 1846 (Herbarium). A tear in the page makes the final word and punctuation mark partially illegible.

was his place in theologically distinctive academic community at Yale. The second source can be traced to *Vestiges* itself.

The author of *Vestiges* concluded his work with a chapter on "The Purpose and General Condition of the Animated Creation." Here he admired the Creator's system that had integrated a "source of gratification" into each of the relations between animals and their environment, and between animals and other animals. Still, the author had to admit that some degree of misery had crept into the system.[121] Misery, he explained, much as Taylor had been doing, must be permitted in order for the system as a whole to attain the highest good. Gravity, for example, could cause injury to a careless boy, but without gravity, the world would be in chaos, causing endless injury even to those who are careful. Likewise, war results from nations that do not get along, but the "tendencies in human nature" that lead to wars, such as "keen assertion of a supposed right" and "resentment of supposed injury" are themselves, like the law of gravitation, benevolent. Indeed, they are indispensably so.[122]

Dana was not the only person at Yale to imbibe this doctrine jointly from the New Haven theologians and from the author of *Vestiges*. Yale philologist James Hedley recorded in his diary in May 1850 that "God does not carry on his work as we should expect. For reasons which we can to some extent divine but never fathom, he works slowly, through long alterations of success and failure."[123] In July Hadley read excerpts in the *New York Tribune* from a new poem, "In Memoriam," by Alfred Lord Tennyson.[124] Inspired by *Vestiges*, the poet had explored in vivid detail the implications of a natural history in which humans emerged through what Hadley had called "long alterations of success and failure":

> Man, her [Nature's] last work, who seem'd so fair,
> Such splendid purpose in his eyes,
> Who roll'd the psalm to wintry skies,

---

121. *Vestiges*, 362. The edition cited is reproduced in Robert Chambers, *Vestiges of the Natural History of Creation and Other Evolutionary Writings, Including Facsimile Reproductions of the First Editions of "Vestiges" and Its Sequel "Explanations,"* ed. with an introduction by James A. Secord (Chicago: Chicago University Press, 1994).

122. *Vestiges* , 364–65.

123. Hadley, *Diary*, 52, entry for 6 May 1850.

124. Hadley, *Diary*, 74, entry for 9 July 1850.

Who built him fanes of fruitless prayer,
Who trusted God was love indeed
And love Creation's final law—
Tho' Nature, red in tooth and claw
With ravine, shriek'd against his creed[125]

Future generations would associate Tennyson's phrase "Nature, red in tooth and claw" with a Darwinian struggle for existence, but Tennyson's source had been *Vestiges*, which he had eagerly purchased immediately upon hearing of its publication in 1844.[126] In both *Vestiges* and "In Memoriam," Yale readers could find a message that resonated with the manner in which their local theologians reconciled the evil manifest in nature with the benevolence of the Creator. Reflecting upon Tennyson's poem and sharing it with others, Hadley wrote, "It is wonderfully full and deep, and to many will seem a revelation of their own inward experience."[127]

Yale readers could find much to appreciate in *Vestiges*, since it emphasized that God had allowed for evil precisely to create a better world. If the human body were more immune to diseases, suggested the author, it also would be less sensitive to tender pleasures. A world without disease would be "a derangement of the whole economy of nature," resulting in "more serious evils."[128] Because the Deity governs nature by the fixed natural laws that He infused into matter at the first moment of creation, humans can discern predictable patterns and thereby form reliable moral laws. Creatures who keep those laws will find Nature's rewards: diminished suffering and increased pleasure.[129]

Finally, the author of *Vestiges* also could be read as a reinforcement of Taylor's argument for eternal salvation. Nature's system as a whole testifies that the Deity is benevolent. Certain parts of that system, however, fail to show the

125. Alfrd Lord Tennyson, "In Memoriam A. H. H." (1850), in *Tennyson: A Selected Edition, Incorporating the Trinity College Manuscripts*, ed. Christopher Ricks (Berkeley, CA: University of California Press, 1989), 331–484, at 399 (Part LVI, Lines 9–16).

126. David Charles Leonard, "Tennyson, Chambers, and Recapitulation," *Victorian Newsletter* 56 (1979): 7–10.

127. Hadley, Diary, 88, entry for 11 Aug. 1850 (reading); 94–95, 100, 111, entries spanning 21 Aug. to 25 Sep. 1850 (sharing).

128. *Vestiges*, 370.

129. *Vestiges* , 377.

DEBATING EVOLUTION BEFORE DARWINISM

Deity's fullest benevolence, since evil and misery still exist in this world. Thus, one may conclude that "the mundane economy might be . . . a portion of some greater phenomena," with an afterlife fulfilling humanity's higher visions for perfect benevolence. "In this faith we may well rest at ease," concluded the author of *Vestiges* with a confidence that his own faith echoed the "doctrines in esteem amongst mankind."[130] Certainly it echoed the chief doctrines esteemed by men at Yale. This may be seen not only in the lives of Dana and Hadley on that campus, but also in the career of a Yale-trained theologian and geologist who presided over the Congregationalist Amherst College in Massachusetts.

Edward Hitchcock: Theologian of Geology

If Dana was Silliman's star pupil at Yale, Silliman also had produced a great man of science whose career took him beyond Yale. Edward Hitchcock, the son of a poor but pious hatter in Deerfield, Massachusetts, became interested in astronomy while at Deerfield Academy in the 1810s. Due to worsening eyesight, he had to abandon that passion, but somehow he began corresponding with Silliman, and his interests turned toward geology.[131] Hitchcock, in fact, was among the few to whom Silliman confided his dream of founding the *American Journal of Science*.[132] Hitchcock also had been nurtured in his family's Congregationalist faith. In 1817 he enrolled for theological training at Yale. Prof. Fitch taught Hitchcock and a handful of other ministerial students in his home, since President Dwight had recently died and Nathaniel Taylor had not yet been hired to found the Yale Theological Department. In Fitch's home, Hitchcock absorbed Stuart's biblical criticism, which had been transmitted via the Yale faculty from Andover; as for dogmatics, he and his classmates felt encouraged by Fitch to push "our discussions on some points to the utmost limits of orthodoxy."[133] A foundation for Taylorite theology, with its willingness to depart from the stricter Old School Presbyterian standards, was thus laid, and in the

130. *Vestiges*, 385–86.

131. Michele L. Aldrich, "Hitchcock, Edward," in *Dictionary of Scientific Biography*, ed. Charles Coulston Gillispie, vol. 6 (New York: Charles Scribner's Sons, 1972), 437–38, at 437.

132. Benjamin Silliman to Edward Hitchcock, 27 Oct. 1817 (Edward Hitchcock Papers, box 3, folder 37).

133. William C. Fowler, "Origin of the Theological School of Yale College," 1869? (Yale Theological Department Miscellaneous Items, Yale University, Yjd 51), 1–3, quoting 3.

decades to come Hitchcock would speak from that foundation when addressing *Vestiges*.

During the period of his theological training, Hitchcock continued to take interest in the natural sciences, particularly geology. He sympathized with, and was encouraged by, Silliman's old-age geology,[134] but also felt a strong vocational alliance to theologians, even if some of them were hesitant to soften their stance on the traditional understanding of Mosaic chronology. In 1822, Hitchcock confided to his geological mentor that he was suffering *"a case of conscience."* The conflict in his heart was not between geology and Genesis, for he and Silliman felt they had resolved that, but between his calling as a pastor and his zeal for "botanical and geological pursuits." To Silliman he confessed a fear that "I might be worshipping idols.—And if these pursuits be the right eye that must be plucked out let them not be spared however painful the effort." Yet the minister hoped, somehow, that his love for geology was not what Christ had in mind when he said, "And if thine eye offend thee, pluck it out: it is better for thee to enter into the kingdom of God with one eye, than having two eyes to be cast into hell fire" (Mark 9:47). He knew Silliman to be a man of faith as well as a professor of geology, so he begged him for advice on how the gospel ministry and geological pursuits might be merged into a single vocation.[135]

Silliman's reply is not extant. Whatever advice the Yale professor may have offered, Hitchcock continued his pastorate in Conway, Massachusetts, for a few more years. Then, in October of 1825, he began auditing Silliman's courses in preparation for a professorship of natural history at Amherst College. His subsequent career in geology was impressive. He served as the state geologist for Massachusetts during the 1830s and helped to found the Association of American Geologists and Naturalists in 1840. Thirty printings of his *Elementary Geology* (1st ed., 1840), a widely used textbook, appeared during his lifetime.[136] Amid these geological pursuits, Hitchcock's passion for religion was never left behind; his career at Amherst was a living vindication of his own career move from the pulpit to the Connecticut River Valley and the classroom: Hitchcock had become America's leading geologist of theology and theologian of geology.

134. Silliman to Hitchcock, 18 Aug. 1820 (Edward Hitchcock Papers, Box 3, Folder 37).

135. Hitchcock to Silliman, 1 Dec. 1822 (Edward Hitchcock Papers, Box 5, Folder 2).

136. Aldrich, 437–38.

"I should feel condemned," said Hitchcock in one of his more memorable sermons before the Amherst student body, "if I did not ... [employ] the discoveries of modern science [to] elucidate and make more impressive the language of Scripture." He was speaking that day on the "Catalytic Power of the Gospel." Christ had said that "the kingdom of heaven is like unto leaven, which a woman took, and hid in three measures of meal, till the whole was leavened" (Matthew 13:33). Hitchcock, as professor of both chemistry and natural theology, explained that it only takes a small amount of leaven to make a big change. As a theologian who sympathized with Taylor's system, he also emphasized that the "kingdom of God" is God's government, through the leaven of the church, just as God's government in nature produces bread through the leavening force of yeast.[137]

Hitchcock's sermons at Amherst reveal the theological heritage he shared with New Haven. Like Fitch and Taylor, he emphasized God's moral perfection as the governor of the universe. Preaching at the college chapel in 1847, he deduced from the perfectibility of God that God must have created the universe with a complete plan. The "general laws" in nature bear evidence to this "preconceived plan," for God has arranged those laws in such a way as to carry out His will. God, who foresaw what would come to pass, must be pleased with it. "If not pleased[,] he would have altered the plan." God included human free agency within this plan, allowing for sin only because such an allowance would work for the greater good.[138] Seen as evidence of God's moral governance, the laws of geology were never far from the chief tenets of Taylorite theology.

In his inaugural address as president of Amherst College in 1845, the Rev. Hitchcock affirmed what he had concluded twenty years prior about the relationship between a career in the natural sciences and a calling in the gospel ministry. "The preacher of the Gospel may consistently devote himself to the work of instructing the young in literature and science" while continuing in the "direct preaching of the gospel occasionally," explained Hitchcock. Educating the young promotes religion, so a minister who leaves the pulpit for the classroom is

137. Edward Hitchcock, "The Catalytic Power of the Gospel," sermon on Matt. 13:33, in his *Religious Truth, Illustrated from Science, in Addresses and Sermons on Special Occasions* (Boston: Phillips, Sampson, and Co., 1857), 223–54, quoting 223.

138. Edward Hitchcock, Sermon on Heb. 3:4, Nov. 1847 (Edward and Orra Hitchcock Papers, Box 9, Folder 18).

not, contrary to some people's views, "abandoning his high calling."[139] Indeed, during his years as a professor of natural history, the Rev. Hitchcock had continued to pray daily in search of God's guidance as to whether he should engage "in any new scientific enterprise."[140] The Amherst presidency, with its attendant duties of preaching in the college chapel and teaching natural theology in the college classroom, satisfied Hitchcock's need to be a minister of the gospel while also keeping him close to research and teaching in the natural sciences.

During his inaugural address, Hitchcock made use of his dual social roles as a geologist and a theologian (or "theologist," as he long had preferred to call himself[141]). In those roles, he brought his newly acquired authority as Amherst College president to bear upon the *Vestiges* controversy. He expressed confidence that theology and natural science could be embraced together, without reservation on either side. "Nevertheless," he admitted, some "unusual interest" has been excited "by the recent work of an English nobleman"—Hitchcock must have had Sir Richard Vyvyan in mind—"entitled, Vestiges of the Natural History of Creation." *Vestiges*, suggested Hitchcock to calm his audience, would not have any serious long-term consequences for religion. Even if its three hypotheses held true—the nebular development of the universe, the spontaneous generation of life, and the transmutation of species—one must still maintain that,

> an intelligent, spiritual, infinite Deity, is quite as necessary to account for existing nature, as on the more common theory, which supposes the universe commanded from nothing at once in a perfect state. Indeed, to endow the particles of matter with the power to form exquisite organic compounds, just at the moment when circumstances are best adapted to their existence, and then to become animated, nay, endowed with instincts, and with lofty intellects; all which results the advocates of these hypotheses must impute to the laws

139. Edward Hitchcock, *The Highest Use of Learning: An Address Delivered at His Inauguration to the Presidency of Amherst College*, published by the Trustees (Amherst: J. S. & C. Adams, 1845), 34.

140. Edward Hitchcock, Private Notes, Dec. 1843–Apr. 1854 (Edward and Orra Hitchcock Papers, Box 19, Folder 4), 87 (entry for Dec. 1843).

141. Edward Hitchcock to Benjamin Silliman, 1 Dec. 1822 (Hitchcock Papers, Box 5, Folder 2); Hitchcock, *The Highest Use of Learning*, 28. In the former instance, "theologist" is apparently synonymous with "theologian," including those theologians who are ignorant of geology; in the latter instance, Hitchcock is speaking of theologians who appreciate the "demonstrations" of God's existence and moral government that geology offers for theology.

impressed upon originally brute matter,—such effects, I say,
demand infinite wisdom, power, and benevolence, even more
imperatively than the common theories of creation.[142]

Writing an undergraduate essay as a junior at Amherst College in 1848, Hitchcock's son, Edward, Jr., echoed his father's evaluation that an account of creation by natural law could reinforce their theological convictions. Geometry, anatomy, and morality were each governed perfectly by the laws God had planned, and if God, in his infinite perfection, planned laws for the government of "inanimate and brute creation," then so much more necessary were God's laws when human minds are involved, so that "there should be uniformity and regularity in all our actions."[143]

Thus, neither the elder Hitchcock nor his son and student, Edward, Jr., had any theological reservations about *Vestiges*. Whereas Dana kept his similar opinion private—expressing it only in the aforementioned letter to Gray—Hitchcock went public in defense of the theological ramifications of *Vestiges*. This did not, however, mean that Hitchcock supported the development hypothesis as a well-grounded natural history. He was a geologist as much, probably more so, than a theologian, and from geology would his disagreements with *Vestiges* arise. His two most widely circulated books, *Elements of Geology* and *Religion of Geology* stressed, respectively, the geological evidence against *Vestiges* and the theological openness that one could have toward *Vestiges*, if only it were not so geologically untenable.

Hitchcock arranged his *Elements of Geology* textbook in a fashion similar to Euclid's *Elements of Geometry*. Yale's students typically studied Euclid's twenty-some-centuries-old text during their sophomore year and afterward buried it, sometimes in a cremated form, in a plot of ground just a stone's throw from Hillhouse Avenue.[144] Whether or not Hitchcock's copy ended up six feet under, he retained much of the Greek sage's pedagogical model. His geological text proceeded by Propositions, giving Definitions where necessary, and then

142. Hitchcock, *The Highest Use of Learning*, 31. This address appeared also in Hitchcock, *Religious Truth*, 9–53.

143. Edward Hitchcock, Jr., "Perfect Order in Nature's Works," essay no. 20, 11 Feb. 1848 (Edward Hitchcock, Jr., Papers, Box 3, Folder 23).

144. James Buchanan Henry and Christian Henry Scharff, *College As It Is: Or, The Collegian's Manual in 1853*, ed. with an introduction by J. Jefferson Looney (Princeton, NJ: Princeton University Libraries, 1996), 144n; Clarence Deming, *Yale Yesterdays* (New Haven: Yale University Press, 1915), 22.

drawing Inferences and offering Proofs. Following each Proof, he usually offered some Inferences and Remarks in order to address the "historical and religious relations" of the geological topics under discussion. By 1854, just fourteen years after the first printing, Hitchcock had expanded *Elementary Geology* into a twenty-fifth edition, which five publishers agreed to issue simultaneously.[145]

One of Hitchcock's "Inferences" in that work pertained directly to the development theory: "It appears that amidst all the diversities of organic life that have existed on the globe, the same general system has always prevailed." He noted that "a late anonymous author very strenuously maintains the doctrine of the creation and gradual development of animals by law, without any special creating agency on the part of the Deity," since no vertebrates have as yet been discovered in the lower Silurian. After citing both *Vestiges* and *Explanations*, Hitchcock objected that "in the whole records of geology, there is not a single fact to make such a metamorphosis probable." The facts, rather, lent support to special creation; vertebrates *had* been found in the upper Silurian, and the upper and lower Silurian periods were virtually continuous. Thus, all the major classes of animals are represented throughout zoological history. "Certainly no geologist will imagine that his science furnishes any evidence of the development theory." Hitchcock warned later in the work that *Vestiges* could be deployed to "annihilate the doctrine of miraculous and special Providence and of prayer."[146] As his later writings emphasized, however, this was not the only theological possibility for *Vestiges*.

In the early 1840s, Hitchcock had delivered several lectures to his students at Amherst, and re-delivered some of them to audiences in New England lyceums, concerning what he called the "religion of geology." In 1851, he prepared those lectures for publication, updating them in order to clarify the theological implications of *Vestiges*. By 1865, Hitchcock's *Religion of Geology* had sold over 12,000 copies in at least eight American and British editions. In that work, Hitchcock emphasized the shared conviction of Silliman, Dana, and the "theologist" himself that natural theology should be entrusted to geologists rather than theologians or philologists, for geology had more to offer on that matter

145. Edward Hitchcock, *Elementary Geology*, 25th ed. (New York: Ivison and Phinney; Chicago: S. C. Griggs, and Co.; Buffalo: Phinney and Co.; Auburn: J. C. Ivison and Co.; Detroit: A. M'Farren; Cincinnati: Moore, Anderson, and Co., 1855), quoting vii.

146. Hitchcock, *Elementary Geology*, 197, 198, 334.

than the other sciences. In common with Stuart, Hitchcock insisted that Scripture should be interpreted, like any other book, according to the rules of grammar and rhetoric, and an historical understanding of the writer's own circumstances. Unlike Stuart, Hitchcock also insisted that geological "discoveries furnish us with another means of its correct interpretation, where it describes natural phenomena." Biblical philology, he reasoned, could allow for either an old or a young earth; thus, only geology could decide what Moses meant.[147]

In promoting geology as the arbiter of Biblical exegesis regarding the history of the earth and its life forms, Hitchcock by no means sought to cut himself loose from his theological moorings. *Religion of Geology* remained firmly anchored in New Haven theology. "Where the existence of evil is concerned," wrote Hitchcock in regard to death and extinction, "we must assume that the present system of the world is the best which infinite wisdom and benevolence could devise." Like Fitch, he asserted that death was not an evil for either animals or humans; death simply was the best means for "removing them to a higher state of existence." Like the author of *Vestiges*, Hitchcock mentioned the law of gravitation as an example of how all natural laws, including those governing death, serve the greatest good, even if occasionally they might cause suffering or pain.[148]

Approaching *Vestiges* from a theological standpoint similar to Dana's, Hitchcock not surprisingly concluded that "the hypothesis of creation by law does not necessarily destroy the theory of religion. . . . [N]o event [caused by a natural law] is any less God's work, then if all [events] were miraculous."[149] So long as he could be clear that he was expanding the religious language of *Vestiges*, rather than following the materialist pathway of Oken, Hitchcock the *theologian* felt quite comfortable with the development hypothesis.[150] Hitchcock the *geologist*, however, could not accept spontaneous generation or species transmutation at a time when all available evidence, as presented by reputable practitioners, such as

147. Edward Hitchcock, *The Religion of Geology and Its Connected Sciences* (Boston: Phillips, Sampson, and Company, 1857), v–xii, 5 (quotation), 45. See also the editions published in Boston: Phillips, Sampson, and Company, 1851, 1852, 1854, 1856; Glasgow and London: William Collins, [1851?], and [1865?]; and, London: James Blackwood, 1859. The title page of the 1857 ed. indicates marked the "twelfth thousand" copy.

148. Hitchcock, *Religion of Geology*, 85 (best system), 105 (higher state), 222 (gravitation).

149. Hitchcock, *Religion of Geology*, 294.

150. Hitchcock, *Religion of Geology*, 297.

Hugh Miller, Richard Owen, and Hitchcock himself, stood against those hypotheses. The nebular hypothesis also had lost support among men of science ever since Lord Rosse's telescope discovered that a supposed nebula was not a nebula at all. Thus, Hitchcock concluded that *Vestiges* "*is not sustained by facts.*"[151]

Not surprisingly, Edward, Jr., likewise was convinced that the scope of natural law—though it could, in keeping with sound theological principles, be extended as far as the author of *Vestiges* had taken it—had not yet been discovered by reputable men of science to account for the creation of life.[152] Creation, therefore, had to remain in the realm of the miraculous. By "miracle" Hitchcock, Sr., did not necessarily mean anything substantially different than a "natural event," since all things were in some way caused by God anyway. The chief distinction for him was epistemological rather than ontological: miracles were those events that men of science could not explain by any known law. Some miracles resulted in the opposite of what a natural law would suggest, while others resulted simply in greater or lesser degrees of the same tendencies produced by natural laws. Indeed, a few events even appeared to be governed entirely by natural laws, but they still were "special providences," thus, more-or-less miracles. Most notably, "conversion takes place in the human heart in perfect accordance with the laws of mind, and could be philosophically explained by them; yet revelation assures us that it is *not of blood,* [natural descent,] *nor of the will of the flesh, nor of the will of man, but of God*" (John 1:13).[153]

It is not accidental that Hitchcock thought of conversion as a "special providence," that is, a "miracle" that otherwise fits his definition of "natural events." New England Congregationalists with affinities for the New Haven theology were convinced, simultaneously, that the Holy Spirit causes conversion

---

151. Hitchcock, *Religion of Geology*, 301–21, 300 (quotation).

152. Edward Hitchcock, Jr., "What Is Life?," essay no. 25, 18 Apr. 1848 (Edward Hitchcock, Jr., Papers, Box 3, Folder 28).

153. Hitchcock, *Religion of Geology,* 330 (insertion original); see also p. 360. For distinctions between "special providence," "miraculous providence," "particular and universal providence," and other subtle varieties under the broad category of "divine providence" see pp. 327–38. For a different reading of Hitchcock, which regards *Vestiges* as "rudely shatter[ing]" his approach to harmonizing geology and theology, and which seeks to account for his definitions of miracles and natural laws without exploring New Haven theology, see Stanley M. Guralnick, "Geology and Religion before Darwin: The Case of Edward Hitchcock, Theologian and Geologist," *Science in America since 1820*, ed. Nathan Reingold (New York: Science History Publications, 1976), 116–30, quoting 120.

and that humans, in keeping with the scientific laws of mental and moral philosophy, exercise free will when being converted. Old School Presbyterians had trouble seeing how these two claims—that God causes, yet man freely chooses—could both be satisfied, and thus they accused New Haven theologians of Arminianism. Neither Taylor nor Fitch owned up to that label; both thought that a person chooses freely within God's system of moral laws and, nevertheless, that the Holy Spirit converts a person's soul. It was precisely this New Haven willingness to speak of the natural and the supernatural in overlapping terms, almost to the point of indistinction, that made *Vestiges* theologically palatable for the very geologists at Yale and Amherst who nonetheless, for natural scientific reasons, rejected the development hypothesis. "I reject them," Hitchcock said regarding the nebular hypothesis, spontaneous generation, and species transmutation during his first official act as college president, "more because they have no solid evidence in their favor, than because I fear that they will ultimately be of much injury to religion."[154]

Theologically, at least, Hitchcock and Dana each felt open to *Vestiges*. Hitchcock ultimately rejected *Vestiges*, since the natural sciences weighed so heavily against the development theory. Dana likewise put his theological openness aside in order to promote a more narrow orthodoxy in the natural sciences. Future generations would view Dana as a great geologist who lived at a time when theology was too conservative to allow for evolutionism. It is, therefore, surprising to learn that, at least for those with New Haven affinities, geology was more conservative than theology as to the question of creation by developmental law. As the next section illustrates, Yale geology also could be more conservative than philology.

The Dana–Lewis Debate of 1856–1857: A Controversy between Two Theological Sciences

Tayler Lewis, a lawyer-turned-philologist, taught "Oriental Languages" (meaning Hebrew and Greek) at Union College, an institution of New School Presbyterians in Schenectady, New York.[155] He had close ties to Yale philology, as

---

154. Hitchcock, *The Highest Use of Learning*, 31.

he occasionally met with Prof. Hadley in New Haven.[156] Like Hadley and Moses Stuart, Lewis regarded the natural, as compared to philological, sciences as impertinent to theology.[157] Like his fellow New Schoolers, he appreciated Taylor's emphasis on God's moral government. His career demonstrates, however, that these elements could be variously combined: either to oppose the development theory or else to sympathize with it, regardless of geological conclusions—that is, if one was, like Lewis, willing to disregard geological conclusions.

In 1845, Lewis reviewed *Vestiges of the Natural History of Creation* for the New York City-based *American Whig Review*. "To style this book infidel," he told his readers, "would be pronouncing upon it too mild a condemnation." The author of *Vestiges* had denied the miraculous, and thereby affirmed atheism —"sheer atheism, dark, chilly, soulless atheism"—even if dressing it in theological language.[158] Three themes, aside from the equation of *Vestiges* with atheism, dominated Lewis's review. First, he emphasized God's moral government, which he felt *Vestiges* had threatened by denying miracles. To support this point he quoted "the ancient Hebrew school of philosophy," according to which God promised, "My salvation shall be forever, and my righteousness (that is my moral government) shall never fail."[159] Second, God's righteousness, which Lewis explained to mean God's moral government, pertained not only to "moral science" but also to "natural [science]." Consequently, the Bible, which testifies of God's moral government, has something to say about both branches of science.[160] Third, when the Bible addresses a topic in the natural sciences, such as can be found in the creation account of Genesis, "interpretation [must be] fairly made on

155. Morgan B. Sherwood, "Genesis, Evolution, and Geology in America before Darwin: The Dana-Lewis Controversy," in *Toward a History of Geology*, Proceedings of the Inter-Disciplinary Conference on the History of Geology, September 7–12, 1967, ed. Cecil J. Schneer (Cambridge, MA: The MIT Press, 1967), 305–16, at 306. The discussion above differs from Sherwood's account in its reference to shared theological commitment between Dana and Lewis to Taylor's moral government theory, but otherwise follows Sherwood's general outline.

156. Hadley, *Diary*, 56 (25 May 1850), 108 (17–18 Sep. 1850).

157. Sherwood, 307.

158. [Tayler Lewis], "Vestiges of the Natural History of Creation," *American Whig Review* 1, no. 5 (May 1845): 525–43, at 527 (condemnation), 537 (atheism).

159. Lewis, "Vestiges," 525, 539 (quotation). Lewis apparently was conflating several verses from Psalm 119 into a concise summary, as no biblical passage precisely matches his wording.

160. Lewis, "Vestiges," 526.

pure philological grounds." Geologists, whether speculative (like the author of *Vestiges*) or more careful (like the author's many geological opponents), had nothing decisive to contribute to such topics as the age of the earth and the length of time God took when creating the earth and its inhabitants.[161]

Ten years later Lewis drew upon his philological expertise to address precisely those exegetical questions, regarding the "days" of creation, that geologists like Silliman and Hitchcock had been trying to answer. His *Six Days of Creation*, published in 1855, defended a day-age theory similar to Silliman's, but rested it upon a philological, rather than geological, foundation. Lewis also repeated in this work his judgment against *Vestiges*: "A development theory which has no divine organization, or acknowledges only one divine organization . . . is as near to atheism as it can be."[162] Lewis did not, however, leave it at that. As Dana had suggested a decade earlier, Lewis proposed:

> A development theory in the sense of species from species, as well as of individual from individual, may be as pious as any other. . . . There . . . is nothing monstrous or incredible in the idea that what had formerly been the residence of an irrational and groveling tenant might now be selected as the abode of a higher life.[163]

America's waters, noted a reviewer of Lewis's *Six Days*, were still disturbed by the "foaming wake in the ocean of thought" left by *Vestiges*.[164] Dana by this time had abandoned his theologically motivated sympathies for the development hypothesis and was seeking to preserve geology and zoology from the spectre of charlatanism surrounding *Vestiges*. He had no tolerance for Lewis's public suggestion that *Vestiges* could be refashioned into pious science, even if it echoed his own private words to Gray from nine years previous. Moreover, Lewis was promoting philology at the expense of geology, and this after faithful Congregationalists like Silliman and Hitchcock had done so much to ensure that geology would enjoy a privileged partnership with theology. It was

161. Lewis, "Vestiges," 528.

162. Lewis, *Six Days of Creation*, 216, quoted in Sherwood, 311.

163. Lewis, *Six Days of Creation*, 248, 249, quoted in Sherwood, 311.

164. Review of *The Six Days of Creation*, by Tayler Lewis, *New York Times*, 16 June 1855, 2.

time for Dana, too, to demonstrate that geology was an indispensable ally to the faith.

Dana submitted a review of Lewis's book to *Bibliotheca Sacra*, a journal edited by Edward Amasa Park of Andover Seminary. "Your article has been received," Park eagerly replied in November 1855, offering to have the printer run as many extra copies as Dana desired.[165] His fifty-page criticism, entitled "Science and the Bible," appeared in the January 1856 issue. Dana began with a quotation from Psalm 19—"The heavens declare the glory of God"—and then proceeded to provide examples of natural patterns, or laws, which demonstrate God's constant workings in nature.[166] His title was meant to imply that geological science and the Bible were united; it was Lewis's philology that stood in opposition to both of them. Dana found it particularly ironic that Lewis had agreed with many of the conclusions geologists had reached, including Silliman's day-age interpretation of Genesis, and yet had ridiculed not only geology but all of the natural sciences, labeling their methods as speculative and their results as unreliable. The greatest irony, however, was that Lewis had affirmed precisely what geologists had, by avoiding speculations, denied, namely, the "infidel geology" of *Vestiges*.[167] Lewis had ostensibly entertained the development hypothesis because the biblical phrases "God created" or "He made" do not in themselves specify the mode or duration for that process.[168] But then Lewis strayed from the plain meaning of the text when he suggested that "let dry land appear" (Genesis 1:9) refers to the formation of mountains and valleys, which are not, Dana emphasized, mentioned in the text.[169] Dana thus could conclude that Lewis had in fact borrowed his exegesis from the very science of geology that he claimed to oppose. Geology, affirmed Dana, was the only reliable source for understanding what Moses meant when he related the creation process.[170] The remainder of Dana's article

---

165. Edward Amasa Park to James D. Dana, 14 Nov. 1855 (Dana Family Papers, Reel 3).

166. James Dwight Dana, "Science and the Bible," 4 pts., *Bibliotheca Sacra* 13, no. 49 (Jan. 1856): 80–129; 13, no. 51 (July 1856): 631–56; 14, no. 56 (Apr. 1857): 388–413; 14, no. 57 (July 1857): 461–524; at pt. 1, p. 80.

167. Dana, "Science and the Bible," pt. 1, pp. 90–93.

168. Dana, "Science and the Bible," pt. 1, p. 98, where he quotes Lewis, *Six Days*, 248–49.

169. Dana, "Science and the Bible," pt. 1, p. 104.

170. Dana, "Science and the Bible," pt. 1, p. 105–6.

presented the harmony between geology and Genesis as to the sequence and duration of various phases of creation.[171]

In Schenectady, Lewis busied himself with a public letter in reply. He denied Dana's charge that he had promoted the infidelity of *Vestiges* and reaffirmed the "style of interpretation" pursued by philologists against the treatment, by geologists, of Moses "as simply expressing scientific or philosophical opinions."[172] Following philological principles, he again insisted that Genesis 1:11–12 teaches that the first plants developed from the earth, rather than being created *ex nihilo*. He clarified that this did not exclude God's direct imposition in nature at the moment of their appearance; thus, in embracing a kind of developmental origin for plants—which he extended also to animals—he was not promoting the deistic or atheistic developmentalism associated with *Vestiges*. Lewis was simply being an honest scholar, admitting that neither *bara* ("create") nor any other Hebrew word, nor any concept in the Semitic mind, meant "creation out of nothing." As a philologist, he felt bound to conclude in favor of some sort of development hypothesis.[173]

Meanwhile, Dana received a flood of private correspondence to herald him as a hero for the natural sciences. Louis Agassiz characterized Dana's review as a "powerful vindication of science versus conceited theology."[174] Several correspondents announced that they were distributing copies of Dana's article, hoping it would dispel the myth propagated by Lewis that geology was out of harmony with Genesis.[175] Even Charles Joy, a faculty member at Lewis's own Union College, wrote to congratulate Dana. He viewed the battle as one between "the orthodoxy of scientific men" like Dana and the prejudices of men like Lewis who held chemistry and geology in low esteem.[176] William Dawson of McGill

---

171. Dana, "Science and the Bible," pt. 1, p. 106–29.

172. Tayler Lewis, "Letter from Professor Lewis," 25 Feb. 1856, published in *Bibliotheca Sacra* 13, no. 50 (Apr. 1856): 471–76, at 472.

173. Lewis, "Letter," 473–75.

174. Louis Agassiz to James D. Dana, 30 Jan. 1856 (Dana Family Papers, Reel 1).

175. A. Caswell to James D. Dana, 11 Feb. 1856 (Dana Family Papers, Reel 1); William McIlvile to James D. Dana, 7 Feb. 1856 (Dana Family Papers, Reel 2); E. B. Andrews to James D. Dana, 6 Mar. 1856 (Dana Family Papers, Reel 1).

176. Charles A. Joy to J. D. Dana, 18 Feb. 1856 (Dana Family Papers, Reel 2).

College in Montreal and Arnold Guyot at Princeton each volunteered to support Dana's cause for geology with further publications.[177]

Dana himself continued the mission of redeeming natural science from the bad name that *Vestiges* and Lewis each had given it. In a rejoinder to Lewis's open letter, Dana quoted from Benjamin Peirce's 1854 presidential address before the American Association for the Advancement of Science. Men of science, argued Peirce and now Dana, held the Bible in high regard. Though the natural sciences necessarily are uncertain in some of their conclusions, Lewis's skepticism toward geology is unwarranted. God himself has made man's mind adapted to nature in order to understand it.[178] After reading this article, Agassiz expressed gratitude that Dana had risen to the occasion since he, unlike Agassiz, was a member of an orthodox church and thus could speak for science without attracting too much suspicion from "clerical ignorance."[179] Dana's friend Charles Bartlett prayed, "God help you & spare your life & health and give you strength & wisdom to battle for the rights, to expose the pseudo Christian philosopher as evil as the open infidel."[180]

Thus, Dana became God's geologist. His calling was to oppose philologists like Lewis who sought to deny the natural sciences their proper role in revealing God in nature, and speculative works such as *Vestiges* that, while ostensibly employing the natural sciences, in fact deviated from the empirical methods of those sciences. Lewis, for his part, was blocked in his attempts to publish in *Bibliotheca Sacra* a full-scale rebuttal to Dana's articles. He did, however, manage to persuade a local printer in Schenectady to issue another book, this time entitled *The Bible and Science; or, The World-Problem* (1856). There Lewis again denied that the physical sciences could ever refute *Vestiges* and reserved for biblical philology the sole authority to determine whether the development theory can be orthodox. This time, however, he more clearly emphasized that he himself did not intend to endorse *Vestiges*. His disagreement with Dana had been over method more than substance: "It is this Word of the Lord, faithfully interpreted, heartily believed, and placed in its proper rank, before all science,

177. J[ohn] W[illiam] Dawson to James D. Dana, 17 Mar. 1856 (Dana Family Papers, Reel 2); Arnold Guyot to James D. Dana, 6 Dec. 1856 (Dana Family Papers, Reel 2).

178. Dana, "Science and the Bible," pt. 2, pp. 631–33.

179. Louis Agassiz to James D. Dana, 6 July 1856 (Dana Family Papers, Reel 1).

180. Ch[arles] M. Bartlett to James D. Dana, [1856] (Dana Family Papers, Reel 1).

and all philosophy—it is this, and this alone, that will effectually slay 'The Vestiges.'"[181]

Dana's "Thoughts on Species" (1857)

As the Dana-Lewis controversy drew to a close, Dana continued his geological, mineralogical, and zoological explorations into what was becoming one of the most vexing questions of the day for naturalists: the concept of "species." At the August 1857 meeting of the American Association for the Advancement of Science in Montreal, Dana presented a paper entitled "Thoughts on Species." *Bibliotheca Sacra* editor Edward Amasa Park was eager to have Dana publish that paper in his journal, so long as he would "preface it by 10 or 20 lines, stating the theological leanings" of the species question. "Some of our ministers might not discern the value of the discussion," explained Park, "for some clergymen are not very deep philosophers." Meanwhile Dana also planned to have his paper published in the *American Journal of Science*, where he could be sure to find a more philosophical readership. Park recommended that neither publication refer to the other, so that each could give the impression of carrying an original contribution.[182]

Dana's introduction for the theologians who subscribed to *Bibliotheca Sacra* spanned six pages, rather than the "10 or 20 lines" for which Park had hoped. For that audience, Dana framed the discussion in terms of ongoing debates about the unity of the human race:

> 1. Is man of one, or of several species[?]
>
> 2. If of one species, was he created on one only, or on different continents, or in other words, was there a plurality of original birth-lands[?]
>
> 3. If of one centre only, was there but one first pair, or a plurality of first pairs[?][183]

181. Tayler Lewis, *The Bible and Science; or, The World Problem* (Schenectady, NY: Van Debogert, 1856), iii (alleging that *Bibliotheca Sacra* would not publish Lewis's articles against Dana), 151, 170–212, quoting 182.

182. Edward Amasa Park to James D. Dana, 18 Sep. 1857 (Dana Family Papers, Reel 3).

183. James D. Dana, "Thoughts on Species," *Bibliotheca Sacra* 14, no. 56 (Oct. 1857): 854–74, at 854.

Dana then proceeded to address the three queries he had presented before the AAAS: "1. What is a species? 2. Are species permanent? 3. What is the basis of variations in species?"[184]

To understand Dana's approach to these questions, one must consider his relationship to two other men of science. First, there was James Davenport Whelpley—Dana's fellow Yale alumnus and Silliman pupil, his partner in prayer on the eve of the 1838 Wilkes Expedition, an across-the-street neighbor, a colleague in both the Yale Natural History Society and the Association of American Geologists and Naturalists, and a fellow charter member of the American Association for the Advancement of Science.[185] In 1845, Dana arranged for Whelpley to publish "Idea of an Atom" in the *American Journal of Science*. There Whelpley argued that atoms lack physical existence, but are simply mathematical points around which various forces of attraction and repulsion act in such a way as to produce the phenomena of mass, heat, electric charge, and so on.[186] When Whelpley reviewed *Vestiges* and *Explanations* for the *American Whig Review* a year later, he drew upon his theory of atomism to evaluate the development theory. Specifically, Whelpley argued that *"Facts . . . press heavily against the progressive hypothesis"* because "there is even no such thing in science as a 'transition.'" He had in mind not only the lack of transitional organic species—for example, something half ape and half human—but also the lack of transitional forms between chemical species—"iron, gold, and oxygen." These species, according to his theory of atomism, were entirely distinct as to the polarities of forces that determined their specific chemical properties. In short, "Nature passes by leaps or intervals, even in motion itself. Even gravity is measured by pulses" (at least according to Whelpley's atomism).[187]

184. Dana, "Thoughts on Species," 859. See also James Dwight Dana, "Thoughts on Species," *American Journal of Science* ser. 2, 24 (Nov. 1857): 305–16, at 305. All future citations will be taken from the latter article.

185. Some biographical information on James Davenport Whelpley is available in an entry for his father: "Whelpley, Philip Melanchthon," in *Appleton's Cyclopaedia of American Biography*, ed. James Grant Wilson and John Fiske, 7 vols. (New York: D. Appleton and Company, 1886–1900), 6:459; for Whelpley's connections with New Haven science, see Yale Natural History Society, Minute Book, 1834–1857 (Records of Clubs, Societies, and Organizations, Box 1, Folder 2), esp. pp. 8, 20, 23, 28, 29, 33, 40.

186. [James Davenport Whelpley], "Idea of an Atom, Suggested by the Phenomena of Weight and Temperature," *American Journal of Science* 48, no. 2 (Jan.–Mar. 1845): 352–68.

187. [James Davenport Whelpley], "A Sequel to 'Vestiges of the Natural History of Creation,'" *American Whig Review* 3 (Apr. 18465): 383–96, quoting 386–87.

A second person who influenced Dana's approach in "Thoughts on Species" was Louis Agassiz. As indicated earlier, Dana corresponded frequently with Asa Gray during the early years of the *Vestiges* debates. By the mid 1850s, however, Gray's nemesis Louis Agassiz had also developed close professional ties with Dana. In particular, the two naturalists exchanged letters in an effort to further Agassiz's idealist metaphysic as applied to the identification of natural kinds: species, genera, and so on.[188]

Echoing the language of Agassiz, Dana addressed the first query of his article—"What is a species?"—by calling for the "true idea of distinction among species," rather than naturalists' own, often arbitrary, classifications. Dana then followed Whelpley's example in first considering "inorganic species," such as chemical elements. Each element had a distinct atomic mass, such as eight units in the case of oxygen, corresponding to its particular qualities of chemical reaction. "Whenever then the oxygen amount and kind of force was concentrated in a molecule, in the act of creation, the species oxygen commenced to exist. And the making of many such molecules instead of one, was only a repetition in each molecule, of the idea of oxygen." In summary, "a *species* corresponds to *a specific amount or condition of concentrated force, defined in the act or law of creation.*"[189] Dana next applied this definition to organic species, which he, like Whelpley, concluded could also be characterized by particular kinds and degrees of forces. Lending support to Agassiz's notion that species were ideas in God's mind, Dana stated that "when individuals multiply from generation to generation, it is but a repetition of the primordial idea-type; and the true notion of the species is not in the resulting group, but in the idea or potential element which is at the basis of every individual of the group."[190] Furthermore, each organic species has an "historical existence," meaning it passes through various stages, from seed to adult, but forever remaining the "type-idea" within this "cycle of evolution."[191]

By "evolution" Dana meant what a later generation of scientists would call "development," that is, the transformation of an embryo into a mature form.

---

188. See, for example, Louis Agassiz to James D. Dana, 16 Feb. 1853, 29 June 1853, 8 July 1853, 7 Aug. 1855, 26 Dec. 1855 (Dana Family Papers, Reel 1).

189. Dana, "Thoughts on Species," 306.

190. Dana, "Thoughts on Species," 307.

191. Dana, "Thoughts on Species," 308.

Meanwhile, "development" in Dana's day—especially in the context of the "development theory" of *Vestiges*—connoted the idea of species transmutation, or what a later generation would call "evolution." Dana turned his audience's attention to that more controversial sense of development, inquiring whether species were permanent. As before, he began with the inorganic world, exemplified by an atom of oxygen. "Is its combining number, its potential equivalent, a varying number,—usually 8, but at times 8 and a fraction, 9, and so on? Far from this; the number is as fixed as the universe. There are no indefinite blendings of elements." Likewise for the organic kingdoms, species are definite. "Were these units capable of blending with one another indefinitely, they would no longer be units, and species could not be recognized. The system of life would be a maze of complexities."[192]

Dana's final subtopic concerned the variations of species. Beginning once more with the inorganic world, he noted that oxygen and iron each vary according to temperature, but always in lawful ways. Likewise, organic species varied, but always within lawful limits. "Variations are not to be arranged under the head of *accidents*; for there is nothing accidental in nature; what we so call, are expressions really of profound law."[193] Dana marveled that zoological specimens from as far away as Japan and the eastern coast of North America could be identified as "identical species," providing "full assurance that species are definite and stable existences."[194] Dana then drew a metaphysical conclusion: species are not merely convenient categories employed by naturalists, but they have real existence in nature. "They are the units fixed in the plan of creation; and individuals are the material expressions of those ideal units."[195]

Though Dana did not here mention God's role in the creation, many of his readers undoubtedly inferred it. In 1855, speaking as the president of the American Association for the Advancement of Science, Dana had argued:

> In thus tracing out the fact, that there has been a plan or
> system of development in the history of this planet, do we
> separate the Infinite Creator from his works? Far from it: no

192. Dana, "Thoughts on Species," 308–9.

193. Dana, "Thoughts on Species," 312.

194. Dana, "Thoughts on Species," 313.

195. Dana, "Thoughts on Species," 315.

more than in tracing the history of a planet. We but study the
method in which Boundless Wisdom has chosen to act in
creation. . . . How vain the philosophy which makes the
creature the God of nature, or nature its own author![196]

Some of those who heard his "Thoughts on Species" presentation at the
1857 meeting or read it in the *American Journal of Science* would have
remembered this. Moreover, those who read "Thoughts on Species" in
Bibliotheca Sacra would have seen this article as a culmination of his two-year
defense in that journal of both geology and God against the latest proponent of
Vestiges of Creation, Tayler Lewis. "Thoughts of Species," therefore, secured
geology on the side of God—against the infidelity found in *Vestiges* and defended
by the renegade philologist Lewis.

But Dana's position did not remain secure for long. Both God and geology
could be regarded in a different manner than he had proposed. Asa Gray sent a
lengthy commentary to Dana on November 7, 1857, just one week after having
read the article. Gray stood unconvinced that inorganic species and organic
species were both "species" in the same sense. "All my experience makes me ~~catio~~
cautious and slow about building too much upon analogies," he wrote, no doubt
thinking of his own experiences in botanical research as well as his social conflict
with Louis Agassiz's idealist zoology. "Until I see further and clearer, I must
continue to think that there is an essential difference between kinds of animals or
plants and kinds of matter." Something subtle but profound was happening in
this sentence. Twelve years earlier Gray had emphasized the distinctions between
living organisms and brute matter; now he was emphasizing a different
distinction, one between *kinds* or *species* of each.

Gray agreed with Dana that species of inorganic matter, such as oxygen,
were permanent; he no longer wanted to be held to such a view regarding species
of organic beings. Agassiz at Harvard and now Dana at Yale, like Whelpley before
him, were insisting that both organic and inorganic species were fixed, but Gray
now thought otherwise. Two months earlier to the day, on September 7, 1857,
Charles Darwin had written a letter to Gray explaining his hypothesis of

196. James D. Dana, *Address before the American Association for the Advancement of Science,
August, 1855* (n.p.: Joseph Lovering, 1855), 30, 36. The sermon-like oratory Dana exhibited in his official
capacity as AAAS president has been overlooked by Sally Gregory Kohlstedt, who portrayed Dana and other
leaders in the AAAS as privately religious but publicly secular, except when an occasional doxology to
nature's God was necessary to appease the public. See Kohlstedt, 6–7, 110.

development by natural selection. "You may be sure," wrote Gray to Dana, "that <before long> there must be one more resurrection of the development theory in a new form, obviating many of the arguments against it, and presenting a more respectable and more formidable appearance than it ever has before."[197]

Ironically, the tables had been turned since 1846, when Gray was opposing the species transmutation theory of *Vestiges* and Dana was suggesting that the theory was perfectly acceptable, at least with respect to theological considerations. Now Dana's God, like Agassiz's, had thought up species as permanent ideas, and so their material instantiations—the individual specimens scrutinized by naturalists—would always remain within fixed limits of variation. This troubled Gray, who had staked so much on the rhetoric of empiricism, even if in substance his own theory of Asian and North American plant origins was predicated upon an unobserved ancient land bridge connecting the continents. At least Gray could point to geological features that plausibly were the remains of that land bridge, but how could Agassiz fill in his own, idealist rhetoric with substance? One cannot help but to suspect that Gray was beginning to worry that Dana was becoming a New Haven Agassiz.

Meanwhile, Edward Hitchcock, who like Dana had once suggested that *Vestiges* posed no religious threat, wrote Dana to congratulate him on the species article. "The finger of Providence it seems to me clearly points to you," he wrote. Hitchcock's primary concern, however, was not species transmutation, but rather polygenism. "I think religion suffers for more in this quarter than for geological difficulties." For Hitchcock, "Thoughts on Species" was nothing short of a message from God that Dana would be the one whom God would send to preserve the natural sciences from Agassiz's polygenism.[198] Dana had introduced the *Bibliotheca Sacra* version by mentioning the debate over the origins of the human races; toward the end of the article in each journal, he concluded that "the barrier against hybridity appears to stand as a physical test of species."[199] That people of African and European descent could have fertile hybrid offspring was well known. Understandably, then, Hitchcock read Dana as a competent man of

197. Asa Gray to James D. Dana, 7 Nov. 1857 (Asa Gray Correspondence, Yale University).
198. Edward Hitchcock to James D. Dana, 4 Feb. 1858 (Dana Family Papers, Reel 2).
199. Dana, "Thoughts on Species," 316.

science who could use Agassiz's own language of "type-ideas" against the heresy of polygenism.

During the era of *Vestiges,* polygenism was, philologically speaking, a more obvious heresy than species transmutation.[200] Moses Stuart, the Andover philologist, not surprisingly shared Hitchcock's concerns about Agassiz, as had Tayler Lewis.[201] When commenting on the Fugitive Slave Law of 1850, Stuart emphasized that a correct interpretation of biblical language would establish that all races had a common origin in Adam and Eve. Here Stuart entreated Louis Agassiz to "investigate the Sacred record with as much impartiality and candor and thoroughness as he has the Book of Nature." He acknowledged "the decisive manner in which [Agassiz] has spoken out against the transmutation theory of the author of the Vestiges of Creation," but he, like Hitchcock, hoped more than anything that the esteemed zoologist would stop misusing his science to endorse polygenism.[202]

These examples may serve as another reminder that evolution *per se* was by no means the main concern of New Haven-affiliated men of science. Species transmutation was simply one topic among many that related to Taylorite theology, which itself was oriented toward God's moral government. As the next section illustrates, Dana himself remained mindful of that theological orientation even after establishing himself as an authority in the natural sciences.

Dana and the Development Theory in 1859

While Gray, Agassiz, Dana, and others contested the meaning of species, Dana continued to view nature and God in Taylor's terms. In May 1859, the *New Englander,* a journal of Yale-affiliated theologians, published an article by Dana concerning "Anticipations of Man in Nature." Dana was responding to a recent book by Taylor's former pupil, Horace Bushnell, entitled *Nature and*

---

200. This becomes clear in the course of Stephen G. Alter, *Darwinism and the Linguistic Image: Language, Race, and Natural Theology in the Nineteenth Century* (Baltimore and London: The Johns Hopkins University Press, 1999).

201. Sherwood, 314.

202. Moses Stuart, *Conscience and the Constitution with Remarks on the Recent Speech of the Hon. Daniel Webster in the Senate of the United States on the Subject of Slavery* (Boston: Crocker & Brewster, 1850), 102. For a broader discussion of Stuart's views on slavery, see Giltner, *Moses Stuart,* 117–29.

Supernature.[203] Bushnell, who by this time had become something of a radical among Congregationalists, sought to reaffirm that there exists a realm not governed by fixed laws, the realm of human free will, spirit, and God himself. Bushnell did not oppose natural science *per se*, but rather cautioned against a growing tendency in the natural sciences to extend the laws of nature also into the realm of "supernature." He wrote:

> a great many of our naturalists, who admit the existence of God, and do not mean to identify his substance with nature, and call him the Creator, and honor him, at least in words, as the Governor of all things, do yet insist that it must be unphilosophical to suppose any present action of God, save what is acted in and through the preordained system of nature. The author of the Vestiges of Creation, for example, (p. 118,) looks on cause and effect as being the eternal will of God, and nature as the all-comprehensive order of his Providence, beside which, or apart from which, he does, and can be supposed to do, nothing.[204]

Dana agreed with Bushnell that naturalists should take care not to exclude the supernatural, and that *Vestiges* could be appropriated for ill ends. A few years earlier Dana himself had concluded his presidential address to the American Association for the Advancement of Science with the reminder: "How vain the philosophy which makes the creature the God of nature, or nature its own author!"[205] Nevertheless, Dana was concerned that Bushnell was drawing the line between the natural and the supernatural in the wrong place. Bushnell had argued that disasters, such as earthquakes and volcanoes, and unfortunate conditions of existence, such as pain and starvation, were sent to earth by God during the geological past in order to prefigure the results of the sins that humans would commit after their creation at some future time. Dana, by contrast, remained consistent with Taylor's theology of nature in asserting that "death is in the system of nature." For example, "the Creator . . . instituted an extensive system of flesh-eaters" for the benevolent purpose of keeping the animal

203. Review of *Nature and the Supernatural,* by Horace Bushnell, *New Englander* 17 (Feb. 1859): 224–58.

204. Horace Bushnell, *Nature and the Supernatural, As Together Constituting the One System of God,* 5th ed. (1st ed., 1858; New York: Charles Scribner, 1860), 60.

205. Dana, *Address,* 36.

population in balance with the plant population. Similarly, ancient geological catastrophes leading to species' extinction, like commonplace circumstances leading to individuals' suffering and death, were "fundamental characteristics of the best system of nature, . . . anticipative of man's nature, but not anticipative distinctly of his sin."[206]

Although Dana denied that God had intervened during the geological past to pass anticipatory judgment on the earth for man's future sins, he did not—like the author of *Vestiges* was reported to have done—exclude God's supernatural action entirely from the history of nature. Dana held that the entirety of geological phenomena, including coal beds, limestone strata, and vegetable and animal remains, "announce, not developing laws, but a creating God." In fact, "science is moving further and further away from any proof of the creation of species through nature's forces."[207]

Dana referred his *New Englander* audience to his recent "Thoughts on Species" articles in *Bibliotheca Sacra* and the *American Journal of Science*. As he had explained there, changes in nature result from the interaction of fixed properties of each specific type, whether inorganic or organic, not from any transformation of those properties themselves.[208] With each new phase of creation—inorganic, plant, invertebrate animal, vertebrate animal, and finally human—God had introduced new kinds into nature, characterized by new laws of properties and their interactions. Whereas death and extinction were part of the natural Providence, these acts of species creation bore the mark of God's supernatural Providence in nature. The clearest example could be seen in the "free will" of man, which itself, Dana agreed with Bushnell, is part of supernature rather than of nature. Human free will could not have arisen from the laws governing lower species, and as everyone who heard Taylor and Fitch preach knew, two things were certain: God governed by means of a moral system, and humans—however they had originated—had enough free will to be responsible for choosing their lot within it.[209]

206. James Dwight Dana, "Anticipations of Man in Nature," *New Englander* 17 (May 1859): 293–334, at 308, 309, 295.

207. Dana, "Anticipations of Man in Nature," 297.

208. Dana, "Anticipations of Man in Nature," 303.

209. Dana, "Anticipations of Man in Nature," 321–26.

Concluding Interpretations: Understanding Vestiges at Yale

Vestiges came to Yale at a time when the academic community was obsessed with a theological rationalization for suffering and imperfection. In addressing this problem, theology professors Taylor and Fitch offered a justification of evil that closely paralleled the final, theodical chapter of *Vestiges*. God governed the world morally, which meant that he allowed evil to occur on the small scale so long as the overall trend in nature worked for the progressive emergence of a greater good; indeed, the allowance for evil was itself a necessary condition for greater good. This moral government theory of New Haven theology filled the Yale campus, and spread far beyond it, prior to the American arrival of *Vestiges* in 1845. When that book became known among Yale-affiliated men of science, Dana and Hitchcock demonstrated that anyone who imbibed Taylor's New Haven theology could welcome *Vestiges* without theological difficulty. All one had to do was emphasize the author's references to divine planning and distance oneself from the materialist versions of the development theory as promoted by Laplace (for nebular development) and Lamarck and Oken (for both spontaneous generation and species transmutation).

But theology was not the only science that had relevance for America's *Vestiges* debates. Both at Yale and beyond, geological consensus weighed heavily against the development theory. Moreover, both the more conservative theologians at Princeton and the more liberal theologians at Harvard rejected *Vestiges* on theological grounds. Yale's Taylorite Congregationalism stood alone among established theological opinion, as sampled in the three case studies, in accommodating *Vestiges*. Nor did Taylor's system have to accommodate *Vestiges*, for it had thrived quite well among New School Calvinists—both Congregationalist and Presbyterian—during the decades before the development theory gained publicity. Thus, although Yale's theology could promote *Vestiges*, it was by no means obvious that it would do so.

Particularly for the geological trio of Silliman, Hitchcock, and Dana, *Vestiges* posed great threats to natural science. It amounted to little consolation, therefore, that *Vestiges* would not be as damaging to the theological sciences as to the natural ones. Furthermore, *Vestiges* could perhaps cause harm theologically. If read as is, the work did seem to invalidate prayer. The God who had sustained Dana and the Wilkes crew in the south seas was a "God that heard prayer." Prayer, in fact, was part of the "moral system" of God's spiritual, moral, and physical laws that governed nature, especially human nature. Silliman, too, prayed earnestly, in agony at times, as did Hitchcock, in search of divine assurance that a life lived for geology also could be a life lived for God. But therein lay a dilemma, for it was not always clear whether geology could be trusted, especially when theology and philology competed for scientific authority and when practitioners in those fields portrayed geology as unstably speculative. It seems, at least in this limited respect, that Silliman and Dana would have been more comfortable at Princeton, where seminary professor Archibald Alexander had said, "Theologians should study geology."[210] *Vestiges* highlighted the disciplinary tensions felt within the Yale community, pitting theologians and philologists and geologists against one another, though not in the manner that a later generation would come to expect.

This case study of *Vestiges* at Yale reveals that when theologians and geologists have been involved in debates concerning evolution, the fulcrum of those debates has not always been evolution itself, as if one must necessarily lean on one side of it or the other. At New Haven, the pivotal question had more to do with the *morality* of God's government, and the *methodology* for exploring God's government, than with the possibly natural and developmental *mode* of various aspects of that government. Thus, Dana and Hitchcock could just as easily support or oppose the development theory of *Vestiges* insofar as it touched upon their theology. Their theology was, both in substance and in implication, distinct from the theologies that structured participation in the sciences at Princeton and Harvard.

Because Silliman, Dana, and Hitchcock championed the empirical methods of geology as superior to those of philology for determining the natural history of the earth and its inhabitants, they chose to reject *Vestiges*. Philology, like theology, could divide on the *Vestiges* question more than geology could. Tayler Lewis suggested that philology could support the development theory, but Moses Stuart, the great pioneer of American philology, rejected *Vestiges*. In Lewis's case, the pivotal issue was not evolution, but rather the authority of philology over geology. In Stuart's case, philological authority also mattered more

210. Allen H. Brown, Lecture Notes on Didactic Theology, by Archibald Alexander, Sep. 1843 (Archibald Alexander Collection, Princeton Theological Seminary, Box 4, Folder 4).

than evolution, but by the 1850s he, like Hitchcock, seems to have been troubled just as much by polygenism.

Most disruptive of all, however, was the widening gulf between Old School Presbyterians and Unitarians. Yale's religious leaders—among whom one may include not only theologians but also geologists who saw themselves as God's spokespersons on natural history—sought to preserve a middle way. Like Presbyterians at Princeton, Yale Congregationalists retained the language of a Trinitarian God who judges sinners, and they preached the gospel not just for the present world, but also for the life to come. Like Unitarians at Harvard, they held God's justice to anthropocentric standards, which required that human initiative be expanded and original sin be diminished. But Yale really belonged to a continuum all its own. New Haven's cross-disciplinary prophets of God's moral government had so distinctly articulated the relationship between God and human nature that *Vestiges*, which was intolerable at both Princeton and Harvard, could at least be entertained at Yale, even if only fleetingly.

Study Questions

1. Identify the unique features of New Haven theology at Yale, explaining Prof. Nathaniel Taylor's emphasis on God's moral government.

2. What is the "design argument" associated with William Paley's *Natural Theology* (1801), and how did New Haven theologians treat this work in distinction to mainstream American Protestants?

3. Contrast Prof. Eleazar Fitch's "gap theory" and Prof. Benjamin Silliman's "day-age theory" for interpreting the chronological aspects of Genesis 1–2. Was this a classic case of theology versus science, or was something more complicated going on?

4. In what sense was Silliman's protégé James Dwight Dana pursuing a career in geology as a theological vocation?

5. What led Dana to conclude that New Haven theology could accommodate the development theory presented in *Vestiges*? Why did Dana nevertheless reject that theory?

6. Which discipline was more conservative at Yale—theology or geology—when it came to the debate concerning evolution? Consider the views of Edward Hitchcock as well as those of Silliman and Dana.

7. What does the debate between Talyer Lewis and James Dana reveal about the rankings of their respective scientific disciplines in American during the 1840s and 1850s?

CONCLUSION

Vestiges of Creation
Leaves a Legacy

*"I suppose now ye are well acquainted
with the* Vestiges of Creation?*"*

a fictional character in William Black's *White Wings*, 1879[1]

Introduction

On November 24, 1859, while *Vestiges* remained a timely topic at Princeton, Harvard, and Yale, London publisher John Murray released a new work on organic species that was, by the day's end, entirely bought up by local booksellers. Two or three dozen of the 1,250 press runs had already been shipped out to naturalists in both Britain and the United States, including Louis Agassiz and Asa Gray at Harvard, and James Dana at Yale.[2] Entitled *On the Origin of Species* and authored by London geologist Charles Darwin, this book soon was recognized throughout America as the most significant evolutionary work since

---

1. William Black, "White Wings: A Yachting Romance," chap. 5 of a serialized novel, *Harper's New Monthly Magazine* 59, no. 352 (Sep. 1879): 555–68, at 565.

2. R. B. Freeman, *Charles Darwin: A Companion* (Folkestone, Kent, England: William Dawson and Sons, 1978), 220.

Vestiges of the Natural History of Creation. Eventually, *Vestiges* became known as a precursor to Darwin's work, a precursor that later could be forgotten, except by historians who mentioned the author of *Vestiges* together with Lamarck. In the early 1860s, however, a reverse emphasis lingered: to some, Darwin's work was a post-*Vestiges* attempt to resuscitate the old "Lamarckian hypothesis," as Henry Darwin Rogers and others still called it.[3] One could hardly discuss Darwin's book without mentioning both Lamarck and the still anonymous author of *Vestiges*, who had popularized "development." For one thing, Darwin's leading British promoter, Thomas Henry Huxley, was suspected by one American writer, as late as 1871, to have written *Vestiges*.[4] Another writer, however, was sure that Lamarck himself had written *Vestiges*. He apparently was ignorant of the fact that the French zoologist had died fifteen years before the work first appeared.[5] The mysterious specter of *Vestiges*, with all that it now represented in the world of American science (both natural and theological), hung thickly in the air as American reviewers introduced the *Origin of Species* to a generation of readers who had been raised on the *Vestiges* debates.

Fifteen years earlier, American reviewers of *Vestiges* had already been speculating about what would come next. A recurring theme was the anticipation that science of the *Vestiges*' sort would eschew the Creator's providence "until the natural becomes everything."[6] Even while avoiding the implication that American reviewers predicted the actual future, it cannot be denied that some of them were foreseeing a potential future. Though oblivious to the London-area geologist who already had sketched out a then private theory of speciation by natural selection,[7]

3. Patsy Gerstner, *Henry Darwin Rogers, 1808–1866: American Geologist* (Tuscaloosa: University of Alabama Press, 1994), 221.

4. See R. R. W., Review of *An Inquiry into the Origin, Development, and Transmission of the Games of Childhood, Galaxy* 11, no. 2 (Feb. 1871): 314–15, at 314. An editor commented, at 315n: "I did not know before that he [Huxley] wrote 'Vestiges of Creation.' However, let it pass—I suppose he did, since it is so stated."

5. " Origin of Darwinism," *Manufacturer and Builder* 8, no. 5 (May 1876): 119.

6. [Tayler Lewis], "Vestiges of the Natural History of Creation," *American Whig Review* 1, no. 5 (May 1845): 525–43.

7. Those Americans who did know of Darwin during the era of *Vestiges* generally did not associate him with organic evolution. Darwin's discovery of new species when visiting the Galapagos Islands, for example, was interpreted as confirmation of God's "creative power" that periodically intervenes in nature and, more tentatively, as evidence suggestive that humans were not the last of God's creation, since the Galapagos species may have been created more recently. See John Augustine Smith, *The Mutations of the Earth; Or An Outline of the More Remarkable Physical Changes, of which, in the Progress of Time, This Earth Has Been*

reviewers in the 1840s worried about whose precursor the author of *Vestiges* might become. As Princeton's Albert Dod confessed, "we have not shown that some other explorer in the same direction may not become more successful."[8]

Sometime after 1859, Darwinism supplanted the *Vestiges* theory. This occurred gradually, and piecemeal.[9] Indeed, for Darwin and a few others, including Asa Gray, the process had begun prior to 1859. As the preceding case studies demonstrated, the era of *Vestiges* was not a monolithic age. Historians of late-nineteenth-century science have revealed that the era of Darwinism was not monolithic either. "Darwinism" could mean a variety of things, including the opposing viewpoints of theistic and materialist evolution. It also could refer to organic evolution by natural selection or else species transmutation by a more Lamarckian means, such as the enhancement and heritability of an organ resulting from its frequent use. Moreover, the late nineteenth century was no less religiously pluralist than the mid part of the century. Given these realizations, it could not have been the case that Darwinism would simply supplant the *Vestiges* theory. Rather, multiple discussions of Darwin's work, conditioned by specific cultural contexts, gradually replaced multiple discussions of *Vestiges*.

This chapter reviews the key findings of the three case studies in terms of the larger methodological stance adopted in this book. In doing so, connections will be suggested between the conclusions reached in those case studies and the conclusions that other scholars have reached in their studies of the Darwinian period. The aim will not be to establish exactly how the era of *Vestiges* evolved

the Subject, and the Theatre: Including an Examination into the Scientific Errors of the Author of the Vestiges of Creation, 1846 Anniversary Discourse, Lyceum of Natural History of New York (New York: Bartlett and Welford, 1846), 38.

8. [Albert B. Dod], "Vestiges of the Natural History of Creation," *Princeton Review* 17, no. 4 (Oct. 1845): 505–57.

9. For examples of how Darwin's theory was integrated into a existing traditions of Christian apologetics that opposed *Vestiges*, see the references to the two works in Josiah Parsons Cooke, Jr., *Religion and Chemistry; or, Proofs of God's Plan in the Atmosphere and Its Elements: Ten Lectures Delivered at the Brooklyn Institute, Brooklyn, N.Y., on the Graham Foundation* (New York: C. Scribner, 1865), esp. 258–59; Albert Barnes, *Lectures on the Evidences of Christianity in the Nineteenth Century, Delivered in the Mercer Street Church, New York, January 21 to February 21, 1867, on the "Ely Foundation" of the Union Theological Seminary* (New York: Harper and Brothers, 1868), 186; Lord Bishop of Carlisle, "The Gradual Development of Revelation," in *Modern Scepticism: A Course of Lectures Delivered at the Request of the Christian Evidence Society* (New York: Anson D. F. Randolph and Co., 1871), 229–364, at 252–53; and, Robert Patterson, *Fables of Infidelity and Facts of Faith* (Cincinnati: Western Tract Society, 1875), 45–48.

into the era of Darwin, but rather to suggest how future researchers might begin to pursue those questions with the same depth of analysis that the *Vestiges* debates have been pursued here. Of particular concern is the degree to which the model of socially situated systematic theologies employed in this study may be useful for the analysis of America's evolution debates in later historical periods. In other words, this chapter explores both the legacy that *Vestiges* left in America and the possible legacy that the foregoing case-study analyses of *Vestiges* might leave for the disciplines of American religious history and the history of science.

Interpreting Vestiges *in Three*
Theologically Shaped Academic Communities

The methodology employed in the case-study chapters proceeded according to the following pattern, which is described more fully in the Methodological Postscript. First, textual relics were, as in historian James Secord's study of *Vestiges'* British receptions, gathered from far beyond the published sources. Second, those relics were rearranged in such away as to piece together a past community's systematic theology, emphasizing, as theologian Paul Tillich has suggested, the relations between components in order to arrive at the whole that is constituted by them. Repeated associations of two ideas in sermon manuscripts, for example, were taken to indicate that those two ideas had a particularly important relationship in the community's systematic theology. Those ideas that serve as the common thread connecting the whole were, in turn, regarded as the central organizing principles that gave meaning to the systematic theology as a whole. Third, that meaning, or "ultimate concern," was interpreted, in the manner of sociologists Max Weber and Robert Merton, as more than a projection of the community's *status quo*. Systematic theology, in this view, also functions dynamically, by structuring social practices and fostering intellectual rationalizations of those practices. This dynamism impacts, as Weber argued, even those individuals who do not themselves assent wholeheartedly to the views predicated by the "ideal type" that the researcher assumes to be applicable to the community being analyzed. Moreover, following Tillich's perspective, the "ultimate concern" toward which theology was oriented was to be read not merely as a cultural condition, but as the existential realization that there existed a gap between the cultural and the transcendent. Theology, viewed as an earnest reaching toward God, thus shapes cultural practices in a spirit of emergency. On this view, to read *Vestiges* within such a community was to re-examine one's relation to God; this could be no casual task. From working models of the three theologically shaped, local academic communities, the case studies allowed for an interpretation of what *Vestiges* meant—that is, how it resonated with the local socio-theological equilibrium—among those who responded to it, and, in turn, how their responses addressed the existential concerns established by that theology.

At Princeton, as was argued in chapter 2, Old School Presbyterians self-consciously organized their theology in terms of God's sovereignty rather than according to His justice or benevolence. This did not mean that they did not believe God to be just or benevolent. It simply meant that God's justice and benevolence took on a different significance in their theology than in the theologies of other communities, because these characteristics were related, at times even subordinated, to collateral doctrines in a distinctive manner. Faithful to their Calvinist heritage, Old Schoolers expressly interpreted God's justice and benevolence in such a way as to reinforce, rather than diminish, their belief in God's sovereign decision to predestine some persons to heaven and others to hell. They did this, chiefly, to refute Arminianism, a deviation from Calvinism that emphasized human free will and thereby compromised the sovereignty of God that ensures that the elect will be saved. But what did this have to do with *Vestiges* in particular or even evolution in general? For Old School Presbyterians, Psalm 97:1 ("The Lord reigns") was the lens through which the creation account in Genesis was read, and it served as the guiding impetus for investigations in the natural sciences. In other words, because the sovereign reign of God over his creation was the central plank of Old School Presbyterians' theology of creation, Princetonians tended to evaluate *Vestiges* in terms of whether or not its developmental view of nature could be accommodated to their understanding of the Creator's sovereignty—the same sovereignty that guaranteed eternal salvation to the elect (among whom they counted themselves). That, more than anything else, was their "ultimate concern," their existential question that imbued their discussions of *Vestiges* with profound theological significance.

Chapter 3 explored a very different theological context, one which predominated at Harvard. There Unitarians boldly proclaimed that God is "the

Creator, not the Sovereign."[10] At Princeton, God was both, but not at Harvard, where Unitarians were concerned, above all, with promoting the view that God had given people the freedom and the ability to discern and live up to an objectively grounded pattern of moral living. They derived their moral philosophy from "internal evidences" (such as the conscience) as well as from "external evidences" (most especially, the New Testament accounts of Christ's moral life, but also including natural history). When people at Harvard addressed *Vestiges*, they emphasized its bearing upon their theological structure of internal and external evidences, and the moral character that they sought to support with that structure. Of course, many critics of *Vestiges* from beyond Harvard also drew attention to "external evidences" from natural history and the Bible, and it would be difficult to find anyone among their contemporaries who did not care about morality. The point, however, is that these empirical stances, and their moral application, functioned differently at Harvard than they did beyond Harvard, because they were linked into a different theological system. In other words, Harvard Unitarians had an "ultimate concern" that others did not share, and it was that ultimate concern that structured their devotion to the empirical sciences and shaped their reactions to *Vestiges*.

The "New Haven theologians" at Yale self-consciously distinguished themselves from both the Presbyterians at Princeton and the Unitarians at Harvard. However petty their concerns may seem by more liberal standards (whether then or now), those concerns powerfully structured the manner in which Yalensians read *Vestiges*. Chapter 4 revealed that neither God's Presbyterian sovereignty, nor a Unitarian moral character, but rather something called God's "moral government," became people's "ultimate concern" in Yale's academic community. Apart from an appreciation of this doctrine, and the comparatively unique importance it had in shaping the systematic theology at Yale, one remains at a loss to explain why both a geologist (James Dwight Dana) and a theologian-geologist (Edward Hitchcock) associated with New Haven theology each expressed full theological openness to *Vestiges*, even though they found it to be faulty on geological terms. The key was the resonance between the theodicy presented toward the end of *Vestiges* and the highly similar notion of

10. Andrew Preston Peabody, *Sermons Connected with the Re-Opening of the Church of the South Parish, in Portsmouth, New Hampshire, preached Dec. 25 & 26, 1858, and Jan. 30 & Feb. 6, 1859* (Portsmouth, NH: James F. Shores, Jr., & Joseph H. Foster; Boston: Crosby, Nichols, and Co., 1859), 8.

God's moral government that structured New Haven theology. Understanding Yale's intellectual climate, therefore, requires that one understand Yale's theology and identify the linkage that members of the Yale academic community experienced between that theology and the other sciences. Out of that theological context, which also was impacted by social friction among geologists, philologists, and theologians, came Yale's characteristic responses to *Vestiges*.

At all three institutions, scientific professionalism remained overtly theological throughout the era of *Vestiges*. Existential theological concerns had become imbedded in the social institutions of scientific education on all three campuses. Not surprisingly these concerns found expression also at joint gatherings, such as the 1845 meeting of the Association of American Geologists and Naturalists, where Benjamin Silliman warned that *Vestiges* promoted "atheism." Joseph Henry, a Princeton Presbyterian, and James Dwight Dana, a Yale Congregationalist, presented their theological frameworks for the natural sciences in their presidential addresses before the American Association for the Advancement of Science during the 1850s. Even at Harvard, which arguably was less sectarian and more secularized than either Princeton or Yale, Asa Gray had immediately characterized Louis Agassiz's first Lowell Lecture on the comparative anatomy as "good natural theology." Throughout the 1850s, and even well beyond, Agassiz continued to speak of "God's thoughts" behind the "ideas" that manifested themselves in nature as distinct species. These, and similar instances that were documented in this book, have lent themselves well to what Paul Tillich envisioned as "a theological history of culture," namely, "the attempt to analyze the theology behind all cultural expressions" in order to better understand the human condition.[11]

Applying a Model of Socially Situated Systematic Theology beyond the Era of Vestiges

As noted in the introductory chapter, historian of science Ronald Numbers has identified four puzzles for which "satisfactory answers" remain wanting: "which American naturalists converted to organic evolution, when [did they do]

---

11. Paul Tillich, *Systematic Theology*, 3 vols. in 1 (1950–1963; rpt., Chicago: University of Chicago Press, 1967), 1:39; see also his *Theology of Culture*, ed. Robert C. Kimball (New York: Oxford University Press, 1959), 7–9, 41.

so, why [did they do] so, . . . and what . . . consequences resulted[?]"[12] Since the case studies in this book, focusing on the period preceding Darwin's *Origin*, identified only a handful of Americans who adopted a theory of organic evolution —Henry Darwin Rogers, Asa Gray, and possibly Samuel Haldeman, Ralph Waldo Emerson, and Henry David Thoreau—little can be said in response to those queries. This study does, however, suggest some answers to a slightly different question, namely, "why did so many naturalists *not* convert to organic evolution during era of *Vestiges*?" Moreover, the answer to that revised question may suggest new ways for researchers to explore the issues outlined by Numbers for the later period, when many Americans did accept evolution.

Extending the Princeton case study into the era of Darwin would likely reveal that Presbyterians on that campus remained focused on God's sovereignty throughout the nineteenth century, and that this unwavering theological commitment continued to structure their engagements with the natural sciences. This seems to have been the case. The British Presbyterian minister James McCosh, who in 1868 became the first president of Princeton College to endorse organic evolution, had written a treatise on God's sovereignty prior to immigrating to America. In that work he opposed *Vestiges*—and, in a revised edition of 1863, he opposed Darwin's theory—for the same reasons that Princetonians had opposed organic evolution during the era of *Vestiges*.[13] But by the late 1860s he became convinced that Asa Gray and other men of science had assembled some compelling empirical evidence supporting a theory of species change. For a Presbyterian like McCosh and his Princeton colleagues, this meant that naturalists had discovered a new mode of God's sovereign action in nature. In view of this new evidence, the same theological system that had barred *Vestiges* from acceptance at Princeton College now could welcome a modified version of Darwin's theory. Likewise, at the seminary, theologian Benjamin Warfield construed Darwin's theory in terms of God's empirically discovered sovereign action in nature. Historians David Livingstone and Mark Noll have interpreted Warfield's evolutionism in the same way that would be expected,

12. Ronald L. Numbers, *Darwinism Comes to America* (Cambridge: Harvard University Press, 1998), 25.

13. James McCosh, *The Method of Divine Government: Physical and Moral*, 13th ed. (1st ed., 1851; London: Macmillan and Company, 1887), 87n (*Vestiges*), 157n (Darwin); notes at vii–viii and 157n, respectively, indicate that the comment concerning *Vestiges* was present at least since the 1855 edition, and that the comment concerning Darwin was inserted in the 1863 edition.

given the findings of the Princeton case study presented in this invesetigation: "the Calvinist conception of sovereignty that governed Warfield's view of human salvation—God as initiator and enabler of human repentance and faith—also governed his view on . . . the physical world."[14] McCosh and Warfield were not simply evolutionists, but *Presbyterian* evolutionists, with the adjective mattering to them much more than the noun.

Continuing the story of Harvard science from the era of *Vestiges* into the era of Darwin would no doubt involve more complications than for the Princeton saga. Already during the 1850s the Harvard community had become divided methodologically between the theological idealism of Louis Agassiz and the theologically respectable probabilistic empiricism of Asa Gray. Moreover, Francis Bowen's inductive empiricism remained a viable alternative during the early debates over Darwin's theory. With Gray endorsing Darwin, and Bowen and Agassiz, each for distinct reasons, opposing both Gray and Darwin,[15] it would seem that any shared theological context that once had shaped Harvard no longer remained effective. In some degree, however, even this fragmentation was itself structured by the domination of Unitarian theology. One should not expect that just because theology unified a community at Princeton, theology necessarily must have functioned in the same manner in all other cases in order to have been relevant in those cases. Unitarian Christology, by separating God's divine nature from Christ's physical nature, encouraged Bowen's moral philosophy and Gray's natural history to go their separate ways. Unitarian moral character required only that the various parties would uphold a rationally defensible view of human nature. The academic freedom that accompanied this theological vision allowed for innovations. By the 1870s, Asa Gray was expressing his evolutionary science in terms that resonated with the Presbyterian Calvinism outlined above. His Congregationalist partner in science, George Frederick Wright, shared with him a

14. David N. Livingstone and Mark A. Noll, "B. B. Warfield (1851–1921): A Biblical Inerrantist as Evolutionist," *Isis* 91 (2000): 283–304, at 290.

15. A. Hunter Dupree, *Asa Gray, 1810–1888* (Cambridge, MA: Harvard University Press, 1959), esp. chaps. 14, 15; Edward Lurie, *Louis Agassiz: A Life in Science* (Chicago and London: University of Chicago Press, 1960), esp. chap. 7; Paul Jerome Croce, "Probabilistic Darwinism: Louis Agassiz vs. Asa Gray on Science, Religion, and Certainty," *The Journal of Religious History* 22, no. 1 (Feb. 1998): 35–58; [Francis Bowen], Review of *On the Origin of Species*, by Charles Darwin, *North American Review* 90, no. 187 (Apr. 1860): 474–506; Francis Bowen, *Remarks on the Latest Form of the Development Theory*, reprinted from *Memoirs of the American Academy*, n.s., vol. 5 (Cambridge: Welch, Bigelow, and Co., 1860).

common vision of seeing organic evolution as God's sovereign action in nature.[16] Meanwhile, Harvard lecturer John Fiske was promoting a "Cosmic Philosophy" that incorporated aspects of Darwin's theory of organic evolution, but drew even more inspiration from Herbert Spencer, who even before Darwin's *Origin* had been sketching out a developmental understanding of the cosmos, human consciousness, and human society.[17] Fiske's "utterly and forever unknowable" God was neither Gray's God nor the God of Unitarianism; if anything, Fiske was closer to Transcendentalism.[18] Thus, the question of whether any theological context at Harvard continued to structure local engagements with science during the Darwinian period requires a full exploration into the range of theological positions that were staked out during that time, and the manner in which the earlier context of Unitarianism gave way to those developments.

At Yale, the central theological concern from the era of *Vestiges*—in this case, God's moral government—seems also to have borne some influence during the era of Darwin. Knowing James Dana's favorable theological reactions to *Vestiges* would suggest that organic evolution would never be of much theological significance to him; his ultimate concerns had already been answered in Taylor's New Haven theology, which would allow for species to evolve if the scientific tide were to change. Perhaps that is why no dramatic moment marks Dana's eventual acceptance of species transmutation. He was not even in a hurry to read or comment upon Darwin's *Origin of Species*.[19] More troubling to him was the extension of naturalism beyond organic evolution to other realms of life, such as when the British physicist John Tyndall raised scientific objections to the efficacy of prayer during the 1860s and 1870s.[20] The resulting debate over prayer even

16. James R. Moore, *The Post-Darwinian Controversies: A Study of the Protestant Struggle to Come to Terms with Darwin in Great Britain and America, 1870–1900* (Cambridge: Cambridge University Press, 1979), 269–80.

17. Charles D. Cashdollar, *The Transformation of Theology, 1830–1890: Positivism and Protestant Thought* (Princeton, NJ: Princeton University Press, 1989), 185–86.

18. Quoted in Jon H. Roberts, *Darwinism and the Divine in America: Protestant Intellectuals and Organic Evolution, 1859–1900*, paperback ed., with a new foreword (Notre Dame, IN: University of Notre Dame Press, 2001), 72.

19. The most careful study of Dana's gradual scientific acceptance of evolution remains William F. Sanford, Jr., "Dana and Darwinism," *Journal of the History of Ideas* 26 (1965): 531–46. Unfortunately, Sanford neglected to consider either Dana's prior responses to *Vestiges* or his ties to New Haven theology.

20. Roberts, *Darwinism and the Divine*, 72–74.

prompted Dana's foe Tayler Lewis to seek reconciliation with the Yale geologist in 1871 in order that the two could defend "the heart of Christian theology," a theology that proclaimed, as Dana had written following the 1839 shipwreck, "a God that heard prayer."[21]

In other respects, however, the importance of Taylor's New Haven theology soon faded at Yale. Theologians Nathaniel Taylor and Eleazar Fitch died, respectively, in 1858 and 1871. Dana's father-in-law and mentor, Benjamin Silliman, died in 1864. James Davenport Whelpley had settled in New York as the editor the *American Whig Review* and then left the United States for a mining expedition in Honduras.[22] The era of Darwin awaited other spokespersons. Noah Porter, a professor of moral philosophy who in 1848 moved into the old Whelpley residence, became the college president in 1871. Eight years into that post, Porter took a stand against political economy professor William Graham Sumner, who was teaching from Herbert Spencer's *Study of Sociology.* Spencer's positivist sociology replaced ethical absolutes, which had been central to God's moral government at the Yale Porter once knew, with the "survival of the fittest." This was Darwinism extended to society, or "social Darwinism," and Sumner was using it to deny that theology had any relevance in the social sciences. But Yale theology itself was changing. A new movement, known as Progressive Orthodoxy, took root in New Haven's First Church during the 1880s, and spread also into the divinity school. Conflating Horace Bushnell's realms of nature and supernature into one, this theology emphasized the immanence of God in nature, and it served as one means by which organic evolution could be assimilated to Protestant theology.[23] Whether this or any other theological movement became sufficiently entrenched in social institutions to shape the careers of late-nineteenth-century

21. Tayler Lewis to James Dana, 13 Mar. 1871; Dana to Robert Bakewell, 24 Mar. 1839 (Dana Family Papers, Yale University).

22. Biographical information on James Davenport Whelpley is scarce. Some data may be found in an entry concerning his father: "Whelpley, Philip Melanchthon," in *Appleton's Cyclopaedia of American Biography,* ed. James Grant Wilson and John Fiske, 7 vols. (New York: D. Appleton and Company, 1886–1900), 6:459. Whelpley has been mentioned in passing by several historians, and his name may be found in the correspondence of several more well-known individuals, including James Dana and J. Peter Lesley, but those sources are too scattered for their citation here to be practical.

23. Bruce Kuklick, *Churchmen and Philosophers: From Jonathan Edwards to John Dewey* (New Haven: Yale University Press, 1985), 217; Cashdollar, 410.

Yale naturalists, such as the self-described "neo-Lamarckian" paleontologist A. S. Packard, remains a question for future research.[24]

Meanwhile, some additional comments can be made concerning a general theme connecting the three case studies featured in this study: the assertion that locally shaped denominational concerns contributed to diverse understandings of *Vestiges* at Princeton, Harvard, and Yale. This claim, if extended into the era of Darwin, would challenge a conclusion reached recently by historian Jon Roberts:

> Within 'mainline' Protestantism, embracing such groups as the Congregationalists, the Presbyterians, the Methodists, the Baptists, the Episcopalians, and the Disciples of Christ, denominational affiliation is not very useful in predicting the responses of religious leaders to the theory of organic evolution [during 1875–1900].[25]

According to Roberts's *Darwinism and the Divine in America* (1989), the results of which he reaffirmed in the 1999 statement just quoted, American Protestants shared a common understanding of organic evolution during the 1860s: most of them rejected it. In the 1870s and 1880s, however, American Protestants diverged into two main groups: some continued to reject evolution, but many others accommodated it. Roberts identified immanentism—as found, for example, in the Progressive Orthodoxy mentioned above—as the crucial theological determinant for many of those Protestant individuals who accepted evolution. Roberts concluded, therefore, that denominational distinctions did not matter, but that a general trend across denominations toward immanentism did matter.[26]

Given that Princeton Presbyterians, Harvard Unitarians, and Yale Congregationalists each rejected organic evolution during the era of *Vestiges* and that members of those communities accepted organic evolution during the final decades of the century, it would seem that Roberts's statement about denominational affiliation not being a useful predictor of people's responses to evolution from 1875 to 1900 also could be applied to the earlier period discussed

24. Peter J. Bowler, *The Eclipse of Darwinism: Anti-Darwinian Evolution Theories in the Decades around 1900* (Baltimore: Johns Hopkins University Press, 1992), 59, 118, 134–36.

25. Jon H. Roberts, "Darwinism, American Protestant Thinkers, and the Puzzle of Motivation," chap. 6 in *Disseminating Darwinism: The Role of Place, Race, Religion, and Gender*, ed. Ronald L. Numbers and John Stenhouse (Cambridge: Cambridge University Press, 1999), 145–172, at 149.

26. Roberts, *Darwinism and the Divine*.

in the case studies of this book. It is important to note, therefore, that the theological dimensions of Numbers's four questions must be explored individually. The predictive questions, "Which American naturalists converted to organic evolution?" and "when?," apparently had little to do with denominationally affiliated theological concerns, just as Roberts has claimed. The case studies explored here have demonstrated, however, that denominational theological concerns can help to explain a nonpredictive question, "why they did so?"—or, as the case was, "why they did not do so?" The final query, "What consequences resulted?," also may have a distinctive answer for each local theological community.

Ultimately, the "why?" question may be more important than the "which Americans?" question. This is especially clear among Presbyterians at Princeton, where the theological context remained more uniform during the nineteenth century than at the other two case-study institutions. Princeton Presbyterians during the era of *Vestiges* opposed organic evolution, and their successors accommodated organic evolution during the era of Darwin. That both of these positions were structured by the same Presbyterian theology suggests something significant about how theology relates to the natural sciences. The correlation of Presbyterianism with both an opposition to, and also a support of, organic evolution does not make denominationally shaped theology less relevant, simply for being less predictive, but rather more relevant, because it shows how thoroughly organic evolution had to be re-packaged in order for it to become acceptable to a theologically stalwart community.

A Word for the Present

This book has argued for more than the claim that *Vestiges* preoccupied theologically minded Americans with sustained contemplations of evolutionary theory prior to the publication of Darwin's evolutionary works, and more than the claim that distinctive local theological contexts profoundly shaped Americans' engagements with *Vestiges*. The present study has offered, above all, the demonstration of a novel means, drawing upon the insights of Weber, Merton, and Tillich, for the historical analysis of how theology has related to the natural sciences. By implication, it may be suggested that future studies, whether examining the reception of a particular scientific theory or else dealing with the broader question of how science and theology interrelate, would benefit from a careful attention to the local social practices in which knowledge is pursued and opinions are formed, and also to the nagging, existential questions that troubled the very persons who found satisfaction in the answers supplied by their theology. Members of the three case-study communities brought to their readings of *Vestiges* concerns that were rooted in American Protestant controversies over Arminianism—a contested theological view of the degree to which human nature has enough good in itself to participate in its own salvation. To investigate the receptions of *Vestiges'* evolutionary natural history in those communities therefore required an exploration into the ways in which humans have understood their own nature—ways that were shaped by their preoccupied grasping toward the divine. This manner of research necessitated a serious engagement with how others have answered one of history's most enduring, and most controversial, questions: "What must I do to be saved?"

No attempt has been made to answer that question here, but only to document how others have answered it, and to explore how their answers shaped their responses to *Vestiges of the Natural History of Creation*. This book's concern with the relation between the salvation question and the evolution question is not, however, idly antiquarian. As cultural anthropologist Clifford Geertz observed:

> From one point of view, the whole history of the comparative
> study of religion . . . can be looked at as but a circuitous, even
> devious, approach to a rational analysis of our own situation,
> an evaluation of our own religious traditions while seeming to
> evaluate only those of exotic others.[27]

This investigation has sought a closer acquaintance with a few "exotic others," who were some of our intellectual ancestors—even if we at times cannot help thinking they were the quirkiest of relatives. Their culture is not ours, but theirs did beget ours, and one or more of their theologies might today still be claimed by at least some of us. Even if not, we owe to them the respect due our elders, for it was they who passed down to us both natural science and theology,

27. Glifford Geertz, *Islam Observed* (Chicago: University of Chicago Press, 1986), 2, quoted in Robert J. Priest, "Christian Theology, Sin, and Anthropology," chap. 2 in *Explorations in Anthropology and Theology*, ed. Frank A. Salamone and Walter Randolph Adams (Lanham, MD: University Press of America, 1997), 23–37, at 24.

two cultural spheres highly valued today, and, for that reason, still highly contested. We might, of course, wish our intellectual ancestors had left a better legacy for us. Before we judge them, however, we may benefit from applying a lesson of their own struggles to our present lives. If they at times spoke with excessive confidence, perhaps so have we. If we can learn to listen historically to them, perhaps we can begin to hear one another with more welcoming ears.

America's debates concerning evolution have seldom been conducted in a spirit of humility that would promote mutual understanding; persuasion and conquest have been the favored means and end for most parties involved. Paradoxically, these disagreements thrive under the very circumstances that should make it impossible for any disagreement to arise. For to disagree with another, in any genuine sense, requires that one first understand the other's position. Failing this, one may ridicule, dismiss, or disrespect, but one has not disagreed, only mischaracterized—or worse, ignored. And that, inescapably, is a far greater offense, for the victim of ignorance is not only the other, but also, and especially, oneself.

Though we may desire a resolution of America's evolution debates through mutual apprehension of the truth, a goal more realistic, but no less noble, also may be suggested. May each person strive to feel another's ultimate concern as that other person feels it, and to communicate one's own ultimate concern in such a way as to help another feel it from one's own perspective. Perhaps a hope for this mutual theological understanding is the most enduring yearning that history can attempt to satisfy for humanity.[28]

28. To satisfy an even deeper human need would, of course, require more than a knowledge of history. It would require an acquaintance with the timeless One who uttered those famous words, "Let there be light . . ." (Genesis 1:3).

Study Questions

1. How and when did Darwin's *Origin of Species* displace *Vestiges of Creation* as the most famous publication advancing a theory of evolution?

2. What was the "ultimate concern" shaping the theology at each of the case-study communities—Princeton, Harvard, and Yale? How did those distinctive systematic theologies, in turn, shape the receptions of *Vestiges* within those communities?

3. What is the difference between *predicting* how people react to a theory of evolution and *explaining* how they react? Can it be the case that this book's concept of "socially situated systematic theologies" is deeply explanatory even though it cannot predict people's reactions? Consider, for example, the suggestion that the Presbyterian emphasis on God's sovereignty explains both opposition to evolution at Princeton (in the era of *Vestiges*) and accommodation to evolution on the same campus (in the era after Darwin).

4. How might the methodology employed in this book be applied to more recent debates involving theology and the natural sciences, such as debates concerning intelligent design theory?

METHODOLOGICAL POSTSCRIPT

Socially Situated Systematic Theologies

Some Lessons from Max Weber and Robert Merton

The approach taken in the case-study chapters draws from the sociological methodology of Max Weber. In his classic study, *The Protestant Ethic and the Spirit of Capitalism* (1904–1905), Weber argued that a distinctively Calvinist understanding of vocation, or one's "calling" in life, fostered the development of the particular version of capitalism that became prevalent in portions of Europe and the United States where Calvinist Reformed believers held positions of social influence during the seventeenth and eighteenth centuries.[1]

---

1. Max Weber, *The Protestant Ethic and the Spirit of Capitalism*, trans. Talcott Parsons, with an introduction by Anthony Giddens (1904–1905; rev. ed., 1920–1921; trans., 1930; rpt., London: Routledge, 2001). The discussion provided above has been informed by Wolfgang Schluchter, *Rationalism, Religion, and Domination: A Weberian Perspective*, trans. Neil Solomon (orig. German essays: 1980, 1985, 1988; Berkeley, CA: University of California Press, 1989). A substantial literature has criticized Weber for inaccurate historical claims concerning the nature of both ascetic Protestantism and capitalism in the societies he surveyed, and for doubtful interpretative claims concerning the relationship between ascetic Protestantism and capitalism. For recent summaries of and contributions to these debates, see Hartmut Lehmann and Guenther Roth, eds., *Weber's "Protestant Ethic": Origins, Evidence, Contexts* (Cambridge: Cambridge University Press, 1993); Mark Valeri, "Religion, Discipline, and Economy in Calvin's Geneva," *Sixteenth Century Journal* 28 (Spring 1997): 123–42; Jacques Delacroix and François Nielsen, "The Beloved Myth: Protestantism and the Rise of Industrial Capitalism in Nineteenth-Century Europe," *Social Forces* 80 (2001): 509–53. For a long, though dated, bibliography, see: "Bibliographie zur Kontroversialliteratur," in *Die Protestantische Ethik II*, ed. Johannes Winckelmann (1978; rpt., Gütersloh, Ger.: Gütersloher Verlagshaus Mohn, 1995), 395–429. The use of Weber in the present investigation does not depend on the

Weber self-consciously argued against an earlier generation of economic history that had been promulgated by Karl Marx and his disciples. Religion for Marx was only an "ideological reflex" of the material relations in a society's economy; neither religion, imagination, metaphysics, or any other expression of human will or sentiment could change those conditions, although they could, and would, be transformed by changes in material conditions.[2] Marx's followers accordingly held that religious doctrines supported existing economic relations by reinforcing a society's class structure. Marx himself memorably called religion the "opium of the people," meaning that by distracting working people's attention with the promised blessings of an afterlife, religious dogma helped them to cope with their oppression under the ruling class and deterred them from rising up in revolution.[3]

In contradiction to Marxist theory, Weber argued that religious beliefs can contribute to profound novelty in social practices, such as by structuring the development of a new kind of capitalist economy. At the core of Weber's argument was the claim that specific religious ideas—his chief example being the Calvinist doctrine that the good works people exhibit through their vocations testify to their divine election for eternal salvation—could structure social and economic reality. In other words, these core beliefs shape the way people organize their subsidiary beliefs, and the structure of that belief system, in turn, patterns their behaviors, transforming their economic attitudes and practices as much as their religious ones. Writing as a sociologist rather than a psychologist, Weber was not concerned with whether each capitalist individual cherished exactly the same Calvinist doctrine of vocation in his heart. It was enough that a system of values, rooted in specific religious doctrines that at least some members of society cherished, was sufficiently pervasive in that society for those values to become integrated into everyday life, thus shaping, and also becoming a

---

specific merits (or demerits) of his own project concerning Protestantism and capitalism, but rather the possible fruitfulness of his vision for the study of how religion and culture interrelate more generally. Moreover, for Weber to have stimulated a long-lasting discussion among scholars, which continues even after many of his original claims have suffered severe criticism, should hardly serve as discouragement for future researchers to borrow selectively from his method.

2. See Karl Marx, "German Ideology, Part I," in *The Marx-Engels Reader*, ed. Robert C. Tucker (New York: Norton, 1972), 118.

3. Karl Marx, "Contribution to the Critique of Hegel's Philosophy of Right," in *Marx and Engels on Religion*, ed. Reinhold Niebuhr (New York: Scribner, 1964), 41–58, at 42.

part of, economic behaviors.[4] Moreover, Weber did not identify the Protestant ethic as a *cause* of the development of capitalism, but rather postulated it as an "ideal type" by which the social scientist can benchmark certain features of a society in order to illuminate the relationship between religion and economics.[5] His approach thus differed from the Marxist tradition not only by conceiving of religion as a cultural force that changes (rather than merely reflects) social conditions, but also by rejecting the possibility of any one-sided history as being complete (whether it be a "materialist" study, like the later writings of Marx and Engels, or even a "spiritualist" one, like Weber's own *Protestant Ethic*, which he regarded as but the beginning of a more comprehensive analysis of culture).[6]

During the 1930s, sociologist Robert Merton applied aspects of Weber's framework to another realm of inquiry: the relations between religion and science. Merton argued that a Puritan *ethos* may be said to have underpinned the social dynamics that fostered the scientific revolution in seventeenth-century England, just as Weber had suggested that such an *ethos*—postulated as an ideal type—could illuminate the development of capitalism.[7] Scholars since Merton have spilled much ink debating the merits of his thesis, just as both critics and supporters have commented laboriously on Weber's own argument ever since it appeared nearly a century ago.[8]

Much of the debate concerning Merton's work has resulted from the implication that modern science originally had (and, some argue, still should have) a foundation in Christian belief. By the middle of the twentieth-century, a

4. See Ralph Schroeder, *Max Weber and the Sociology of Culture* (London: SAGE, 1992), esp. 145.

5. Weber, *Protestant Ethic*, 33, 56, 149n28. See also Schluchter, 17–27; L. M. Lachmann, *The Legacy of Max Weber: Three Essays* (Berkeley, CA: Glendessary Press, 1971), 26–28.

6. See Karl Löwith, *Max Weber and Karl Marx*, ed. with an introduction by Tom Bottomore and William Outhwaite, with a new preface by Bryan S. Turner (London: Routledge, 1993), chap. 4.

7. Robert K. Merton, *Science, Technology, and Society in Seventeenth Century England* (1938; rpt., New York: H. Fertig, 1970).

8. On Weber, see the sampling of literature cited at note 1, above. On Merton, see esp. the twenty-four essays, spanning from 1938 to 1989, compiled in I. Bernard Cohen, ed., *Puritanism and the Rise of Modern Science: The Merton Thesis*, with the assistance of K. E. Duffin and Stuart Strickland (New Brunswick: Rutgers University Press, 1990). It is important to recognize that much of the debate over "Merton's thesis" has been a debate over the claim, inaccurately attributed to Merton, that Puritanism somehow caused or explains the scientific revolution. See Steven Shapin, "Understanding the Merton Thesis," *Isis* 79 (Dec. 1988): 594–605; and, H. Floris Cohen, *The Scientific Revolution: A Historiographical Inquiry* (Chicago: University of Chicago Press, 1994), 314–21.

sizable "harmony" literature had developed, arguing that good science (epitomized by Newton and so forth) and orthodox religion (variously defined) *have* (historically) or *can* (philosophically) or *should* (politically) be harmonious partners.[9] At times these three arguments were merged almost into one.[10] This "harmony" literature, whether scholarship or propaganda or both, came as a backlash against a "warfare" historiography spawned during the late nineteenth century by those advocating a more secular ideal for the academy.[11] Unfortunately, neither the intellectual heirs of the "warfare" histories nor the "harmony" scholars who critiqued them adequately considered the ways in which Merton's own approach (which was often taken as a contribution to the "harmony" side) creatively transcended the diametrical opposition that separate those interpretations.

Just as Weber was not claiming that all Calvinists and only Calvinists become modern capitalists, neither was Merton claiming that only Puritans and all Puritans become modern scientists. Nor were they claiming that the best Calvinists and Puritans become the best capitalists and scientists. Each scholar had a much more profound contribution to offer, a contribution that can be recognized only if the subtleties of their arguments be granted proper attention. Weber focused on a particular version of the doctrine of vocation that was rooted in medieval theology, repackaged and emphasized by the reformer Martin Luther, and ultimately transposed by Calvinists into a Puritan framework that was distinct from both its medieval and Lutheran predecessors. Weber then suggested a resonance in *ethos* between this Puritan understanding of vocation and the attitudes that promoted the emergence of a particular form of capitalism he studied, as distinct from the social patterns of economic relations found elsewhere, even in other capitalist environments.

9. For recent reflections by a scholar who has engaged these issues for over a generation, see Ian G. Barbour, *Religion in an Age of Science: The Gifford Lectures, 1989–1991, Vol. 1* (San Francisco: Harper and Row, 1990).

10. For example, Nancy R. Pearcey and Charles B. Thaxton, *The Soul of Science: Christian Faith and Natural Philosophy* (Wheaton, IL: Crossway Books, 1994).

11. For an overview of the "warfare" and "harmony" literature, see David C. Lindberg and Ronald L. Numbers, eds., "Introduction," in *God and Nature: Historical Essays on the Encounter between Christianity and Science* (Berkeley: University of California Press, 1986), 1–18. The classic "conflict" studies are John William Draper, *History of the Conflict between Religion and Science* (New York: D. Appleton and Company, 1874); and, Andrew Dickson White, *The Warfare of Science* (New York: D. Appleton and Company, 1876).

Similarly, Merton concerned himself not with the relationship between science and religion *in abstracto*, but rather with a concrete variety of each and the mindsets that resonated between them both. Moreover, neither scholar treated religion as a set of dogmas that either corroborated or contradicted new knowledge claims from science, but rather each of them regarded religion as a values-complex that structured the very ways in which various activities—social, economic, or scientific—were pursued. Thus, Merton and Weber cannot rightfully be classified anywhere along a warfare-conflict continuum; theirs was an entirely different project, as is the present study.[12]

During the late twentieth century, historians roundly rejected both "harmony" and "warfare" models, suggesting that these tended toward gross over-generalizations and gave unwarranted credence to partisan *cliché* conclusions ("the founding fathers of modern science were men of God," or, in retort, "the Church has usually opposed scientific progress"). The recent literature also has faulted both "warfare" and "harmony" studies with the failure to provide historical explanations that addressed the changing nature of both natural science and religion over the centuries. These criticisms, though appropriate, lack a coherent sense of direction. Some scholars have suggested that historians should consider the role of "religion *in* science," noting that many scientists under historical investigation imbued their work with religious motives, assumptions, and reactions.[13] Others scholars have questioned whether it is possible even to make a responsible historical distinction between science and religion, given that so many historical persons intertwined the two quite closely (or refused to subdivide the one, if it be granted that no distinction can be made).[14] Amid these and other sundry suggestions, two late-twentieth-century surveyors of the literature could not do better than suggest that historians study the "engagement" between science and religion, with "engagement" having some seven distinct meanings, including the marital connotation of the two becoming

12. For a helpful corrective of these misconceptions of Merton's project, see Shapin, esp. 596–97, 604.

13. Charles Coulston Gillispie, *Genesis and Geology: A Study in the Relations of Scientific Thoguht, Natural Theology, and Social Opinion in Great Britain, 1790–1850*, with a foreword by Nicolaas A. Rupke and a new preface by the author (Cambridge, MA: 1996), xxix.

14. David B. Wilson, "On the Importance of Eliminating *Science* and *Religion* from the History of Science and Religion: The Cases of Oliver Lodge, J. H. Jeans, and A. S. Eddington," chap. in 2 in *Facets of Faith and Science*, ed. Jitse M. van der Meer, 4 vols. (Lanham, MD: University Press of America, 1996), 1:27–47.

one.[15] However theoretically astute, such suggestions offer little practical guidance for a scholar desiring to spin a body of historical research into a usable narrative thread.[16]

A return to Weber may be of service here. Weber, and Merton after him, anticipated many of the useful insights of the late-twentieth-century turn away from the harmony-warfare dichotomy, while also offering a constructive narrative framework that the more recent literature has overlooked. In common with Weber, Merton, and the post-dichotomy historiography, this book will not be concerned with questions of whether religion in general (or even Protestantism in general) supports or subverts science in general (or even evolutionary theory in general). Applying Weber's notion of an "ideal type" as an analytical tool for grasping how socially pervasive attitudes shape individuals' behaviors around common practices, the case studies in chapters 2 through 4 sought, first, to identify particular core beliefs of specific varieties of Protestantism that performed a regulative function in the arrangement of subsidiary beliefs. The notion of an "ideal type" permitted specific examples of individuals' theological expressions to be analyzed methodologically as manifestations of a singular theological construct that presumably gave rise to those expressions. Then, the second phase of analysis examined how that socially shared system of beliefs channeled individuals' responses to the distinctive version of evolutionary theory presented in *Vestiges of the Natural History of Creation* into discernible patterns characteristic of the communities that adhered to those core beliefs. This approach also considered exceptions within the case-study communities in effort to determine the limits at which the "ideal" constructs remain explanatory. The next section outlines an explicitly theological method for pursuing these stages of analysis.

15. John Brooke and Geoffrey Cantor, *Reconstructing Nature: The Engagement of Science and Religion*, Glasgow Gifford Lectures (Edinburgh: T and T Clark, 1998), 6–8.

16. Those scholars who have made positive contributions to the post-dichotomy historiography generally have done so by suggesting complex ways in which the theological dispositions of *individual* scientific thinkers (such as Kepler and Newton) shaped their work, rather than exploring, as this present study seeks to do, the role of theology in orienting a *community's reception* of scientific ideas. See, for example, the articles in John Hedley Brooke, Margaret J. Osler, and Jitse M. van der Meer, eds., *Science in Theistic Contexts: Cognitive Dimensions, Osiris* 16 (2001). A recent community-study project by David Livingstone will be discussed below, in the section entitled "From 'Common Context' to Theological Specificity."

Paul Tillich's Existential Systematic Theology as a Hermeneutic for Historical Analysis

This study has enhanced the interpretive framework of Weber's and Merton's sociology with an explication, gleaned from theologian Paul Tillich, of the way in which a community's systematic theology functions to prioritize doctrines that shape exegetical and rhetorical practices. Tillich discovered that "the smallest problem, if taken seriously and radically, drove me to all other problems and to the anticipation of a whole in which they could find their solution."[17] For Tillich's notion of "systematic theology," all parts of a theology fit together as a whole, the parts deriving their significance from their relation to that larger picture. Applied to the historical study of the theologies that shaped reactions to *Vestiges* at Princeton, Harvard, and Yale, this means that it cannot suffice simply to note that two Harvard theologians, for example, objected that *Vestiges* undermined the authority of Scripture. One must also determine where scriptural authority fit within the larger picture of Harvard theology, and how that theology related to Harvard natural science; failing these goals, one cannot claim to have understood the Harvard position.[18]

Tillich further conceived of theology as a system oriented toward an "ultimate concern," by which he meant an existential question that cuts to the very heart of life: the difference between being and not-being.[19] The case studies in this book have revealed that botanists, zoologists, and geologists, no less than moral philosophers, philologists, and theologians, were intensely concerned about the theological implications of *Vestiges*. Arguably, theology mattered to them for the same reason it mattered to Tillich: theology centered on fundamental questions concerning the problem of their own existence.

An explicitly *theological* understanding of theology calls into question the naturalistic perspective of cultural anthropologists, such as Clifford Geertz, who generally have viewed theology as a formalization of religion, and religion as a projection of cultural conditions that, in turn, interprets and reshapes those

17. Paul Tillich, *Systematic Theology*, 3 vols. in 1 (1950–1963; rpt., Chicago: University of Chicago Press, 1967), 1:xi.

18. Unitarians' biblical concerns, as has been revealed in chapter 3, had nothing to do with a literal interpretation of Genesis and much to do with a preoccupation with Transcendentalism.

19. Tillich, *Systematic Theology*, 1:11.

conditions. Tillich, in contrast, regarded theology as a culturally informed statement of the fundamental, and transcultural, human condition itself.[20] The reader is not required to share Tillich's essentially theological view of culture or his universalist claim that the same, ultimately spiritual, existential questions preoccupy persons in all cultures; all that is requested of the reader is a willingness to consider this perspective as potentially indicative of the self-understanding of the historical individuals encountered in this book. A careful use of Tillich's perspective offers, in other words, an opportunity to come to know those persons more approximately as they themselves would have wanted to be known, namely, as people whose relationship with God was the existential root of their cultural engagements with one another—and with the author of *Vestiges*.

This is not, however, to suggest that Tillich's analysis should be applied wholesale to the nineteenth-century persons in this study; such a course would violate historical responsibility. For one thing, Tillich himself denied a chief claim made by many of the theologians in the case-study communities, namely, that theology can be logically grounded upon inductive summaries of Biblical evidence or universally shared rational assumptions. Consequently, an unqualified application of Tillich's approach would obscure an understanding of those historical persons' own perceptions.[21] Second, that which gave wholeness to Tillich's existentialist Protestant theology of the mid-twentieth century is irrelevant to the era of *Vestiges*; of relevance here is only that which gave wholeness to the theologies at Princeton, Harvard, and Yale when *Vestiges* was being discussed there.[22] Finally, it should be emphasized that although some scholars have suggested ways in which Tillich's theological program could be developed into a normative conceptualization of how theology and the natural sciences *ought* to relate, the use of Tillich's program in this investigation is only as an analytical tool aimed at describing and explaining how theology and the

20. Clifford Geertz, "Religion As a Cultural System," chap. 4 in *The Interpretation of Cultures: Selected Essays* (New York: Basic Books, 1973), 87–125; Paul Tillich, *Theology of Culture*, ed. Robert C. Kimball (New York: Oxford University Press, 1959), esp. chaps. 1 and 4.

21. Tillich, *Systematic Theology*, 1:8–9, 34–36.

22. Tillich, *Systematic Theology*, 1:66. For further clarification, it may be noted that Tillich's own "method of correlation" (which he sets forth at pp. 59–66) has not been employed in this study, since that method was his proposal for developing a twentieth-century theology, rather than a method applicable to historical persons remote from Tillich's own situated encounter with God.

natural sciences *did* relate.[23] The goal is to offer what Tillich called "a theological history of culture," that is, "the attempt to analyze the theology behind all cultural expressions" in order to better understand the human condition.[24]

The evidence presented in the three case studies has suggested a tightly woven relationship between two questions that members of the local academic communities asked themselves: "What must I do with *Vestiges*?" and "What must I do to be saved?" The first question was the initial motivator for this study. It is drawn from people's concrete experiences of reading and discussing the first comprehensive theory of evolution to gain wide circulation in America. The second question arguably was so closely intertwined with Americans' receptions of *Vestiges* that it, too, deserved special focus. For that reason, the originally planned thesis question—something like "What did people think of *Vestiges*, and how did religion factor into that?"—has been replaced with a more refined question: "How did the relationship between 'What must I do with *Vestiges*?' and 'What must I do to be saved?' shape the receptions of *Vestiges* in the three case study communities?"

Like "What must I do with *Vestiges*?," "What must I do to be saved?" also is drawn from concrete human experience. But it reaches, according to Tillich, beyond personal experience for an answer that transcends human experience. The pleading questioner also hopes that the answer will enter into human experience for the purpose of rescuing humans from the kind of existential plight that makes one ask the question in the first place.[25] "What must I do to be saved?" was the question that a jailor in Philippi asked when an earthquake shook open the prison gate. As his hymn-singing missionary prisoner, Paul of Tarsus, was liberated by the quake, the jailor drew his sword in preparation to take his own life before his superiors charged him with unpardonable negligence (Acts 16:29). "What must I do to be saved?" also was a question asked, under very different circumstances, by the theologians and naturalists encountered in the three case studies. They belonged to distinct Protestant denominations precisely because they thought that a carefully articulated answer to that question

23. See Donald E. Arther, "Paul Tillich's Perspectives on Ways of Relating Science and Religion," *Zygon* 36, no. 2 (June 2001): 261–67; Robert John Russell, "The Relevance of Tillich for the Theology and Science Dialogue," *Zygon* 36, no. 2 (June 2001): 269–308.

24. Tillich, *Systematic Theology,* 1:39; see also his *Theology of Culture,* 7–9, 41.

25. Tillich, *Systematic Theology,* 1:40–43, 57.

mattered, and they cherished their distinctive answers to that question because they belonged to distinct Protestant denominations that taught them to do so. This was even true among the comparatively liberal religious leaders at Harvard who shied away from specific doctrinal formulas, since this laxity was prompted by the firm conviction that human-made creeds distanced people from the true God. It was not only the answer to, but also the implications of, the salvation question that mattered. "To be saved from what?" was answered differently by Princeton Presbyterians, who emphasized the total depravity of each human person, than by Harvard Unitarians, who viewed human nature far more optimistically. At Yale the system was oriented in yet a third direction. Consequently, "What must I do to be saved?" resonated in different ways with "What must I do with *Vestiges*?" on those three campuses.

It is, moreover, both the guiding assumption and the documented conclusion of this study that to understand the place of *Vestiges* in American history requires that one understand the place of distinctive theologies in American history, and, simultaneously, that an understanding of the receptions of *Vestiges* helps one to understand the theologies that shaped those receptions. The methodology is therefore constructively circular at two levels. At the inner level, *Vestiges* is deployed to elucidate a community's systematic theology, even as each community's systematic theology is deployed to account for that community's response to *Vestiges*. At an outer level, the claim that such circularity can contribute to historical understanding, rather than reduce itself to the vacuity of a tautology, is a claim that serves as both a method and a conclusion in this investigation. This notion of constructive circularity has been derived from Weber's approach of employing an "ideal type" (in his case, the Puritan work ethic) as both a product of and a necessary precondition for social science analysis.[26] In this case, the ideal types are the systematic theologies of the

---

26. Weber, notably, argued that "the spirit of capitalism" could not be defined at the onset of his study, except provisionally, since it was the object of his study not only to demonstrate the importance of that *Geist* in shaping the development of capitalism, but also to discover what the ideal type of that *Geist* was. See esp. Weber, *Protestant Ethic*, 13–14. For further commentary on Weber's idealism, and its service as both method and result, see Talcott Parsons, "Introduction," to Max Weber, *The Sociology of Religion*, trans. and intro. by Talcott Parsons, with a new foreword by Ann Swindler (German orig., 1922; English trans., 1963; rpt., Boston: Beacon, 1993), xxix–lxxvii, esp. at xxx–xxxiv. For additional discussion of this point, together with an identification of some Kantian assumptions shared by Weber and Tillich, see Keith Tester, "Between Sociology and Theology: The Spirit of Capitalism Debate," *The Sociological Review* 48, no. 1 (Feb. 2000): 43–57. For qualifications concerning the limit to which Weber may be justly considered Kantian, see Schluchter, 31.

three case-study communities. In the very process of arguing that these theologies had a profoundly significant impact upon the receptions of *Vestiges*, the theologies themselves became revealed with increasing clarity, for the ultimate concerns of *Vestiges'* readers have been identified not only from the archival records of their writings that did not address *Vestiges*, but more especially from those that did. Theology, in other words, was so inextricably wrapped up with the *Vestiges* debates that it cannot be isolated as an independent variable, but rather must serve both as a an explanatory method and a phenomenon to be explained.

One might, of course, object that something other than theology was central to the communities under investigation. The persuasiveness of the account offered here may be measured, therefore, by the degree to which that objection loses force after this book has been given a full read. Weber, contrary to frequent misreadings of his work, did not claim to have explained all of European capitalism solely by appeal to Calvinism. He openly conceded that his account remained incomplete without particular insights that could be gained only through alternative methodologies, including the methodology of his rival school, the Marxists. In fact, Weber proposed his Puritan *ethos* model merely as a propaedeutic for future inquiry; it was to be a tool for ongoing scholarship, not an established conclusion that would bring scholarly debate to a close.[27] Likewise, Merton never wrote that Puritanism accounted for the rise of modern science; he claimed, more modestly, that Puritanism helped to explain some of the social structures that surrounded the practices of science as those practices emerged in seventeenth-century England.[28] The crucial question now, therefore, is not whether systematic theology explains everything about the receptions of *Vestiges*, but whether it explains enough to deserve a prominent role among other

27. Weber, *Protestant Ethic*, 125. This oft-overlooked passage deserves to be quoted in full: "Here we have only attempted to trace the fact and the direction of its [the Puritan ethos'] influence to their [the capitalists'] motives in one, though very important point. But it would also further be necessary to investigate how Protestant Asceticism was in turn influenced in its development and its character by the totality of social conditions, especially economic. The modern man is in general, even with the best of will, unable to give religious ideas a significance for culture and national character which they deserve. But it is, of course, not my aim to substitute for a one-sided materialism [i.e., Marxist sociology] an equally one-sided spiritualistic causal interpretation [i.e., Weber's own research program] of culture and history. Each is equally possible, but each, if it does not serve as the preparation, but as the conclusion of an investigation, accomplishes equally little in the interest of historical truth."

28. Shapin, 596–97.

historical narratives that likewise seek to guide present-day readers through the textual artifacts of nineteenth-century natural science.

The strong focus on theology in this book may be encouraged not only by the fruitfulness of Weber's and Merton's studies, but also by a conclusion reached by historian Paul Conkin in his study of nineteenth-century American Reformed theology. Conkin suggested that inter-denominational rivalry, at times leading also to intra-denominational doctrinal disputes, preoccupied theologians more than the tensions that were arising between theology and the natural sciences.[29] In narrating some of those debates, Conkin highlighted the significance of Arminianism, which, as has been explained more fully in the case-studies, was a critique of the Calvinist doctrines of total depravity and predestination. For the present it suffices to recall that the distinctive theological systems at Princeton, Harvard, and Yale each took a different position in the debates over Arminianism, and that those positions shaped the ways in which *Vestiges* was understood on the three campuses.

Even with the importance of religion for nineteenth-century American academic life conceded, however, objections still could be raised that "systematic theology" does not provide the best avenue for investigating responses to *Vestiges* at Princeton, Harvard, and Yale. Anthropologists—who, more frequently than historians, have at least considered using such a model—have raised concerns that systematic frameworks of analysis fail to capture the religious experiences of people whose beliefs are in fact non-systematic. Systematic theology may well be the goal and product of academic theologians, but for everyday believers, and especially for the merely occasional participants in a community's religious rites, does faith ever construct a world picture according to the coherence predicated in a formal system of theology?[30] Among historians, similar methodological concerns have motivated the rise of scholarship on "the religion of the people," which is assumed to be quite different than the theology of the religious leaders.[31] Thus, it should also be clear that this study has adopted

29. Paul K. Conkin, *The Uneasy Center: Reformed Christianity in Antebellum America* (Chapel Hill, NC: University of North Carolina Press, 1995), 210.

30. Douglas J. Davies, *Anthropology and Theology* (Oxford: Berg, 2002), 20–22.

31. See, for example, Thomas A. Tweed, ed., *Retelling U.S. Religious History* (Chicago: University of Chicago Press, 1997); Robert Orsi, *Thank You, St. Jude: Women's Devotion to the Patron Saint of Hopeless Causes* (New Haven, CT: Yale University Press, 1996). The contributors to Tweed's volume argue that American historians should depart from the meta-narratives that previous generations of scholars based

the burden of demonstrating that formalized theology does in fact have significant ramifications throughout the local social community in which, and on behalf of which, that theology is articulated, despite the presence of idiosyncrasies, or even resistance, among the community members themselves.

A second, and closely related, objection to a methodological emphasis on theology is that religion (meaning a complex of attitudes, beliefs, and behaviors pertaining to the supernatural) encompasses much more than theology (meaning doctrines and creeds that formalize religious beliefs), and is therefore the more historically significant factor.[32] In reply, it must be emphasized that the approach taken in this book does look beyond formal theology in order to discern what may be termed a *socially situated systematic theology*, that is, a manner in which formal doctrinal expressions were integrated into the daily life of a community, which was united by specific social practices that cohered, in part at least, precisely because those formal doctrinal expressions rationalized those social practices. Such practices included classroom instruction, chapel sermons, and public discourses on natural science (a topic that during the era of *Vestiges* involved frequent theological references). By considering these practices both with respect to the intellectual claims made and the social functions of those claims, the case studies explored the sociological dimensions of a community's systematic theology—all with the expectation that responses to *Vestiges* were formed by those dimensions.

largely upon New England Puritan theology and instead pursue themes of social analysis, viewing religion in terms of its bearing upon issues of gender, race, and cross-cultural contact; none of the contributors recommends a closer attention to theology. Orsi borrows from cultural anthropology and psychoanalysis to explore the varied private meanings that women attached to their devotion to St. Jude; he argues, in effect, that formal theology has little relevance to the historical experience of religion. John McGreevy, standing out as an exception to recent scholarship, encouraged American religious historians to explore "the theology of everyday life," which he defines as the manner in which "theological beliefs and practices shape understandings of such perennial themes as suffering, nationalism, migration, and the market, even as they undergird faith communities." See John T. McGreevy, "Faith and Morals in the United States, 1865–Present," *Reviews in American History* 26, no. 1 (Mar. 1998): 239–49, at 244.

32. Clinton Bennett, *In Search of the Sacred: Anthropology and the Study of Religions* (London: Cassell, 1996), 17. For the definition of "religion" employed above, see Gary Farraro, *Cultural Anthropology: An Applied Perspective* (St. Paul: West, 1992), 262; James Peoples and Garrick Bailey, *Humanity: An Introduction to Cultural Anthropology* (St. Paul: West, 1988), 318.

From "Common Context" to Theological Specificity

Until recently, most treatments of natural science and religion in the nineteenth century adopted what historian Robert Young called a "Common Context" approach.[33] Young's study of Victorian British periodicals indicated that common tropes of natural theology, shared across regional, social, political, and religious boundaries, provided a broadly shared intellectual framework in which matters of faith and natural science were harmonized. Scholars of the American scene have likewise pursued this line of argument. For example, Theodore Bozeman emphasized the Scottish Realist ideal of inductive natural, moral, and theological sciences at Princeton during the decades preceding the Civil War. His title, *Protestants in an Age of Science* (1977), suggested that this portrait of science-religion harmony at Princeton was indicative of a much larger picture. Indeed, one year later Herbert Hovenkamp revealed just that, when he published *Science and Religion in America, 1800–1860* (1978). Hovenkamp, incidentally, dealt only with Protestants, but, like Bozeman, he employed a title that suggested a much broader representation than his evidence presented. Thus, both explicitly and implicitly, these pioneering studies—which for twenty-five years remained the standard survey works—portrayed a generally uniform view of science-religion relations in pre-Darwinian America. Hovenkamp's account, since it included voices beyond the Princeton of Bozeman's focus, did highlight some exceptions to the dominance of Scottish Realism in American intellectual life, such as instances of German Idealism at Harvard and Yale, but otherwise he preserved an image of common context—the perfect set-up for a demolition by Darwinism in 1859.[34]

New scholarship, focused on the British scene, has challenged both the claim of a common pre-Darwinian context, and the supposed shattering of it by Darwinism. Secord's study, as mentioned earlier, revealed that *Vestiges* preempted much of the excitement that has, he argued, mistakenly been

33. Robert M. Young, "Natural Theology, Victorian Periodicals, and the Fragmentation of a Common Context," in *Darwin's Metaphor: Nature's Place in Victorian Culture* (Cambridge: Cambridge University Press, 1985), 126–63.

34. Theodore Dwight Bozeman, *Protestants in an Age of Science: The Baconian Ideal and Antebellum American Religious Thought* (Chapel Hill: University of North Carolina Press, 1977), xiv, 163–65; Herbert Hovenkamp, *Science and Religion in America, 1800–1860* (Philadelphia: University of Pennsylvania Press, 1978), chap. 10.

attributed to Darwin's *Origin of Species*. Secord further argued that, far from everyone reading *Vestiges* through a common interpretative context of British natural theology, readers approached the text in remarkably varied ways.[35] Likewise, Jonathan Topham's groundbreaking study of the Bridgewater Treatises —a series of natural theological works from the 1830s that Young regarded as exemplary of a supposed "common context"—has revealed that those texts did not mean the same thing to everyone. For example, William Buckland's Bridgewater Treatise concerning geology, though dedicated to the proposition that the Creator's wisdom, power, and glory were manifest in nature, nevertheless was read by at least one atheist reader as a source book for his own pre-Darwinian theory of naturalistic evolution.[36] Whereas Topham and Secord each focused upon readership practices, David Livingstone has promoted a new school, called "the geography of science." In a challenge to the common context concept even for members of a single denomination—Presbyterianism in this case —Livingstone concluded that distinct local factors prompted the near-simultaneous rejection of organic evolutionism at Belfast, toleration for it at Princeton, and accommodation of it at Edinburgh. At the heart of Livingstone's argument is the contention that evolutionism was not a straightforward "ism" at all, but rather a cultural encounter that could, and did, mean quite different things when encountered in distinctive local settings.[37]

Argument, of course, is not the same as persuasion. Have these recent studies accomplished their goal of invalidating the common context narratives? The results, it seems, are mixed. Secord, by far, provides the greatest depth of documentary evidence. If there was a scrap of paper with the word "Vestiges" on it, it seems that Secord managed to find it, and clearly he found enough scraps to show that not even those figures known well to historians of science were always reading *Vestiges* through the lens of the Bridgewater Treatises' natural theology.

35. James A. Secord, *Victorian Sensation: The Extraordinary Publication, Reception, and Secret Authorship of "Vestiges of the Natural History of Creation"* (Chicago: University of Chicago Press, 2000), *Victorian Sensation*, esp. 3, 138, 299, 333–34, 532.

36. Jonathan R. Topham, "Beyond the 'Common Context': The Production and Readings of the Bridgewater Treatises," *Isis* 89 (1998), 233–62.

37. David N. Livingstone, "Science, Region, and Religion: The Reception of Darwinism in Princeton, Belfast, and Edinburgh," chap. 1 in *Disseminating Darwinism*, ed. Numbers and Stenhouse, 7–38; also published as "Situating Evangelical Responses to Evolution," chap. 8 in *Evangelicals and Science in Historical Perspective*, ed. David N. Livingstone, D. G. Hart, and Mark A. Noll (New York and Oxford: Oxford University Press, 1999), 193–219.

Topham's study, drawn from a step-by-step analysis of the "publication circuit" linking authors to printers to booksellers to readers, and so on, likewise provides some compelling cases in which something other than the words on the page had a strong impact upon the meanings people associated with those words. Livingstone's argument, however, has a much thinner evidential base. The "context" he revealed was derived from but a few lectures, books, and newspaper reports. His claim to have determined "not only what could be *said*, but what could be *heard*, about evolution in these three different localities,"[38] makes for a potentially enlightening objective, but its attainment remains out of reach. Until diaries, sermon manuscripts, correspondence and other such sources can be located to bring present readers closer to the historical action, one must suspend judgment as to how decisively the local contexts were able to divide those three Presbyterian communities.

Meanwhile, those desiring to learn about natural science and theology in pre-Darwinian America are left with the conviction that Baconian Scottish Realism, the design argument, and biblicism carried a lot of weight. Bozeman's and Hovenkamp's emphasis on Scottish Realism was mentioned above. Chapter 2 of this book, which focused on Princeton, has demonstrated that a Presbyterian understanding of God's sovereignty mattered more than Scottish Realism, and in fact shaped the version of Scottish Realism that Princetonians articulated. That chapter and chapter 4, which focused on Yale, suggested that God's sovereignty, at Princeton, and God's moral government, at Yale, also mattered more than the "design argument" of William Paley. Historians often have portrayed Paley's argument—roughly, that design in nature proves a divine Designer—as central to pre-Darwinian American relations between theology and the natural sciences. Hovenkamp, for example, singled out Paley's *Natural Theology* (1801) as the most influential work for American natural theology.[39] Daniel Walker Howe's subsequent study of Harvard moral science concluded, however, that Paley's works received significant criticism at that school.[40] Numerous examples of pre-Darwinian dissatisfaction with Paley also can be found at Yale, both in the lecture manuscripts of professors and in the notebooks and diaries of students, as

38. Livingstone, "Situating Evangelical Responses to Evolution," 196.

39. Hovenkamp, 219.

40. Daniel Walker Howe, *The Unitarian Conscience: Harvard Moral Philosophy, 1805–1861*, with a new introduction (Middletown, CT: Wesleyan University Press, 1988), 65.

documented in chapter 4. That chapter indicates that the same organizing principles that constituted the systematic theology espoused by members of the Yale community also influenced how they applied the so-called "design argument." In other words, apart from an understanding of Yale's systematic theology, the fact that all students were required to read Paley conveys little historical significance.

Today's readers may be surprised to learn that the problem American critics generally had with *Vestiges* was not that it denied Paley's identification of design in nature, nor that it attempted to have design without a divine Designer, for *Vestiges* in fact did neither of these things. Occasionally reviewers did charge *Vestiges* with banishing the Creator from His creation, but these statements must be understood in terms of *their* conceptualization of *their* debates, not *today's* conceptualizations of *today's* debates. It will at times be difficult for this present audience to view *Vestiges*, the first popular presentation of a comprehensive theory of evolution in the English language, as a work that both affirmed and indeed was founded upon a notion of divine design. Many of today's readers have been socialized to suspect that the recent craze over "intelligent design theory" is merely another form of "creation science" in disguise and that "evolution," if it is to be "scientific," is incompatible with any "intelligent design theory."[41]

Debates during the era of *Vestiges* were conducted in different terms. What was at stake was not always creation versus evolution, nor design versus chance; moreover, it *never* was creationism *versus* science. Rather, the debate centered over what kind of Designer-Creator science should reveal, a question laden with theological implications and often answered from theological presuppositions. Generally, it was supposed that those presuppositions in no way impinged upon (but in fact made possible) a scientific approach to the problem.

The "design argument" is thus best analyzed as an adaptable doctrine that could, depending upon which theological system appropriated it, be used to support a variety of natural histories, including the natural history in *Vestiges* itself. Particularly chapters 2 (regarding Princeton) and 4 (regarding Yale) have underscored that *Vestiges* did promote a vision of divine design. At Princeton, it was rejected for not fitting with Presbyterian understandings of God's

41. For a discussion of the cultural battles over whether intelligent design theory should be classed as creation science, theistic evolution, or some other category, see Ronald L. Numbers, *Darwinism Comes to America* (Cambridge: Harvard University Press, 1998), 15–21.

sovereignty; at Yale, it was welcomed as a reinforcement of the local "New Haven theology." Only by understanding "design" as a component integrated into a larger rationalization of the world, such as a community's systematic theology, can its significance for that community be grasped.

A similar methodological task concerns the topics of biblical literalism and biblical inerrancy. Contrary to common perceptions of historical debates, and of contemporary American life, what people think of the Bible's reliability and relevance in matters of natural science and what people think about evolution are not that closely correlated. Whether discussing *Vestiges of Creation* or *Origin of Species*, historians frequently have assumed that Protestant oppositions to evolution could be explained in terms of conflicts that "fundamentalists" or proto-fundamentalists perceived between Scripture and science, particularly, between the Book of Genesis and geology.[42] This investigation has shown that such conflicts were relatively unimportant in motivating people's actual responses to *Vestiges* and, more particularly, that historians' references to "biblical literalism," "scriptural geology," and related catch-phrases have provided but a shallow explanation of why anyone expressed discomfort or disagreement with the theories they encountered in *Vestiges*.

Few persons who articulated a discrepancy between the Bible and *Vestiges* could themselves be characterized as biblical literalists. Most were willing, for example, to interpret the "days" in Genesis 1 as geological eons of innumerable millennia or to regard the six days of creation as merely the last of several series of creations God made on this earth, with His reported making of the sun on day three referring symbolically to the reappearance of the sun after the earth's atmosphere thinned. These and other hermeneutical maneuvers enabled Christians to regard as harmonious witnesses of God's work both the Mosaic record and the geological record, which during the 1830s and 1840s seemed increasingly indicative of an ancient past populated with species unlike those that

42. Hofstadter, 14; Thomas Jefferson Wertenbaker, *Princeton, 1746–1896* (Princeton, NJ: Princeton University Press, 1946), 232–33; John H. Giltner, "Genesis and Geology: The Stuart-Silliman-Hitchcock Debate," *Journal of Religious Thought* 23, no. 1 (1966–1967): 3–13, rpt. in John H. Giltner, *Moses Stuart: The Father of Biblical Science in America* (Atlanta: Scholars Press, 1988), 66–74; Hovenkamp, 209; Louise L. Stevenson, *Scholarly Means to Evangelical Ends: The New Haven Scholars and the Transformation of Higher Learning in America, 1830–1890* (Baltimore and London: Johns Hopkins University Press, 1986), 27; Rachel Laudan, "The History of Geology, 1780–1840," in *Companion to the History of Modern Science*, ed. R. C. Olby, *et al.* (London: Routledge, 1990), 314–25, at 323; David R. Oldroyd, *Thinking about the Earth: A History of Ideas in Geology* (Cambridge: Harvard University Press, 1996), 6.

presently exist. Christians who insisted upon a literal interpretation of the Mosaic chronology during the era of *Vestiges* generally were concerned more with the future return of Christ than with the past creation, and it was their eschatological visions that required both a literalist hermeneutic (most notably for the "thousand years" in Revelation 20) and a defense of a recent *fiat* creation.[43] Others, however, could and did take a high view of Scripture's wording in a different direction. As historians Mark Noll and David Livingstone have emphasized, it is surprising, and therefore instructive, to realize that Princeton theologian Benjamin Warfield was both the father of "biblical inerrancy," as it became codified and coopted by fundamentalists in the early twentieth century, and also one of America's major pioneers of theistic evolutionism.[44] The words recorded in the Bible were by no means irrelevant to the diverse ways in which nineteenth-century Christians related science and religion, but neither was the sacred text anything close to determinative of the positions they articulated.[45]

These examples direct historians of religion away from the Bible itself and toward the deeper organizing principles according to which different groups interpreted what the Bible taught and around which they arranged those teachings into what they regarded as a cohesive whole. In other words, one is led to consider the importance of systematic theology, for even those who were not themselves systematicians nonetheless had a sense that some religious

43. Rodney L. Stiling, "Scriptural Geology in America," chap. 7 in *Evangelicals and Science in Historical Perspective*, ed. Livingstone, Hart, and Noll, 177–92; Robert Whalen, "Genesis, Geology, and Jews: The New York Millenarians of the Antebellum Era," *American Presbyterians* 73, no. 1 (Spring 1995): 9–22; Robert Whalen, "Calvinism and Chiliasm: The Sociology of Nineteenth Century American Millenarianism," *American Presbyterians* 70, no. 3 (Fall 1992): 163–72.

44. David N. Livingstone and Mark A. Noll, "B. B. Warfield (1851–1921): A Biblical Innerantist as Evolutionist," *Isis* 91 (2000): 283–304.

45. This also holds true for the most widely known instance of religious opposition to evolutionism in the twentieth century, the 1925 Scopes trial. Though much of the rhetoric was couched in terms of biblicism, the cross-examination of anti-evolutionist attorney William Jennings Bryan by Clarence Darrow revealed that Bryan did not interpret Genesis literally, even if he was insistent that Genesis refuted evolutionism. On this historical point, both Edward Larson, a historian who accepts evolutionary theory, and Henry Morris, a pioneer in young-earth creation science, agree. See Edward J. Larson, *Summer for the Gods: The Scopes Trial and America's Continuing Debate over Science and Religion* (Cambridge: Harvard University Press, 1998), 182–191; Henry M. Morris, *History of Modern Creationism*, new updated ed., with a foreword by John C. Whitcomb (Santee, CA: Institute for Creation Research, 1993), 73. Surveying the fundamentalist movement more generally, George Marsden has noted that even conservative Baptists themselves at times denied holding a biblical literalist posture. See George M. Marsden, *Fundamentalism and American Culture: The Shaping of Twentieth Century Evangelicalism, 1870–1925* (New York: Oxford University Press, 1980), 107.

convictions carried more weight than others. It would have been these priorities, according to Tillich's "theological history of culture," that dictated how the rest of one's convictions, including one's interpretations of Scripture passages, fit together.[46]

As was revealed in this study, Harvard Unitarians did not care for a literal interpretation of Genesis 1–2, but they did push for a literal interpretation of any New Testament passage that arguably portrayed Christ as inferior to the Father, for this reinforced their conviction that only the Father, and not Christ, is God.[47] For their own part, Princeton-affiliated Old School Presbyterians found themselves holding to the literal sense of Scripture when it best fit their doctrines of limited atonement and predestination, but they departed from such a strict interpretation in cases where the bare words of Scripture could be construed as contrary to Presbyterian Calvinism.[48]

The case studies have revealed that Unitarians and Presbyterians each were caught in what Tillich called a "theological circle," even if each group believed that it was simply drawing inductions from the plain facts of Scripture.[49] Neither of their hermeneutic circles centered around biblical literalism, nor was any other exegetical model determinative, but rather their exegesis was "theological" in Tillich's sense. That is, it flowed from the entirety of their systematic theologies as governed by the peculiar doctrines that each group, having found a particular answer to its existential question, treasured as most important. In other words, *socially situated systematic theologies shaped how people were reading Scripture.* For many of the persons in this study, such exegetical practices—perhaps perceived at the time as objective generalizations of

46. Tillich, *Systematic Theology*, 1:47–52.

47. James Walker, *A Discourse on the Deference Paid to the Scriptures by Unitarians* (Boston: American Unitarian Association, 1832); Andrew Preston Peabody, *Sermons Connected with the Re-Opening of the Church of the South Parish, in Portsmouth, New Hampshire, Preached Dec. 25 & 26, 1858, and Jan. 30 & Feb. 6, 1859* (Portsmouth, NH: James F. Shores, Jr., & Joseph H. Foster; Boston: Crosby, Nichols, and Co., 1859), esp. 26.

48. "The Extent of the Atonement," 2 pts., *Presbyterian Magazine* 4, no. 11 (Nov. 1854): 496–502, 4, no. 12 (Dec. 1854): 533–40.

49. Tillich, *Systematic Theology*, 8–11. For Tillich, the circle is not merely hermeneutic, but "theological," that is, existential—connected to God who is the source of all existence. God revealed himself in Scripture precisely for the purpose of answering the most deeply felt need that any human being can have —the need for reunion with one's Creator, from Whom one has been self-alienated by the tragedy that human existence itself has become as a result of sin.

the plain facts in Scripture, but analyzable historically as self-confirming hermeneutic circles[50]—were not peripheral to the rest of life, but rather shaped one's entire rationalization of the world and thus held sway over how one regarded *Vestiges*.

Those rationalizations, according to this model of analysis, were neither shared across the boundaries of geography and religious denomination, nor so idiosyncratic as to leave each person with nothing more than a personal faith. Rather, the cultural rationalizations amid which *Vestiges* was interpreted were shaped by convictions that developed within a local community of fellow believers. The concrete experiences of a lived relationship with one's co-religionists shaped the significance of commonplace terms, such as "design," "nature," and "God," making the significance of those terms community-specific. Admittedly, writers from the three academies under study often employed nearly identical verbal expressions, such that at first glance a review of *Vestiges* published at Harvard may not seem so distinct from one published at Princeton, for example. Nevertheless, if the model of analysis employed in this present study has merit, then it reveals that the meaning of those reviews is not to be found in the literal verbal formulations that appeal, almost invariably, to pervasive Anglo-American tropes about God and nature, but rather in the relationship between those formulations and other elements that are prevalent in each local community's socially situated systematic theology.

In order to discover what a particular verbal expression meant to the subjects of this study, a close reading of the archival documents has served as a surrogate for what anthropologists call "participant observation." The aim was to explore each case-study community's interpretations of *Vestiges* in the same

---

50. Such an analysis does not necessarily lead to truth relativism. Some recent authors, accepting much of the postmodernist corpus of critiques against modernist assumptions about "scientific" or "objective" hermeneutics, have nonetheless found ways to support their conviction that absolute truth both exists and can be known, at least in increasingly refined approximations, by humans. One such approach is to replace Tillich's circle with a spiral: as the text of Scripture and the systematic theology being deployed in an exegesis of Scripture engage one another in dialogue, the exchange may lead to a refinement of the systematic theology employed, such that the reader may asymptotically approach the true interpretation of Scripture—with "true" understood as a universal, rather than contingent, quality. Thus, although the analysis employed in the case studies for this book may lend itself well to relativistic theologies, it by no means precludes a more universalist application. See Grant R. Osborne, *The Hermeneutical Spiral: A Comprehensive Introduction to Biblical Interpretation* (Downers Grove, IL: InterVarsity Press, 1991); and James W. Voelz, *What Does This Mean? Principles of Biblical Interpretation in the Post-Modern World* (St. Louis: Concordia, 1995).

manner that one might, for example, interpret what Charlotte Elliott's hymn, "Just As I Am," means to Baptists and Lutherans today. Initially, it would seem that the two religious groups understand the hymn in the same way, since they both sing the same lyrics to the same musical setting, namely, William Bradbury's *Woodworth*. But upon closer analysis, it becomes obvious that the hymn has different, even opposing, meanings for the two groups. Baptists typically employ this hymn for altar calls. The piano music begins softly toward the end of the sermon, and then the lyrics are heard as people who have not previously professed their faith in Christ are invited to step forward to the altar and invite Jesus into their heart as personal Savior: "Just as I am, without one plea, / but that Thy blood was shed for me / . . . O, Lord, I come, I come." In the case of Lutherans, however, only those who previously have professed their faith publicly through the rite of confirmation are invited forward to the altar when this otherwise "same" hymn is heard. There they kneel to receive the Blood of Christ, which was shed on Calvary as mentioned in the song and also mysteriously becomes present in the eucharistic wine. Lutherans do not invite Jesus to enter their hearts, but rather find Him hidden in bread and wine in the Sacrament of Holy Communion. In fact, Lutherans regard the Baptist conversion theology of "inviting Jesus into one's heart" to be heretical. Baptists, meanwhile, find heretical the Lutheran teaching of Christ's physical presence in the Sacrament of Holy Communion and the Lutheran denial that conversion requires the degree of human initiative suggested by the Baptist altar call. Though Baptists and Lutherans share much in common, their "ultimate concerns" differ sufficiently to create entirely different experiences of what otherwise is the "same" hymn.[51] The case studies in this investigation seek to read textual relics of Americans' past experiences with *Vestiges* in the manner that an anthropologist might "read" Baptists' and Lutherans' present experiences with "Just As I Am." This requires

---

51. As for the hymn's writer, Charlotte Elliott, she was an invalid unable to assist her fellow believers in the physical aspects of church work, so, on one sleepless night in 1834, she pleaded that Christ would nonetheless accept her, "Just as I am." Her own preferred interpretation is, of course, irrelevant to the fact that Baptists and Lutherans have each imbued her words with meanings that resonate with their own respective systematic theologies. See Evangelical Lutheran Synod, *Evangelical Lutheran Hymnary* (St. Louis: MorningStar Music Publishers, 1996), hymn 319 (which, not incidentally, is found in the Holy Communion section), and pp. 10, 35–36, 38–39 (for a reprint of sixteenth-century Lutheran confessional writings concerning the real presence); Fred L. Precht, *Lutheran Worship Hymnal Companion* (St. Louis: Concordia Publishing House, 1992), 378 (for an account of Elliott); both the Baptist and Lutheran usages of this hymn have been observed in person.

that the investigator enter into the social practice of the three case-study communities' systematic theologies, rather than simply take their common source texts (the Bible, William Paley's *Natural Theology*, etc.) at face value.

The published reviews surveyed in chapter 1, therefore, embodied merely a hint of the larger dynamic shaping Americans' encounters with *Vestiges*. That larger dynamic centered—if the core proposition of this book is true—upon the socially situated systematic theologies by which the development theory was enveloped, and through which it acquired a unique set of meanings in each of the three case-study communities. Somewhere in the resonance between a published objection to *Vestiges* and a sermon preached in the college chapel, somewhere between a letter written to a fellow botanist and a public lecture concerning the development theory, somewhere between the ministerial constituency on the board of trustees and what one's father-in-law thought of the natural sciences— *somewhere within one's socially situated theological condition* emerged what Tillich would call an "ultimate concern," a deeply felt conviction that one had better say something about *Vestiges*, and say it well, because one's relationship to God depends upon it.[52]

By now it should be clear that the goal of this study was not to identify strict causal relationships between theological beliefs and the acceptance or rejection of evolutionary theory. Nor is the goal to state, according to a looser standard of causality, which groups were *more likely* to accept or reject evolution. Rather, the goal has been to identify what evolution *had to mean* for each group, if that group were to *understand it as something acceptable*, and likewise, what evolution *had to mean* if the group were to *understand it as something it unacceptable*. The answers to those questions have been sought largely in terms

52. It is significant that the climatic phrase in that sentence is "one's relation to God"; historians, sociologists, and anthropologists generally view the importance of religion to be how it shapes one's relation to the state (Marx), how it creates the relationships that constitute a society (Durkheim), or how it maps out the meanings of cultural experience (Geertz). The people in this study likely would not recognize themselves in any such portrayal. Rather than try to explain why they believed in God (which too often results in explaining away that belief, e.g., as a tool for social negotiation, or for psychological coping), this study accepts their stated beliefs as given, and tries to explain how those beliefs impacted other aspects of life, especially the reading of *Vestiges*. Nevertheless, the "givenness," which is predicated by Tillich as an inescapable "ultimate concern," is not to be confused with transcultural or transhistorical permanence, since, as Tillich argued, "Every religious act, not only in organized religion, but also in the most intimate movement of the soul, is culturally formed." Just as Tillich viewed Christ as not only fully God but also fully human, and therefore both fully eternal and fully historical, so also religion in this book will be understood as both ultimate in its concern and yet historical in its form. See Tillich, *Theology of Culture*, 42.

of the local community's theology, and from these answers an explanation emerged as to why people responded to *Vestiges* in the ways they did.

Study Questions

1. Max Weber argued that specific religious beliefs can shape social and economic practices that give structure to a people's culture. Merton extended Weber's analysis to the culture of modern science in particular. How do their perspectives assist in interpreting the distinctive amalgamations of theology and the natural sciences at the three case-study communities explored in this book?

2. Distinguish between the "harmony" and the "warfare" schools of thought within the history of science and religion. Identify examples that support each of these perspectives. Does the history of Americans' reactions to the development theory in *Vestiges* fit with either of these models?

3. What insights has this book borrowed from Paul Tillich's notion of "systematic theology"?

4. What was the "Common Context" approach to the history of nineteenth-century science, and what are the limitations of that approach?

5. On what basis does the author conclude that the first two chapters of Genesis did not have a strong influence upon the relationship between theology and the natural sciences in nineteenth century America? What, instead, was the driving influence upon the relationship between theology and the natural sciences?

6. Define "socially situated systematic theology" and evaluate how useful this concept is for explaining how nineteenth-century Americans responded to the development theory in *Vestiges*.

Appendices

Appendix A: Reviews of Vestiges *in American Periodicals and Pamphlets*[1]

A[llen], J[oseph] H[enry]. "Vestiges of Creation and Sequel." *Christian Examiner and Religious Miscellany* 4th ser., 5, no. 3 (May 1846): 333–49.

[Allen, William H.] "Vestiges of the Natural History of Creation; Explanations: A Sequel to 'Vestiges of the Natural History of Creation.'" *Methodist Quarterly Review* 3rd ser., 6 (Apr. 1846): 292–327.

"The Author of the *Vestiges of the Natural History of Creation*." *American Whig Review* 3 (Feb. 1846): 168–79.

[Bowen, Francis]. "A Theory of Creation." *North American Review* 60 (Apr. 1845): 426–78.

Bowen, Francis. *A Theory of Creation.* Rpt. from the *North American Review.* Boston: n.p., 1845.

[Brewster, David]. "Vestiges of the Natural History of Creation." *Living Age* 6 (20 Sep. 1845): 564–82.

[Dana, James Dwight]. "Vestiges of the Natural History of Creation." *American Journal of Science* 48, no. 2 (Mar. 1845): 395.

[———]. "Sequel to the Vestiges of Creation." *American Journal of Science* 2d ser., 1, no. 2 (Mar. 1846): 250–54.

[Dod, Albert B.]. "Vestiges of the Natural History of Creation." *Princeton Review* 17, no. 4 (Oct. 1845): 505–57.

"Explanations: A Sequel to 'Vestiges of the Natural History of Creation,' by the Author of That Work." *(New York) Churchman* 15 (7 Feb. 1846): 195.

"Explanations: A Sequel to 'Vestiges of the Natural History of Creation,' by the Author of That Work." *Southern Literary Messenger* 12, no. 3 (Mar. 1846): 191.

"Explanations: A Sequel to 'Vestiges of the Natural History of Creation.'" *Living Age* 8 (7 Mar. 1846): 442–45.

[Gray, Asa]. "Explanations: A Sequel to 'Vestiges of the Natural History of Creation.'" *North American Review* 62 (Apr. 1846): 465–506.

---

1. This list of American reviews is more comprehensive than that found in Robert Chambers, Vestiges of the Natural History of Creation, and Other Writings, ed. James Secord (Chicago: University of Chicago Press, 1992), appendix C. Secord's appendix also lists many British reviews. No American reviews have appeared in modern reprint, though several British reviews may be found in John M. Lynch, ed., "Vestiges" and the Debate before Darwin, 7 vols. (Bristol: Thoemmes, 2000). The Making of America Project, an online database of nineteenth century American literature hosted by the University of Michigan (http://moa.umdl.umich.edu/) and Cornell University (http://cdl.library.cornell.edu/moa/) provided an efficient means for quickly locating a large number of references to Vestiges in periodicals, pamphlets, and books (accessed throughout 2001–2003). Those cited here, and in the bibliography (which is restricted to works cited in this book), represent only a sampling of the known literature that addressed Vestiges during the nineteenth century.

[Lewis, Tayler]. "Vestiges of the Natural History of Creation." *American Whig Review* 1, no. 5 (May 1845): 525–43.

[Mantell, Gideon Algernon]. "Vestiges of the Natural History of Creation." *American Journal of Science* 49, no. 1 (June 1845): 191.

"Monthly Literary Bulletin." *U.S. Democratic Review* 16, no. 80 (Feb. 1845): 202–7.

Remarks upon a Recent Work Published by Wiley and Putnam of New York, Entitled The Natural History of the Vestiges of Creation [sic]. Philadelphia: Carey and Hart, 1846.

Review of *Explanations: a Sequel to the Vestiges of the Natural History of Creation*, by the author of that work, *Nassau Monthly* 6, no. 7 (May 1847): 217–28.

Review of *Vestiges of the Natural History of Creation, American Whig Review* 1, no. 2 (Feb. 1845): 215.

Smith, John Augustine. *The Mutations of the Earth; Or An Outline of the More Remarkable Physical Changes, of which, in the Progress of Time, This Earth Has Been the Subject, and the Theatre: Including an Examination into the Scientific Errors of the Author of the Vestiges of Creation*, 1846 Anniversary Discourse, Lyceum of Natural History of New York. New York: Bartlett and Welford, 1846.

[Strong, Edward]. "Vestiges of Creation and Its Reviewers." *New Englander* 4 (Jan. 1846): 113–27.

Taylor, George. "Theories of Creation and the Universe," review of *Vestiges* and *Explanations, Debow's Review* 4, no. 2 (Apr. 1847): 177–94.

"Vestiges of the Natural History of Creation." *Living Age* 4, no. 34 (4 Jan. 1845): 60–64.

"Vestiges of the Natural History of Creation." *New York Journal of Medicine and the Collateral Sciences* n.v. (Mar. 1845): 269.

"Vestiges of the Natural History of Creation." *Southern Literary Messenger* 11, no. 4 (Apr. 1845): 255.

"Vestiges of the Natural History of Creation." *United States Catholic Magazine and Monthly Review* 6, no. 5 (May 1847): 229–57.

Whewell, William. *Indications of the Creator: Extracts, Bearing upon Theology, from the History and the Philosophy of the Inductive Sciences*. Philadelphia: Carey & Hart, 1845.

[Whelpley, James Davenport]. "A Sequel to 'Vestiges of the Natural History of Creation.'" *American Whig Review* 3 (Apr. 1846): 383–96.

Appendix B: Nineteenth-Century American Reprints of Vestiges

Year	City	Publisher	Editions[2]	Supplements
1845	New York	Wiley & Putnam	V1	
1845	New York	Wiley & Putnam	V3	Introduction by George Cheever
1845	New York	Wiley & Putnam	V3	*North British Review* article, by David Brewster
1845	Philadelphia	Carey & Hart	E1	*Indications of the Creator*, by William Whewell
1846	New York	Wiley & Putnam	V	
1846	New York	Wiley & Putnam	V2	
1846	New York	Wiley & Putnam	V3	*North British Review* article, by David Brewster
1847	New York	Harper & Bros.	V	
1847	New York	Harper & Bros.	V, E	
1848	New York		V	
1848	New York	Coyler	V3, E	*North British Review* article, by David Brewster
1848	New York	Wiley & Putnam	V3, E	*North British Review* article, by David Brewster
1852	Cincinnati	James	V, E	
1853	New York		V	
1854	New York	Harper & Bros.	V, E	
1857	New York	Harper & Bros.	V, E	
1858	Cincinnati	James	V, E	
1858	New York	Harper & Bros.	V, E	
1859	New York	Harper & Bros.	V, E	
1862	New York	Harper & Row	V, E	
1868	New York	Harper & Row	V, E	
1872	New York	Harper & Row	V, E	
1875	New York		V	
1884	New York	R. Worthington	V12	Introduction by Alexander Ireland, identifying Robert Chambers as the author
1890	New York	G. Routledge	V2	Introduction by Henry Morely

2. V=*Vestiges*; E=*Explanations*, the sequel volume; numerals indicate the British edition, if known, that was reprinted; all American reprints until the 1884 ed. are believed to have been unauthorized.

Appendix C: Manuscript Conventions

Quotations from manuscripts preserve original capitalization, underlining, and spelling, with nonstandard spelling often being affirmed by the use of "[*sic*]," lest it be confused for a typographical error. Original punctuation has also been retained, with the exception that dashes of varying lengths have been replaced by a standard em dash ("—"). Superscripted abbreviations, such as "M^r," are reproduced with normal characters, thus: "Mr." Interlined text is surrounded by angled brackets ("<" and ">"), and deletions are indicated by a strikethrough line ("~~deleted text~~"). When appropriate, comments in the body of this book or else in a footnote indicate that some insertions or deletions were made by a second hand or by the same hand but in a different ink. An ellipsis (". . .") indicates that a portion of text present in the manuscript has been omitted in the quotation; in no cases did a manuscript itself have an ellipsis. Unless indicated in the footnotes to be otherwise, text appearing between square brackets ("[" and "]") has been inserted to clarify or else to expound upon the original text in the manuscript.

Archival organization practices vary widely, even among branch libraries at a single university. The citation format employed in this book represents a compromise between the peculiar citation formats preferred by each archival library and that recommended in *The Chicago Manual of Style*.

> **Letters:** Sender's Name to Recipient's Name, Date (Collection Title, Library Name, Box, Folder, and Item Numbers [if applicable]).

> **Other Documents:** Author's Name, "Document Title" or Description, Date (Collection Title, Library Name, Box, Folder, and Item Numbers [if applicable]), Page Number [if applicable].

Subsequent citations of the same document will be abbreviated when this can be done unambiguously. Since most of the manuscripts bear no clear title themselves, the descriptive title supplied by the archivists generally will be used; in that case, it will not appear in quotation marks. Dates appearing in brackets indicate that either an archivist or the researcher has dated a manuscript that itself contains no date. Further information on that dating process is provided in the notes when applicable. Collection numbers (when applicable) are specified in the bibliography under "Archives Consulted."

BIBLIOGRAPHIES

Archival Collections Consulted

Academy of Natural Sciences at Philadelphia
- William Parker Foulke Papers
- Samuel S. Haldeman Papers
- Joseph Leidy Papers

Amherst College: Robert Frost Library
- Edward and Orra White Hitchcock Papers
- Edward, Jr., and Mary Judson Hitchcock Papers

Academy of Natural Sciences at Philadelphia Library
- William Parker Foulke Papers
- Samuel S. Haldeman Papers
- Joseph Leidy Papers

American Philosophical Society Library
- Louis Jean Randolph Agassiz Papers (B/Ag12.ms)
- Broadside Collection
- Joseph Carson Papers (B/C239)
- Edward Waller Claypole Papers pertaining to Geology (557.3 C57)
- Darwin–Lyell Correspondence (B/D25.L)
- Darwin–Lyell Papers (B/D25.L1)
- William Parker Foulke Papers (B/F826)
- John Fries Frazer Papers (B/F865)
- Asa Gray Papers (B/G78)
- John Thomas Gulick Papers (B/G96)
- Arnold Henry Guyot Papers (B/G98)
- Samuel S. Haldeman Papers (B/H 129)
- Robert Hare papers (B/H22)
- LeConte Family Papers (B/L493f)
- John L. LeConte Papers (B/L493)
- J. Peter Lesley Papers (B/L56 and H.S. Film 12)
- Leo Lesquereux Autobiography (B/L567)

- Letters of Scientists Papers (509/L56)
- Sir Charles Lyell Papers (B/L981)
- Miscellaneous Manuscript Collection
- Maria Mitchell Papers (HS Film 9)
- Moore Autograph Collection (B/M781)
- Samuel George Morton Papers (B/M843)
- Notes taken from Professor Olmsted's lectures on natural philosophy (B/OL5)
- Ord-Waterton Correspondence (B/Or2)
- Sir James Paget Collection (B/P212)
- Caspar Wistar Pennock Papers (B/P3825)
- Benjamin Silliman Papers (B/Si4)
- John Torrey Papers (B/T63)
- John Warner Papers (B/W243)
- Windsor Literary Society Papers (806/W72)
- Joseph Winlock Letterbook (B/721)
- Chauncey Wright Papers (B/W933)
- John Howard Wurtz [sic: Wurts], Lectures on natural history (504/W95)
- Aaron Young, Jr., Papers (B/Y81)

Harvard University: Andover-Harvard Divinity School Library
- Pamphlets Collection
- Edward Amasa Park Lectures (bMS 406, bMS 466)
- Faculty Writings File (bMS 13001)
- Theodore Parker Papers (bMS 101)
- Records relating to the Dudleian Lectures (bMS 523)

Harvard University: Asa Gray Herbarium
- Asa Gray Papers

Harvard University: Countway Library of Medicine
- Jeffries Wyman Papers (H MS c12)

Harvard University: Harvard University Archives at Pusey Library
- Biographical Files (HUG 300)
- Divinity School Essays (HUG 88xx)
- Dudlein Lectures (HUC 5340)
- Francis Bowen Papers (HUG 1232)
- Jason Martin Gorham, Diary (HUD 848.34)
- Asa Gray Papers (HUG 1435)
- Harvard Natural History Society Papers (HUD 3599)
- T. Wentworth Higginson, Notes in Divinity School (HUC 8847.386)
- Lecture Notes (HUG 86xx)
- Letters to the Treasurer (UA I 50.8.VT)
- Theodore Lyman, A Digest of Two Courses and a Half of Lectures (HUC 8857.398)
- John Mead, John Mead's Journal (HUD 848.54.5)

- Records of the Theological School (UA V 328.4 VT)
- Reports to the Board of Overseers (UA II 10.5)
- Senior Exhibitions and Commencement Exercises (HUC 68xx)
- Edmund Quincy Sewall, Diary of Edmund Quincy Sewall (HUD 944.80)
- Richard Alan Shapiro, "The Rumford Professorship," B.A. thesis (HU 92 85 787)
- John Langdon Sibley, Private Journal (HUG 1791.72 VT)
- John Austin Stevens, Letters as a Student (HUD 842.83)
- Themes (HUC 88xx)
- Papers of James Walker: Sermons (UAI 15.888)

Harvard University: Houghton Library
- Louis Agassiz Papers (bMS Am 1419)
- Francis Bowen Correspondence (bMS Am 972)
- North American Review Papers (bMS Am 1704.10)
- Benjamin Peirce Papers
- Henry David Thoreau, Extracts mostly upon Natural History (HEW 12.7.10)

Harvard University: Museum of Comparative Anatomy, Ernst Mayr Library
- Louis Agassiz Correspondence and Manuscripts (bAg 15)

Historical Society of Pennsylvania
- Samuel George Morton Manuscripts (Yi 2, 7388–7390)

New York Botanical Garden Library
- John Torrey Papers

Presbyterian Historical Society
- George Musgrave Giger Papers (RG 285)

Princeton Theological Seminary: Luce Library
- Archibald Alexander Manuscript Collection
- Archibald Alexander Hodge Manuscript Collection
- Charles Hodge Manuscript Collection
- William H. Green Collection
- James [=Joseph] Addison Alexander Collection
- Samuel Miller Manuscript Collection
- Samuel Miller, Jr., Manuscript Collection

Princeton University: Dept. of Manuscripts and Archives at Firestone Library
- John R. Buhler, Diary, General Manuscript Collection (CO#199)
- Cameron Family Papers (CO#355)
- Eli Field Cooley Papers (CO#410).
- Duffield Family Papers (CO#421)
- General Manuscripts, Miscellaneous Collection (CO#140)
- Ashbel Green Collection (CO#257)

- John Miller Papers (CO#632)
- Charles Woodruff Shields Papers (CO#343)
- Theodore W. Tallmadge Papers (CO#353)

Princeton University: Mudd Manuscript Library
- Autograph Collection (AC#040)
- Board of Trustees Minutes and Records
- Faculty Files
- Lecture Notes Collection (AC#052)
- Office of the President Records (AC#117)
- Offprint Collection (AC#121)
- Scrapbook Collection (AC#026)
- Undergraduate Alumni Collection (AC#104)

Smithsonian Institution Archives
- Spencer Fullerton Baird Papers (RU 7002)
- Joseph Henry Papers (RU 7001)
- Office of the Secretary Papers (RU 43)

Yale University: Cushing Medical Historical Library
- Yale Medical Dept. M.D. Dissertations (T 113 Y11)

Yale University: Divinity School Library
- Horace Bushnell Papers (RG 39)
- Eleazar Thompson Fitch Papers (RG 93)

Yale University: University Archives at Sterling Library
- Records of the Calliopean Society (YRG 40, RU 857)
- Records of Clubs, Societies, and Organizations (YRG 40, RU 56)
- Commencement Orations and Poems (YRG 47, RU 140)
- Collection of Compositions (YRG 47, RU 331)
- Course Lectures (YRG 47, RU 159)
- Dana Family Papers (MGN 164 and HM 160 microfilm)
- Yale Student Diaries (YRG 41, RU 861)
- Gibbs Family Papers (MGN 236)
- Goodrich Family Papers (MGN 242)
- Edward C. Herrick Papers (MGN 691)
- Linonian Society Records (YRG 40, RU 206)
- Elias Loomis Family Papers (MGN 331)
- Natural Science Manuscripts (MGN 583)
- John Pitkin Norton Papers (MGN 367)
- Papers on Yale Collections (YRG 47–F, RU 103)
- Play Collection (YRG 47, RU 119)
- Poetry Collection (YRG 47, RU 377)
- Noah Porter Papers (MGN 1131)

- Rhetorical Society of the Theological Dept. Papers (YRG 40, RU 40)
- Silliman Family Papers (MGN 450 and HM 140 microfilm)
- Wayland Family Papers (MGN 1067)

Articles Cited: Unknown Authors[1]

"Agassiz's Natural History." *Atlantic Monthly* 1, no. 3 (Jan. 1858): 320–33.
"Agreement of Science and Revelation." *Nassau Monthly* 6, no. 6 (Apr. 1847): 181–85.
"The American Journal of Science." *Southern Literary Messenger* 9, no. 2 (Feb. 1853): 126.
"Association of American Geologists and Naturalists." *New York Herald*, 9 May 1845, 1.
"The British House of Commons." *New York Times*, 9 Dec. 1852, 4.
"Darwin on the Origin of Species." Review of *On the Origin of Species*, by Charles Darwin. *North American Review* 90, no. 187 (Apr. 1850): 474–507.
"Darwin on the Origin of Species." Review of ten works, rpt. from the *Edinburgh Review*. *Living Age* 66, no. 840 (7 July 1860): 3–26.
"Errata." *North American Review* 87, no. 181 (Oct. 1858): 572.
"European Correspondence." *Southern Literary Messenger* 11, no. 5 (May 1845): 323–26.
"The Evils of Unsanctified Literature." *Princeton Review* 15, no. 1 (Jan. 1843): 65–77.
"The Extent of the Atonement." 2 pts. *Presbyterian Magazine* 4, no. 11 (Nov. 1854): 496–502, 4, no. 12 (Dec. 1854): 533–40.
"Fish and Fishermen." *Debow's Review* 15, no. 2 (Aug. 1853): 143–60.
"Geological Paper." *Nassau Monthly* 8, no. 3 (Nov. 1848): 81–88.
"Joseph Henry Allen." *Boston Transcript*, 21 Mar. 1898.
"Journalism in the United States." *Southern Quarterly Review* 3, no. 6 (Apr. 1851): 500–18.
"Literary and Other Items." *New York Times*, 14 Apr. 1859, 12.
"Miscellaneous." *New York Times*, 14 Apr. 1859, 12.
"The New Movement in Philosophy." *New York Herald*, 1 Nov. 1847, 2.
"New Publications Received." *North American Review* 82, no. 490 (Aug. 1898): 288–92.
"Origin of Darwinism." *Manufacturer and Builder* 8, no. 5 (May 1876): 119.
"Papers, Critical, Exegetical, and Philosophical." *Ladies' Repository* 17, no. 1 (Jan. 1857): 53–55.
"Personal." *New York Times*, 8 Jan. 1859, 1.
"Personal." *New York Times*, 23 Apr. 1859, 11.
"Philosophy of Natural History." *North American Review* 19, no. 45 (Oct. 1824): 395–411.
"Potato Jelly," *Boston Daily Evening Transcript*, 2 Dec. 1845, n.p.
"The Present State of Oxford University." *Princeton Review* 26, no. 3 (July 1854): 409–36.
Review of *Chambers' Miscellany of Useful and Entertaining Knowledge*, ed. by William Chambers. *Debow's Review* 5, no. 4 (Apr. 1848): 399.
Review of *A Discourse on the Studies of the University of Cambridge*, by Adam Sedgwick. *American Journal of Science* 2d ser., 11 (May 1851): 144–46.
Review of *Four Lectures on Spiritual Christianity*, by Isaac Taylor. *North American Review* 61, no. 128 (July 1845): 159–81.
Review of *The Lives of the Brothers Humboldt, Alexander and William,* by Juliette Bauer. *Princeton Review* 25, no. 2 (Apr. 1853): 324–26.
Review of *Nature and the Supernatural,* by Horace Bushnell. *New Englander* 17 (Feb. 1859): 224–58.
Review of *On the Origin of Species*, by Charles Darwin. *Dial* 1, no. 3 (Mar. 1860): 196–97.
Review of *On the Relation between the Holy Scriptures and Some Parts of Geological Science*, by John Pye Smith, *Princeton Review* 13, no. 3 (July 1841): 368–94.

1. For reviews of *Vestiges* in American periodicals, see Appendix A.

Review of *The Six Days of Creation,* by Tayler Lewis. *New York Times,* 16 June 1855, 2.
Review of *The Soul,* by George Bush. *Princeton Review* 18, no. 2 (Apr. 1846): 219–60.
Review of *The True Intellectual System of the Universe,* by Ralph Cudworth. *Christian Examiner* 27, no. 3 (Jan. 1840): 289–319.
"The Study of Nature: Lecture by Prof. Agassiz." *Boston Weekly Courier* (26 Mar. 1859).
"The Truth of Christianity." *North American Review* 63, no. 133 (Oct. 1846): 382–432.
"The Value of History." *Nassau Monthly* 8, no. 1 (Sep. 1848): 12–14.
[Untitled]. *New York Times,* 21 Jan. 1859, 1.
"Webber, Samuel." *Appleton's Cyclopaedia of American Biography.* Ed. James Grant Wilson and John Fiske, vol. 6, p. 405. 7 vols. New York: D. Appleton and Company, 1886–1900.
"Whelpley, Philip Melanchthon." In *Appleton's Cyclopaedia of American Biography.* Ed. James Grant Wilson and John Fiske, vol. 6, p. 459. 7 vols. New York: D. Appleton and Company, 1886–1900.

Articles Cited: Known Authors

A[gassiz], L[ouis]. "Contemplations of God in the Kosmos." *Christian Examiner* 4th ser., 15 (Jan. 1851).
Aldrich, Michele L. "Hitchcock, Edward." In *Dictionary of Scientific Biography.* Ed. Charles Coulston Gillispie, vol. 6, pp. 437–38. New York: Charles Scribner's Sons, 1972.
[Allen], J. S. Review of *An Essay On the Philosophy of Medical Science,* by Elisha Bartlett. *Southern Literary Messenger* 11, no. 6 (June 1845), 330–340.
Appel, Toby A. "Jeffries Wyman, Philosophical Anatomy, and the Scientific Reception of Darwin in America." *Journal of the History of Biology* 21, no. 1 (Spring 1988): 69–94.
Arther, Donald E. "Paul Tillich's Perspectives on Ways of Relating Science and Religion." *Zygon* 36, no. 2 (June 2001): 261–67.
Baatz, Simon. "'Squinting at Silliman': Scientific Periodicals in the Early American Republic, 1810–1833." *Isis* 82 (1991): 223–44.
Black, William. "White Wings: A Yachting Romance." Chap. 5 of a serialized novel. *Harper's New Monthly Magazine* 59, no. 352 (Sep. 1879): 555–68.
[Bowen, Francis]. Review of *Ancient Sea-Margins, or Memorials of Changes in the Relative Level of Sea and land,* by Robert Chambers, and *A Memoir upon the Geological Action of the Tidal and Other Currents of the Ocean,* by Charles Henry Davis. *North American Review* 69, no. 144 (July 1849): 246–69.
[– – –]. Review of *On the Origin of Species,* by Charles Darwin. *North American Review* 90, no. 187 (Apr. 1860): 474–506.
Bowler, Peter J. "The Changing Meaning of 'Evolution.'" *Journal of the History of Ideas* 36, no. 1 (1975): 95–114.
Brooke, John Hedley. "Richard Owen, William Whewell, and the *Vestiges.*" *British Journal for the History of Science* 10, no. 35 (1977): 132–45.
Brown, Sanborn C. "Thompson, Benjamin (Count Rumford)." *Dictionary of Scientific Biography.* Ed. Charles Coulston Gillispie, vol. 13, pp. 350–51. New York: Charles Scribner's Sons, 1976.
Cabot, J. Elliot. "The Life and Writings of Agassiz." *Massachusetts Quarterly Review* 1, no. 1 (Dec. 1847): 96–119.
Clarke, Thomas William. "The Unity of the Human Race." *Harvard Monthly* 1, no. 7 (July 1855): 337–40.
Croce, Paul Jerome. "Probabilistic Darwinism: Louis Agassiz vs. Asa Gray on Science, Religion, and Certainty." *The Journal of Religious History* 22, no. 1 (Feb. 1998): 35–58.
Dana, James Dwight. "Anticipations of Man in Nature." *New Englander* 17 (May 1859): 293–334, at 308, 309, 295.
– – –. "Science and the Bible." 4 pts. *Bibliotheca Sacra* 13, no. 49 (Jan. 1856): 80–129; 13, no. 51 (July 1856): 631–56; 14, no. 56 (Apr. 1857): 388–413; 14, no. 57 (July 1857): 461–524.

– – –. "Thoughts on Species." *American Journal of Science* 2d ser., 24 (Nov. 1857): 305–16.

– – –. "Thoughts on Species." *Bibliotheca Sacra* 14, no. 56 (Oct. 1857): 854–74.

Delacroix, Jacques, and Nielsen, François. "The Beloved Myth: Protestantism and the Rise of Industrial Capitalism in Nineteenth-Century Europe." *Social Forces* 80 (2001): 509–53.

Farrelly, Maura Jane. "'God Is the Author of Both': Science, Religion, and the Intellectualization of American Methodism." *Church History* 77, no. 3 (Sep. 2008): 659–87.

Giltner, John H. "Genesis and Geology: The Stuart–Silliman–Hitchcock Debate." *Journal of Religious Thought* 23, no. 1 (1996–1967): 3–13.

Gladstone, W. E. "Ecce Homo." Review of *Ecce Homo: A Survey of the Life and Work of Jesus Christ. Living Age* 96, no. 1235 (1 Feb. 1868): 259–66.

Gray, Julian C. "Finding the Right Mineralogy Text: Dana's System of Mineralogy," *Tips and Trips*, 28, no. 4 (Apr. 1999): 7. Available online: *www.gamineral.org/Dana-system.htm.*

Habich, Robert D. "Emerson's Reluctant Foe: Andrews Norton and the Transcendental Controversy." *New England Quarterly* 65, no. 2 (June 1992): 208–37.

Harrison, James. "Erasmus Darwin's Views of Evolution." *Journal of the History of Ideas* 32 (1971): 247–64.

Hayden, C. B. "A Resume of Geology." *Southern Literary Messenger* 12, no. 11 (Nov. 1846): 658–671.

Hilton, Boyd. "The Politics of Anatomy and an Anatomy of Politics c. 1825–1850." In *History, Religion, and Culture: British Intellectual History 1750–1950.* Ed. Stefan Collini, Richard Whatmore, and Brian Young, 179–97. Cambridge: Cambridge University Press, 2000.

Hodge, M. J. S. "The Universal Gestation of Nature: Chambers' *Vestiges* and *Explanations.*" *Journal of the History of Biology* 5, no. 1 (1972): 127–51.

Howe, Daniel Walker. "The Cambridge Platonists of Old England and the Cambridge Platonists of New England." *Church History* 57, no. 4 (Dec. 1988): 470–85.

Hunt, W. Holman. "The Pre-Raphaelite Brotherhood: A Fight for Art." *Living Age* 170, no. 2195 (17 July 1886): 131–39.

Johns, Adrian. "Science and the Book in Modern Cultural Historiography." *Studies in History and Philosophy of Science* 29, no. 2 (1998): 167–94.

Kendall, Martha B. "Lesley, J. Peter." *Dictionary of Scientific Biography.* Ed. Charles Coulston Gillispie. Vol. 8, pp. 260–61. New York: Charles Scribner's Sons, 1973.

King, Horatio C. "Dickinson College," *The American University Magazine*, 6, no. 1 (Apr.-May 1897), 4–33.

LeConte, Joseph. "Evolution in Relation to Materialism." *Princeton Review* 1 (Jan.-June 1881): 149–74.

Leonard, David Charles. "Tennyson, Chambers, and Recapitulation." *Victorian Newsletter* 56 (1979): 7–10.

Livingstone, David N., and Noll, Mark A. "B. B. Warfield (1851–1921): A Biblical Inerrantist as Evolutionist." *Isis* 91 (2000): 283–304.

Looney, J. Jefferson. "'An Awfully Poor Place': Edward Shippen's Memoir of the College of New Jersey in the 1840s," *Princeton University Library Chronicle* 59 (Autumn 1997): 9–14.

[Lord, David N.] "A Letter to the Corporation of Yale College, on the Doctrines of the Theological Professors in that Institution." *Views in Theology* 4, no. 16 (May 1835): 291–341.

MacPherson, Ryan Cameron. "When Evolution Became Conversation: *Vestiges of Creation,* Its Readers, and Its Respondents in Victorian Britain." Essay review of *Victorian Sensation,* by James A. Secord, and *"Vestiges" and the Debate before Darwin,* ed. John M. Lynch. *Journal of the History of Biology* 34, no. 3 (2001): 565–79.

– – –. "Natural and Theological Science at Princeton, 1845–1859: *Vestiges of Creation* Meets the Scientific Sovereignty of God," *Princeton University Library Chronicle* 65, no. 2 (2004): 184–235.

Mayr, Ernst. "Agassiz, Darwin, and Evolution." *Harvard Library Bulletin* 13 (1959): 165–94.

McGreevy, John T. "Faith and Morals in the United States, 1865–Present." *Reviews in American History* 26, no. 1 (Mar. 1998): 239–49.

[Minor, Benjamin Blake]. "Notice of New Works." *Southern Literary Messenger* 12, no. 3 (Mar. 1846): 189–92.

Moorhead, James H. "The 'Restless Spirit of Radicalism': Old School Fears and the Schism of 1837," *Journal of Presbyterian History* 78, no. 1 (Spring 2000): 19–33.

DEBATING EVOLUTION BEFORE DARWINISM

Nartonis, David K. "Locke–Stewart–Mill: Philosophy of Science at Dartmouth College, 1771–1854." *International Studies in the Philosophy of Science* 15, no. 2 (2001): 167–75.

Noll, Mark A. "Common Sense Traditions and American Evangelical Thought." *American Quarterly* 37, no. 2 (Summer 1985): 216–38.

Priest, Robert J. "Christian Theology, Sin, and Anthropology." Chap. 2 in *Explorations in Anthropology and Theology*. Ed. Frank A. Salamone and Walter Randolph Adams. Lanham, MD: University Press of America, 1997.

Rand, Benjamin. "Philosophical Instruction in Harvard University from 1636 to 1906." *Harvard Graduates' Magazine* 37 (Dec. 1928): 188–200.

Reese, Ronald Lane, Everett, Steven M., and Craun, Edwin D. "'In the Beginning . . .': The Ussher Chronology and Other Renaissance Ideas Dating the Creation." *Archaeoastronomy* 5, no. 1 (1982): 20–23.

Robinson, David. "The Road Not Taken: From Edwards, through Chauncy, to Emerson." *Arizona Quarterly* 48, no. 1 (Spring 1992): 45–60.

Russell, Robert John. "The Relevance of Tillich for the Theology and Science Dialogue." *Zygon* 36, no. 2 (June 2001): 269–308.

Schweitzer, Ivy. "Transcendental Sacramentals: 'The Lord's Supper' and Emerson's Doctrine of Form." *New England Quarterly* 61, no. 3 (Sep. 1988): 398–418.

Shapin, Steven. "Understanding the Merton Thesis." *Isis* 79 (Dec. 1988): 594–605.

Sherwood, Morgan B. "Genesis, Evolution, and Geology in America before Darwin: The Dana-Lewis Controversy." In *Toward a History of Geology*, Proceedings of the Inter-Disciplinary Conference on the History of Geology, September 7–12, 1967. Ed. Cecil J. Schneer, pp. 305–16. Cambridge, MA: The MIT Press, 1967.

Shippen, Edward. "Some Notes about Princeton," ed. J. Jefferson Looney, *Princeton University Library Chronicle* 59 (Autumn 1997): 15–57.

Sloan, Phillip R. "Lamarck in Britain: Transforming Lamarck's Transformism." In *Jean-Baptiste Lamarck, 1744–1826*. Ed. Goulven Laurent, 677–687. Paris: CTHS Publications, 1997.

– – –. "Whewell's Philosophy of Discovery and the Archetype of the Vertebrate Skeleton: The Role of German Philosophy of Science in Richard Owen's Biology." *Annals of Science* 60 (2003): 39–61.

Smucker, Isaac. "Who Were the Aboriginals of North America?" *Ladies' Repository* 22, no. 4 (Apr. 1862): 239–43.

Stanton, William. "Dana, James Dwight," in *Dictionary of Scientific Biography*. Ed. Charles Coulston Gillispie, vol. 3, pp. 549–54. New York: Charles Scribner's Sons, 1971.

Tester, Keith. "Between Sociology and Theology: The Spirit of Capitalism Debate." *The Sociological Review* 48, no. 1 (Feb. 2000): 43–57.

Thigpen, Thomas Paul. "Bushnell's Rejection of Darwinism." *Church History* 57 (1988): 499–513.

Topham, Jonathan R. "Beyond the 'Common Context': The Production and Readings of the Bridgewater Treatises." *Isis* 89 (1998): 233–62.

Turner, James. "Le concept de science dans l'Amérique du XIXe siècle." *Annales* 57, no. 3 (May–June 2002): 753–72.

Tyndall, John. "Inaugural Address of Professor John Tyndall." Rpt. from *Nature*. *Living Age* 122, no. 1581 (26 Sept. 1874): 802–24.

Valeri, Mark. "Religion, Discipline, and Economy in Calvin's Geneva." *Sixteenth Century Journal* 28 (Spring 1997): 123–42.

W., R. R. Review of *An Inquiry into the Origin, Development, and Transmission of the Games of Childhood*. *Galaxy* 11, no. 2 (Feb. 1871): 314–15.

Wallace, Peter, and Noll, Mark. "The Students of Princeton Seminary, 1812–1929: A Research Note." *American Presbyterians* 72, no. 3 (Fall 1994): 203–15.

Walls, Laura Dassow. "Textbooks and Texts from the Brooks: Inventing Scientific Authority in America." *American Quarterly* 49, no. 1 (1997): 1–25.

Whalen, Robert. "Calvinism and Chiliasm: The Sociology of Nineteenth Century American Millenarianism." *American Presbyterians* 70, no. 3 (Fall 1992): 163–72.

–––. "Eleazar Lord and the Reformed Tradition: Christian Capitalist in the Age of Jackson." *American Presbyterians* 72, no. 4 (Winter 1994): 219–28.

–––. "Genesis, Geology, and Jews: The New York Millenarians of the Antebellum Era." *American Presbyterians* 73, no. 1 (Spring 1995): 9–22.

Wedgwood, Julia. "John Ruskin." *Living Age* 225, no. 2909 (7 Apr. 1900): 1–8.

Welch, R. B. "The Modern Theory of Forces." *Princeton Review* 4, no. 13 (Jan. 1875): 28–38.

[Whelpley, James Davenport]. "Idea of an Atom, Suggested by the Phenomena of Weight and Temperature." *American Journal of Science* 48, no. 2 (Jan.–Mar. 1845): 352–68.

[Wilder, Daniel Webster]. "Anonymous Books." *Harvard Magazine* 1, no. 1 (Dec. 1854): 22–24.

Williams, Henry Smith. "The Century's Progress in Biology." *Harper's* 95, no. 570 (Nov. 1897): 930–42.

Wilson, Philip K. "Arnold Guyot (1807–1884) and the Pestalozzian Approach to Geology Education." *Eclogae Geologicae Helvitiae* 92 (1999): 321–25.

Winchell, Alexander. "Voices from Nature." *Ladies' Repository* 23, no. 10 (Oct. 1863): 625–28.

Woodall, Guy R. "George Barrell Cheever." In *Dictionary of Literary Biography*. Ed. John W. Rathbun and Monica M. Grecum, vol. 59, pp. 72–79. Detroit: Bruccoli Clark Layman Book, 1987.

Wright, Conrad. "The Religion of Geology." *New England Quarterly* 14 (June 1941): 334–58.

Youmans, E. L. "Spencer's Evolution Philosophy." *North American Review* 129, no. 275 (Oct. 1879): 389–404.

Books and Pamphlets Cited: Unknown Authors[2]

Abstract of the Proceedings of the Sixth Annual Meeting of the Association of American Geologists and Naturalists, Held in New Haven, Conn., April, 1845. New Haven: B. L. Hamlen, 1845.

Addresses at the Inauguration of the Professors in the Theological Department of Yale College, September 15, 1861. New Haven: E. Hayes, 1861.

Catalogue of Books Belonging to the Linonian Society, Yale College, November 1846. New Haven: J. H. Benham, 1846.

Catalogue of the College of New Jersey, for 1840–41 through *Catalogue . . . 1858–59*. Princeton: John T. Robinson, 1858–1859.

Catalogue of the Library of the Society of Brothers in Unity, Yale College, April, 1846. New Haven: B. L. Hamden, 1846.

Catalogue of the Officers and Students of the Theological Seminary at Princeton, New Jersey, 1834–1835 through *Catalogue . . . 1850–1851*. Princeton: John T. Robinson, 1835–1851.

Catalogue of the Officers and Students in Yale College, 1852–53. Rev. ed. Springfield, MA: B. L. Thunderclap, 1852.

The Confession of Faith; the Larger and Shorter Catechisms, with the Scripture-Proofs at Large, Together with the Sum of Saving Knowledge. Edinburgh: D. Hunter Blair, 1845.

Discourses and Addresses at the Ordination of the Rev. Theodore Dwight Woolsey, LL.D., to the Ministry of the Gospel, and His Inauguration as President of Yale College, October 21, 1846. New Haven: Yale College, 1846.

Essays, Theological and Miscellaneous, Reprinted from the Princeton Review, Second Series, Including the Contributions of the Late Rev. Albert B. Dod, D.D. New York and London: Wiley and Putnam, 1847.

---

2. For American pamphlets that responded directly to *Vestiges* in the mid 1800s, see Appendix A.

Books and Pamphlets Cited: Known Authors

Agassiz, Louis. *Essay on Classification*. Ed. Edward Lurie. Cambridge, MA: Harvard University Press, 1962.

– – –. *An Introduction to the Study of Natural History, in a Series of Lectures Delivered in the Hall of the College of Physicians and Surgeons, New York*. New York: Greeley & McElrath, Tribune Building, 1847.

– – –. *Life, Letters, and Works of Louis Agassiz*. Ed. Jules Marcou. 2 vols. New York: Macmillan, 1896.

– – –. *Twelve Lectures on Comparative Embryology, Delivered before the Lowell Institute, in Boston, December and January, 1848–9*. Stenographic report by James W. Stone, originally for the *Boston Traveller*. Boston: Redding & Co.; Gould, Kendall, & Lincoln; James Munroe & Co.; New York: Dewitt & Davenport; Tribune Buildings; Philadelphia: G. B. Zieber & Co., 1849.

Ahlstrom, Sidney E., and Carey, Jonathan S., eds. *An American Reformation: A Documentary History of Unitarian Christianity*. Middletown, CT: Wesleyan University Press, 1985.

Alexander, Archibald. *A Brief Outline of the Evidences of the Christian Religion* Philadelphia: American Sunday School Union, 1825.

Allen, Joseph Henry. *Our Liberal Movement in Theology: Chiefly as Shown in Recollections of the History of Unitarianism in New England, Being a Closing Course of Lectures Given in the Harvard Divinity School*, 3d ed. Boston: Roberts Brothers, 1892.

Alter, Stephen G. *Darwinism and the Linguistic Image: Language, Race, and Natural Theology in the Nineteenth Century*. Baltimore and London: The Johns Hopkins University Press, 1999.

American Institute of the City of New York. *Alphabetical and Analytical Catalogue of the American Institute Library. With Rules and Regulations*. New York: W. L. S. Harrison, 1852.

Appel, Toby A. *The Cuvier-Geoffroy Debate: French Biology in the Decades before Darwin*. New York: Oxford University Press, 1987.

Bacon, Francis. *Novum Organon, with Other Parts of the Great Instauration*. Trans. and ed. Peter Urbach and John Gibson. Chicago: Open Court, 1994.

Barbour, Ian G. *Religion in an Age of Science: The Gifford Lectures, 1989–1991, Vol. 1*. San Francisco: Harper and Row, 1990.

Barnes, Albert. *Lectures on the Evidences of Christianity in the Nineteenth Century, Delivered in the Mercer Street Church, New York, January 21 to February 21, 1867, on the "Ely Foundation" of the Union Theological Seminary*. New York: Harper and Brothers, 1868.

Barton, William E. *The Soul of Abraham Lincoln*. New York: George H. Doran Company, 1920.

Bennett, Clinton. *In Search of the Sacred: Anthropology and the Study of Religions*. London: Cassell, 1996.

Blake, John Lewis. *A Biographical Dictionary* [etc.]. Philadelphia: H. Cowperthwait and Company, 1859.

Bowler, Peter J. *The Eclipse of Darwinism: Anti-Darwinian Evolution Theories in the Decades around 1900*. Baltimore: Johns Hopkins University Press, 1992.

– – –. *Evolution: The History of an Idea*. Rev. ed. Berkeley, CA: University of California Press, 1989.

– – –. *Fossils and Progress: Paleontology and the Idea of Progressive Evolution in the Nineteenth Century*. New York: Science History Publications, 1976.

Bowen, Francis. *Critical Essays, on a Few Subjects Connected with the History and Present Condition of Speculative Philosophy*. 2d ed. Boston: James Munroe and Company, 1845.

– – –. *The Principles of Metaphysical and Ethical Science Applied to the Evidences of Religion*. New ed., rev. and annotated, for the use of colleges. Boston: Brewer and Tileston, 1855.

– – –. *Remarks on the Latest Form of the Development Theory*. Rpt. from *Memoirs of the American Academy*, n.s., vol. 5. Cambridge: Welch, Bigelow, and Co., 1860.

Bozeman, Theodore Dwight. *Protestants in an Age of Science: The Baconian Ideal and Antebellum American Religious Thought*. Chapel Hill: University of North Carolina Press, 1977.

Brooke, John Hedley. *Science and Religion: Some Historical Perspectives*. Cambridge: Cambridge University Press, 1991.

Brooke, John, and Geoffrey Cantor. *Reconstructing Nature: The Engagement of Science and Religion*. Glasgow Gifford Lectures. Edinburgh: T and T Clark, 1998.

Brooke, John Hedley, Margaret J. Osler, and Jitse M. van der Meer, eds. *Science in Theistic Contexts: Cognitive Dimensions. Osiris* 16 (2001).

Brown, Chandos Michael. *Benjamin Silliman: A Life in the Young Republic.* Princeton, NJ: Princeton University Press, 1989.

Brown, Jerry Wayne. *The Rise of Biblical Criticism in America, 1800–1870: The New England Scholars.* Middletown, CT: Wesleyan University Press, 1969.

Bruce, Robert V. *The Launching of Modern American Science 1846–1876.* The Impact of the Civil War series. Ed. by Harold M. Hyman. New York: Alfred A. Knopf, 1987.

Buckland, William. *Geology and Mineralogy Considered with Reference to Natural Theology.* 2 vols. London: William Pickering, 1836.

Buffon, Georges Louis Leclerc, comte de. *Las épocas de la naturaleza.* Ed. and trans. Antonio Beltrán Marí. Madrid: Alianza, 1997.

———. *Les époques de la nature.* Crit. ed., with an introduction and notes by Jacques Roger. Paris: Éditions du Muséum, 1962.

Bushnell, Horace. *Christian Nurture.* Introduced by Luther A. Weigle. 1849. Rpt., New Haven: Yale University Press, 1967.

———. *God in Christ: Three Discourses Delivered at New Haven, Cambridge, and Andover.* Hartford, CT: Brown and Parsons, 1849.

———. *Nature and the Supernatural, As Together Constituting the One System of God,* 5th ed. 1st ed., 1858. New York: Charles Scribner, 1860.

Calhoun, Daniel, ed. *The Educating of Americans: A Documentary History.* Boston: Houghton Mifflin Company, 1969.

Calhoun, David B. *Princeton Seminary. Vol. 1: Faith and Learning, 1812–1868; vol. 2: The Majestic Testimony: 1869–1929.* Carlisle, PA: The Banner of Truth Trust, 1994.

Capper, Charles, and Conrad Edick Wright, eds. *Transient and Permanent: The Transcendentalist Movement and Its Contexts.* Boston: Massachusetts Historical Society, 1989.

Carlisle, Lord Bishop of. "The Gradual Development of Revelation." In *Modern Scepticism: A Course of Lectures Delivered at the Request of the Christian Evidence Society.* New York: Anson D. F. Randolph and Co., 1871.

Cashdollar, Charles D. *The Transformation of Theology, 1830–1890: Positivism and Protestant Thought.* Princeton, NJ: Princeton University Press, 1989.

Chambers, Robert. *Vestiges of the Natural History of Creation and Other Evolutionary Writings, Including Facsimile Reproductions of the First Editions of "Vestiges" and Its Sequel "Explanations."* Ed. with an introduction by James A. Secord. Chicago: Chicago University Press, 1994.

[———]. *Vestiges of the Natural History of Creation.* 2nd ed. from the 3rd London ed.[3] Intro by George B. Cheever. New York: Wiley and Putnam, 1845.

Conkin, Paul K. *The Uneasy Center: Reformed Christianity in Antebellum America.* Chapel Hill: University of North Carolina Press, 1995.

Cohen, H. Floris. *The Scientific Revolution: A Historiographical Inquiry.* Chicago: University of Chicago Press, 1994.

Cohen, I. Bernard, ed. *Puritanism and the Rise of Modern Science: The Merton Thesis.* With the assistance of K. E. Duffin and Stuart Strickland. New Brunswick: Rutgers University Press, 1990.

Cooke, Josiah Parsons, Jr. *Religion and Chemistry; or, Proofs of God's Plan in the Atmosphere and Its Elements. Ten Lectures Delivered at the Brooklyn Institute, Brooklyn, N.Y., on the Graham Foundation.* New York: C. Scribner, 1865.

Corsi, Pietro. *The Age of Lamarck: Evolutionary Theories in France, 1790–1830.* Rev. and updated. Trans. Jonathan Mandelbaum. Berkeley, CA: University of California Press, 1988.

Crowe, Michael J. *Modern Theories of the Universe from Herschel to Hubble.* New York: Dover, 1994.

3. A listing of nineteenth-century American reprints may be found in Appendix B.

Crouthamel, James L. *Bennet's New York Herald and the Rise of the Popular Press*. Syracuse, NY: Syracuse University Press, 1989.

Dana, James D. *Address before the American Association for the Advancement of Science, August, 1855*. n.p.: Joseph Lovering, 1855.

Dalzell, Jr., Robert F. *Enterprising Elite: The Boston Associates and the World They Made*. Cambridge, MA: Harvard University Press, 1987.

Daniels, George H. *American Science in the Age of Jackson*. New York: Columbia University Press, 1968.

Darwin, Erasmus. *Zoonomia; or, The Laws of Organic Life*. Facsimile rpt., with a new preface by Throm Verhave and Paul R. Bindler. 2 vols. 1794–1796. Rpt., New York: AMS Press, 1974.

Davies, Douglas J. *Anthropology and Theology*. Oxford: Berg, 2002.

Dawkins, Richard. *The Blind Watchmaker: Why the Evidence of Evolution Reveals a Universe without Design*. New York: Norton, 1986.

Deming, Clarence. *Yale Yesterdays*. New Haven: Yale University Press, 1915.

Desmond, Adrian. *Archetypes and Ancestors: Palaeontology in Victorian London, 1850–1875*. Chicago: University of Chicago Press, 1984.

–––. *The Politics of Evolution: Morphology, Medicine, and Reform in Radical London*. Chicago: University of Chicago Press, 1989.

Desmond, Adrian, and James Moore. *Darwin*. New York: Norton, 1992.

Douglas, Ann. *The Feminization of American Culture*. 2d ed., with a new introduction. New York: Knopf, 1977.

Draper, John William. *History of the Conflict between Religion and Science*. New York: D. Appletons and Company, 1874.

Dulles, Joseph H., ed. *Princeton Theological Seminary Biographical Catalogue, 1909*. Trenton, NJ: MacCrellish and Quigley Printers, 1909.

Dupree, A. Hunter. *Asa Gray, 1810–1888*. Cambridge, MA: Harvard University Press, 1959.

Dwinelle, John Whipple. *Address on the Acquisition of California by the United States*. San Francisco: Sterett and Cubery, 1866.

Elliott, Clark A., and Rossiter, Margaret W., eds. *Science at Harvard University: Historical Perspectives*. Bethlehem: Lehigh University Press, 1992.

Emerson, Ralph Waldo. *The Collected Works of Ralph Waldo Emerson, Vol. 1: Nature, Addresses, and Lectures*. Introductions and notes by Robert E. Spiller. Text established by Alfred R. Ferguson. Cambridge, MA: Harvard University Press, 1971.

–––. *The Journals and Miscellaneous Notebooks of Ralph Waldo Emerson*. Vol. 9 (1843–1847) ed. Ralph H. Orth and Alfred R. Ferguson. 16 vols. Cambridge, MA: The Belknap Press of Harvard University Press, 1960–1982.

–––. *The Letters of Ralph Waldo Emerson*. Ed. Ralph L. Rusk (vols. 1–6) and Eleanor M. Tilton (vols. 7–10). 10 vols. New York: Columbia University Press, 1939–1999.

–––. *Miscellanies*. Ed. J. E. Cabot. Boston: Houghton, Mifflin, and Company, 1884.

–––. *Representative Men: Seven Lectures*. Text established by Douglas Emory Wilson. Introduced by Andrew Delbanco. Cambridge, MA: Harvard University Press, 1996.

Evangelical Lutheran Synod. *Evangelical Lutheran Hymnary*. St. Louis: MorningStar Music Publishers, 1996.

Everett, Charles Carroll. *Memoir of Joseph Henry Allen*. Rpt. from *The Publications of the Colonial Society of Massachusetts*, vol. 6. Cambridge: John Wilson and Son, 1902.

Farraro, Gary. *Cultural Anthropology: An Applied Perspective*. St. Paul: West, 1992.

Fish, Henry Clay. *History and Repository of Pulpit Eloquence*. 2 vols. in 1. New York: Dodd and Mead, 1850.

Fitch, Eleazar T. *Sermons, Practical and Descriptive, Preached in the Pulpit of Yale College*. New Haven: Judd and White, 1871.

Freeman, R. B. *Charles Darwin: A Companion*. Folkestone, Kent, England: William Dawson and Sons, 1978.

Furness, William H. *Nature and Christianity: A Dudleian Lecture Delivered in the Chapel of the University at Cambridge [i.e., Harvard], Wednesday, May 12, 1847*. Boston: William Crosby and H. P. Nichols, 1847.

Fry, C. George, and Fry, Jon Paul. *Congregationalists and Evolution: Asa Gray and Louis Agassiz*. Lanham, MA: University Press of America, 1989.

Gabriel, Ralph Henry. *Religion and Learning at Yale: The Church of Christ in the College and University, 1757–1957*. New Haven, CT: Yale University Press, 1958.

Garraty, John A., and Carnes, Mark C., eds. *American National Biography*. 24 vols. New York: Oxford University Press, 1999.

Geertz, Clifford. *The Interpretation of Cultures: Selected Essays*. New York: Basic Books, 1973.

Gerstner, Patsy. *Henry Darwin Rogers, 1808–1866: American Geologist*. Tuscaloosa: University of Alabama Press, 1994.

Gillispie, Charles Coulston. *Genesis and Geology: A Study in the Relations of Scientific Thoguht, Natural Theology, and Social Opinion in Great Britain, 1790–1850*. With a foreword by Nicolaas A. Rupke and a new preface by the author. Cambridge, MA: 1996.

Gilman, Daniel C. *The Life of James Dwight Dana: Scientific Explorer, Mineralogist, Geologist, Zoologist, Professor in Yale University*. New York: Harper and Brothers, 1899.

Greenspan, Ezra. *George Palmer Putnam: Representative American Publisher*. University Park, PA: Pennsylvania State University Press, 2000.

Giltner, John H. *Moses Stuart: The Father of Biblical Science in America*. Atlanta: Scholars Press, 1988.

Glick, Thomas F., ed. *The Comparative Reception of Darwinism*. Austin: University of Texas Press, 1974.

Gould, Stephen Jay. *The Mismeasure of Man*. New York: W. W. Norton, 1981.

Gray, Asa. *Darwiniana: Essays and Reviews Pertaining to Darwinism*. 1876. Rpt., New York: D. Appleton and Company, 1884.

———. *Natural Science and Religion: Two Lectures Delivered to the Theological School of Yale College*. New York: Charles Scribner's Sons, 1880.

Guelzo, Allen C. *Abraham Lincoln: Redeemer President*. Grand Rapids, MI: William B. Eerdmans, 1999.

———. *Edwards on the Will: A Century of Theological Debate*. Middletown, CT: Wesleyan University Press, 1989.

Gundlach, Bradley J. *Process and Providence: The Evolution Question at Princeton, 1845–1929*. Grand Rapids: William B. Eerdmans, 2013.

Gura, Philip F., and Joel Myerson. *Critical Essays on American Transcendentalism*. Boston: G. K. Hall and Company, 1982.

Guyot, Arnold. *The Earth and Man: Lectures on Comparative Physical Geography, in Its Relation to the History of Mankind*. Trans. from the French by C. C. Felton. 1849. Rpt., Boston: Gould and Lincoln, 1858.

Hadley, James. *Diary (1843–1852) of James Hadley*. Edited with a foreword by Laura Hadley Moseley. New Haven: Yale University Press, 1951.

Haven, Joseph. *A History of Philosophy*. New York: Sheldon & Company, 1876.

Henry, James Buchanan, and Christian Henry Scharff. *College As It Is: Or, The Collegian's Manual in 1853*. Ed. with an introduction by J. Jefferson Looney. Princeton, NJ: Princeton University Libraries, 1996.

Henry, Joseph. *The Papers of Joseph Henry*. Ed. Nathan Reingold (vols. 1–5) and Marc Rothenberg (vols. 6–) *et al.* 8 vols. to date. Washington, DC: Smithsonian Institution Press, 1972–1998.

Herndon, William H., and Jesse W. Weik. *Herndon's Life of Lincoln: The History and Personal Recollections of Abraham Lincoln as Originally Written by William H. Herndon and Jesse W. Weik*. With introduction and notes by Paul M. Angle. 2 vols. Cleveland and New York: The World Publishing Company, 1949.

Hill, Thomas. *The Annual Address before the Harvard Natural History Society, Delivered by Rev. Thomas Hill, Thursday, May 19, 1853*. Cambridge, MA: J. Bartlett, 1853.

Hofstadter, Richard. *Social Darwinism in American Thought*. With a new introduction by Eric Foner. Boston: Beacon Press, 1992.

Hirrel, Leo P. *Children of Wrath: New School Calvinism and Antebellum Reform*. Lexington, KY: The University Press of Kentucky, 1998.

Hitchcock, Edward. *Elementary Geology*. 25th ed. New York: Ivison and Phinney; Chicago: S. C. Griggs, and Co.; Buffalo: Phinney and Co.; Auburn: J. C. Ivison and Co.; Detroit: A. M'Farren; Cincinnati: Moore, Anderson, and Co., 1855.

———. *The Highest Use of Learning: An Address Delivered at His Inauguration to the Presidency of Amherst College*. Published by the Trustees. Amherst: J. S. & C. Adams, 1845.

———. *The Religion of Geology and Its Connected Sciences*. Boston: Phillips, Sampson, and Company, 1857.

Hodge, Arhibald Alexander. *The Confession of Faith: A Handbook of Christian Doctrine Expounding the Westminster Confession*. 1869. Rpt., London: Billing and Sons, 1978.

Hodge, Charles. *A Brief Account of the Last Hours of Albert B. Dod: Nov. 20th, 1845*. Princeton: John T. Robinson, [1845?].

Hodge, Charles. *Systematic Theology*. 3 vols. 1871–1872. Facsimile rpt., Peabody, MA: Hendrickson, 2001.

Hofstadter, Richard. *Social Darwinism in American Thought*. With a new introduction by Eric Foner. 1st ed., 1944. New ed., Boston: Beacon Press, 1992.

Hovenkamp, Herbert. *Science and Religion in America, 1800–1860*. Philadelphia: University of Pennsylvania Press, 1978.

Howe, Daniel Walker. *The Unitarian Conscience: Harvard Moral Philosophy, 1805–1861*. With a new introduction. Middletown, CT: Wesleyan University Press, 1988.

Jardine, N., Secord, J. A., and Spary, E. C., eds. *Cultures of Natural History*. Cambridge: Cambridge University Press, 1996.

Johnston, James F. W. *Notes on North America: Agricultural, Economical, and Social*. 2 vols. Boston: Charles C. Little and James Brown; Edinburgh and London: William Blackwood and Sons, 1851.

Keller, Charles Roy. *The Second Great Awakening in Connecticut*. New Haven: Yale University Press, 1942.

Kohlstedt, Sally Gregory. *The Formation of the American Scientific Community: The American Association for the Advancement of Science 1848–60*. Urbana: University of Illinois Press, 1976.

Kuklick, Bruce. *Churchmen and Philosophers: From Jonathan Edwards to John Dewey*. New Haven: Yale University Press, 1985.

———, ed. *The Unitarian Controversy, 1819–1823*. 2 vols. New York: Garland Publishers, 1987.

Lachmann, L. M. *The Legacy of Max Weber: Three Essays*. Berkeley, CA: Glendessary Press, 1971.

Larson, Edward J. *Summer for the Gods: The Scopes Trial and America's Continuing Debate over Science and Religion*. Cambridge, MA: Harvard University Press, 1998.

LeConte, Joseph. *The Autobiography of Joseph LeConte*. Ed. William Armes. New York: D. Appleton and Company, 1903.

Lehmann, Hartmut, and Guenther Roth, eds. *Weber's "Protestant Ethic": Origins, Evidence, Contexts*. Cambridge: Cambridge University Press, 1993.

Lindberg, David C., and Ronald L. Numbers, eds. *God and Nature: Historical Essays on the Encounter between Christianity and Science*. Berkeley: University of California Press, 1986.

Livingstone, David N. *Darwin's Forgotten Defenders: The Encounter between Evangelical Theology and Evolutionary Thought*. Grand Rapids, MI: William B. Eerdmans, 1987.

Livingstone, David N., D. G. Hart, and Mark A. Noll, eds. *Evangelicals and Science in Historical Perspective*. New York and Oxford: Oxford University Press, 1999.

Lord, David N. *Geognosy; Or, The Facts and Principles of Geography Against Theories*. New York: Franklin Knight, 1855.

Lord, Eleazar. *The Epoch of Creation; The Scripture Doctrine Contrasted with the Geological Theory*. New York: Scribner, 1851.

Löwith, Karl. *Max Weber and Karl Marx*. Ed. with an introduction by Tom Bottomore and William Outhwaite. With a new preface by Bryan S. Turner. London: Routledge, 1993.

Lurie, Edward. *Louis Agassiz: A Life in Science*. Chicago and London: University of Chicago Press, 1960.

Lyell, Charles. *Principles of Geology*. Facsimile rpt. with a new introduction by Martin J. S. Rudwick. 3 vols. 1830–1833. Rpt., Chicago: University of Chicago Press, 1990.

Lynch, John M., ed. *"Vestiges" and the Debate before Darwin*. 7 vols. Bristol: Thoemmes, 2000.

Maclean, John. *History of the College of New Jersey, from Its Origin in 1746 to the Commencement of 1854*. 2 vols. Philadelphia: J. B. Lippincott and Co., 1877.

Marsden, George M. *The Evangelical Mind and the New School Presbyterian Experience: A Case Study of Thought and Theology in Nineteenth-Century America*. New Haven: Yale University Press, 1970.

———. *Fundamentalism and American Culture: The Shaping of Twentieth Century Evangelicalism, 1870–1925*. New York: Oxford University Press, 1980.

———. *The Soul of the American University: From Protestant Establishment to Established Nonbelief*. New York and Oxford: Oxford University Press, 1994.

McCosh, James. *The Method of Divine Government: Physical and Moral*, 13th ed. 1st ed., 1851. London: Macmillan and Company, 1887.

Menand, Louis. *The Metaphysical Club*. New York: Farrar, Straus, and Giroux, 2001.

Mercantile Library Association. *Catalogue of the Mercantile Library in New York*. New York: Baker, Godwin & Co., 1850.

Mercantile Library Association. *Supplement to the Catalogue of the Mercantile Library in New York, Containing the Additions Made to August 1856*. New York: Baker, Godwin & Co., 1856.

Merton, Robert K. *Science, Technology, and Society in Seventeenth Century England*. 1938. Rpt., New York: H. Fertig, 1970.

Milhauser, Milton. *Just Before Darwin: Robert Chambers and Vestiges*. Middletown, CT: Wesleyan University Press, 1959.

Miller, Glenn T. *Piety and Intellect: The Aims and Purposes of Ante-Bellum Theological Education*. Atlanta, GA: Scholars Press, 1990.

Miller, Perry. *The New England Mind: From Colony to Province*. 1953. 9th printing. Cambridge: Harvard University Press, 1998.

———. *The New England Mind: The Seventeenth Century*. 1953. 2d printing. Cambridge, MA: Harvard University Press, 1954.

Mitchell, Samuel S. *Joseph Henry. In Memoriam. Funeral Address, by His Pastor, Rev. Samuel S. Mitchell, D.D., May 16, 1878*. Washington, DC: Thomas McGill and Co., 1878.

Mivart, St. George Jackson. *On the Genesis of Species*. New York: D. Appleton, 1871.

Moore, James R., ed. *History, Humanity, and Evolution*. Cambridge: Cambridge University Press, 1989.

———. *The Post-Darwinian Controversies: A Study of the Protestant Struggle to Come to Terms with Darwin in Great Britain and America, 1870–1900*. Cambridge: Cambridge University Press, 1979.

Morison, Samuel Eliot. *Three Centuries of Harvard, 1636–1936*. Cambridge, MA: Harvard University Press, 1936.

Morrell, Z. N. *Flowers and Fruits from the Wilderness*. Boston: Gould and Lincoln, 1872.

Morris, Henry M. *History of Modern Creationism*. New updated ed., with a foreword by John C. Whitcomb. Santee, CA: Institute for Creation Research, 1993.

Myerson, Joel, ed. *Emerson and Thoreau: The Contemporary Reviews*. Cambridge: Cambridge University Press, 1992.

New York State Library. *Catalogue of the New York State Library, January 1, 1850*. Albany: C. Van Benthuysen, 1850.

New York State Library. *Catalogue of the New York State Library, 1855: General Library*. Albany: C. Van Benthuysen, 1856.

Niebuhr, Reinhold, ed. *Marx and Engels on Religion*. New York: Scribner, 1964.

Noll, Mark A. *Princeton and the Republic: The Search for a Christian Enlightenment in the Era of Samuel Stanhope Smith*. Princeton, NJ: Princeton University Press, 1989.

———, ed. *The Princeton Theology, 1812–1921: Scripture, Science, and Theological Method from Archibald Alexander to Benjamin Breckinridge Warfield*. Grand Rapids, MI: Baker Book House, 1983.

Nott, J. C., Gliddon, George R., eds. *Types of Mankind; or, Ethological Researches, Based upon the Ancient Monuments, Paintings, Sculptures, and Crania of Races, and upon the Natural, Geographical, Philological, and Biblical History*. Philadelphia: Lippincott, Grambo, & Co., 1855.

Numbers, Ronald. *Creation by Natural Law: Laplace's Nebular Hypothesis in American Thought*. Seattle: University of Washington Press, 1977.

———. *Darwinism Comes to America*. Cambridge: Harvard University Press, 1998.

Numbers, Ronald L., and Stenhouse, John, eds. *Disseminating Darwinism: The Role of Place, Race, Religion, and Gender*. Cambridge: Cambridge University Press, 1999.

Olby, R. C., *et al.*, eds. *Companion to the History of Modern Science*. London: Routledge, 1990.

Olmsted, Denison. *The Student's Common-Place Book, on a New Plan; Uniting the Advantages of a Note Book and Universal Reference Book. Applied Alike to the College Student, and to the Professional Man*. New Haven: S. Babcock, 1838.

Orsi, Robert. *Thank You, St. Jude: Women's Devotion to the Patron Saint of Hopeless Causes*. New Haven, CT: Yale University Press, 1996.

Osborne, Grant R. *The Hermeneutical Spiral: A Comprehensive Introduction to Biblical Intepretation*. Downers Grove, IL: InterVarsity Press, 1991.

Paine, Martyn. *Physiology of the Soul and Instinct, as Distinguished from Materialism*. New York: Harper & Brothers, 1872.

Paley, William. *Natural Theology; or, Evidences of the Existence and Attributes of the Deity, Collected from the Appearances of Nature*. 12th ed. 1st ed., 1801. Weybridge, Eng.: S. Hamilton, 1809.

Patrides, C. A., ed. *The Cambridge Platonists*. Cambridge, MA: Harvard University Press, 1970.

Patterson, Robert. *Fables of Infidelity and Facts of Faith*. Cincinnati: Western Tract Society, 1875.

Peabody, Andrew P. *The Analogy of Nature and the Bible: The Dudleian Lecture Delivered in the Chapel of Harvard University, May 14, 1856*. Cambridge, MA: Metcalf and Co., 1856.

Peabody, Andrew Preston. *Sermons Connected with the Re-Opening of the Church of the South Parish, in Portsmouth, New Hampshire, Preached Dec. 25 & 26, 1858, and Jan. 30 & Feb. 6, 1859*. Portsmouth, NH: James F. Shores, Jr., & Joseph H. Foster; Boston: Crosby, Nichols, and Co., 1859.

Pearson, Thomas. *Infidelity; Its, Aspects, Causes, and Agencies: Being the Prize Essay of the British Organization of the Evangelical Alliance*. New York: R. Carter & Brothers, 1854.

Peoples, James, and Garrick Bailey. *Humanity: An Introduction to Cultural Anthropology*. St. Paul: West, 1988.

Potter, Alonzo, ed. *Lectures on the Evidence of Christianity, Delivered in Philadelphia, by Clergymen of the Protestant Episcopal Church, in the Fall & Winter of 1853–4*. Philadelphia: E. H. Butler & Co., 1855.

Precht, Fred L. *Lutheran Worship Hymnal Companion*. St. Louis: Concordia Publishing House, 1992.

Quincy, Josiah. *Speech of Josiah Quincy, President of Harvard University, before the Board of Overseers of That Institution, February 25, 1845, on the Minority Report of the Committee of Visitation, Presented to That Board by George Bancroft, Esq., February 6, 1845*. Boston: Charles C. Little and James Brown, 1845.

Rappaport, Rhoda. *When Geologists Were Historians, 1665–1750*. Ithaca: Cornell University Press, 1997.

Reingold, Nathan, ed. *Science in America since 1820*. New York: Science History Publications, 1976.

Richards, Robert J. *The Meaning of Evolution: The Morphological Construction and Ideological Reconstruction of Darwin's Theory*. Chicago: The University of Chicago Press, 1992.

Roberts, Jon H. *Darwinism and the Divine in America: Protestant Intellectuals and Organic Evolution, 1859–1900*. Paperback ed., with a new foreword. Notre Dame, IN: University of Notre Dame Press, 2001.

Roberts, Jon H., and Turner, James. *The Sacred and the Secular University*. Introduced by John F. Wilson. Princeton, NJ: Princeton University Press, 2000.

Robinson, David. *Apostle of Culture: Emerson as Preacher and Lecturer*. Philadelphia: University of Pennsylvania Press, 1982.

Rogers, William Barton. *Life and Letters of William Barton Rogers*. Ed. Emma Rogers. 2 vols. Boston and New York: Houghton, Mifflin, and Company, 1896.

Rossi, Paolo. *The Dark Abyss of Time: The History of the Earth and the History of Nations from Hooke to Vico*. Trans. Lydia G. Cochrane. Chicago: University of Chicago Press, 1984.

Rudolph, Frederick. *Curriculum: A History of the American Undergraduate Course of Study since 1636*. 1977. Rpt., San Francisco: Jossey–Bass Publishers, 1992.

Rudwick, Martin J. S. *The Great Devonian Controversy: The Shaping of Scientific Knowledge among Gentlemanly Specialists*. Chicago: University of Chicago Press, 1985.

———. *The Meaning of Fossils: Episodes in the History of Palaeontology*. 2d ed. Chicago: University of Chicago Press, 1985.

Rupke, Nicholaas A. *The Great Chain of History: William Buckland and the English School of Geology (1814–1849)*. Oxford: Clarendon Press, 1983.

Schluchter, Wolfgang. *Rationalism, Religion, and Domination: A Weberian Perspective*. Trans. Neil Solomon. Orig. German essays: 1980, 1985, 1988. Berkeley, CA: University of California Press, 1989.

Schroeder, Ralph. *Max Weber and the Sociology of Culture*. London: SAGE, 1992.

Secord, James A. *Controversy in Victorian Geology: The Cambrian-Silurian Dispute*. Princeton, NJ: Princeton University Press, 1986.

———. *Victorian Sensation: The Extraordinary Publication, Reception, and Secret Authorship of "Vestiges of the Natural History of Creation."* Chicago: University of Chicago Press, 2000.

Selden, William K. *Princeton Theological Seminary: A Narrative History, 1812–1992*. Princeton, NJ: Princeton University Press, 1992.

Silliman, Benjamin. *An Address Delivered before the Association of the Alumni of Yale College, in New Haven, August 17, 1842*. New Haven: B. L. Hamlen, 1842.

———. *Suggestions Relative to the Philosophy of Geology, as Deduced from the Facts and to the Consistency of Both the Facts and Theory of This Science with Sacred History*. New Haven: B. L. Hamden, 1839.

Smith, Hariette Knight. *The History of the Lowell Institute*. Boston: Lamson, Wolffe, and Co., 1898.

Southall, James P. C. *The Recent Origin of Man, As Illustrated by Geology and the Modern Science of Pre-Historic Archaeology*. Philadelphia: J. B. Lippincott & Co., 1875.

Stanton, William. *The Leopard's Spots: Scientific Attitudes toward Race in America, 1815–1859*. Chicago: University of Chicago Press, 1960.

Stevenson, Louise L. *Scholarly Means to Evangelical Ends: The New Haven Scholars and the Transformation of Higher Learning in America, 1830–1890*. Baltimore and London: Johns Hopkins University Press, 1986.

Stewart, John W. *Mediating the Center: Charles Hodge on American Science, Language, Literature, and Politics*. Studies in Reformed Theology and History series, vol. 3, no. 1. Princeton, NJ: Princeton Theological Seminary, Winter 1995.

Stewart, John W., and James H. Moorhead, eds. *Charles Hodge Revisited: A Critical Appraisal of His Life and Work*. Grand Rapids, MI: William B. Eerdmans, 2002.

Stokes, Anson Phelps. *Memorials of Eminent Yale Men: A Biographical Study of Student Life and University Influences during the Eighteenth and Nineteenth Centuries*. 2 vols. New Haven: Yale University Press, 1914.

Story, Ronald. *The Forging of an Aristocracy: Harvard and the Boston Upper Class, 1800–1870*. Middletown, CT: Wesleyan University Press, 1980.

Stuart, Moses. *Conscience and the Constitution with Remarks on the Recent Speech of the Hon. Daniel Webster in the Senate of the United States on the Subject of Slavery*. Boston: Crocker & Brewster, 1850.

Sweeney, Douglas A. *Nathaniel Taylor, New Haven Theology, and the Legacy of Jonathan Edwards*. Oxford: Oxford University Press, 2003.

Taylor, Nathaniel W. *Essays, Lectures, Etc. upon Select Topics in Revealed Theology*. Ed. Noah Porter. New York: Clark, Austin, & Smith, 1859.

Tennyson, Alfrd Lord. "In Memoriam A. H. H." 1850. In *Tennyson: A Selected Edition, Incorporating the Trinity College Manuscripts*. Ed. Christopher Ricks, 331–484. Berkeley, CA: University of California Press, 1989.

Thoreau, Henry David. *Journal*. Gen. ed. John C. Broderick. 6 vols. Princeton, NJ: Princeton University Press, 1981–2000.

–––. *Walden and Other Writings*. New York: Barnes and Noble, 1993.

Tillich, Paul. *Systematic Theology*. 3 vols. in 1. 1950–1963. Rpt., Chicago: University of Chicago Press, 1967.

–––. *Theology of Culture*. Ed. Robert C. Kimball. New York: Oxford University Press, 1959.

Tucker, Robert C., ed. *The Marx-Engels Reader*. New York: Norton, 1972.

Turner, James. *Language, Religion, Knowledge: Past and Present*. Notre Dame, IN: University of Notre Dame Press, 2003.

–––. *The Liberal Education of Charles Eliot Norton*. Baltimore: The Johns Hopkins University Press, 1999.

Tweed, Thomas A., ed. *Retelling U.S. Religious History*. Chicago: University of Chicago Press, 1997.

United States Military Academy Library. *Catalogue of the Library of the U.S. Military Academy, West Point, N.Y.* New York: J. F. Trow, 1853.

Ussher, James. *The Annals of the World: James Ussher's Classic Survey of World History*, rev. and updated by Larry and Marion Pierce. Green Forest, AK: Master Books, 2003.

van der Meer, Jitse M., ed. *Facets of Faith and Science*. 4 vols. Lanham, MD: University Press of America, 1996.

Vander Stelt, John C. *Philosophy and Scripture: A Study of Old Princeton and Westminster Theology*. Marlton, NJ: Mack Publishing Company, 1978.

Voelz, James W. *What Does This Mean? Principles of Biblical Interpretation in the Post-Modern World*. St. Louis: Concordia, 1995.

Wacker, Grant. *Religion in Nineteenth Century America*. New York: Oxford University Press, 2000.

Walker, James. *A Discourse on the Deference Paid to the Scriptures by Unitarians*. Boston: American Unitarian Association, 1832.

–––. *Unitarianism Vindicated against the Charge of Not Going Far Enough*, 3d ed. Boston: James Munroe & Co., 1832.

–––. *Unitarianism Vindicated against the Charge of Skeptical Tendencies*. Boston: James Munroe & Co., 1832.

Walker, James B. *God Revealed in the Process of Creation, and by the Manifestation of Jesus Christ; Including an Examination of the Development Theory contained in the "Vestiges of the Natural History of Creation."* Boston: Gould and Lincoln; New York: Sheldon and Co.; Cincinnati: George S. Blanchard, 1870.

Wayland, John T. *The Theological Department in Yale College, 1822–1858*. American Religious Thought of the 18th and 19th Centuries series. Ed. Bruce Kuklick. Rpt., New York and London: Garland Publishing, 1987.

Weber, Max. *The Protestant Ethic and the Spirit of Capitalism*. Trans. Talcott Parsons. With an introduction by Anthony Giddens. 1904–1905. Rev. ed., 1920–1921. Trans., 1930. Rpt., London: Routledge, 2001.

–––. *The Sociology of Religion*. Trans. and intro. by Talcott Parsons. With a new foreword by Ann Swindler. German orig., 1922. English trans., 1963. Rpt., Boston: Beacon, 1993.

Wedgwood, Julia. *John Ruskin*. New York: Tucker, 1900.

Weeks, Edward. *The Lowells and Their Institutes*. Boston: Little, Brown, and Co., 1966.

Wellek, René. *Confrontations: Studies in the Intellectual and Literary Relations between Germany, England, and the United States during the Nineteenth Century*. Princeton, NJ: Princeton University Press, 1965.

Wells, David F., ed. *Reformed Theology in America: A History of Its Modern Development*. Grand Rapids, MI: Baker, 1997.

Wertenbaker, Thomas Jefferson. *Princeton, 1746–1896*. Princeton, NJ: Princeton University Press, 1946.

West, Nathaniel. *The Corruption of Established Truth and Responsibility of Educated Men. An Address before the Alumni of the University of Michigan, June 27, 1856*. Detroit, MI: n.p., 1856.

Wharton, Francis. *A Treatise on Theism, and on the Modern Skeptical Theories*. Philadelphia: J. B. Lippincott & Co., 1859.

White, Andrew Dickson. *The Warfare of Science*. New York: D. Appleton and Company, 1876.

Williams, George Huntston, ed. *The Harvard Divinity School: Its Place in Harvard University and American Culture*. Boston: The Beacon Press, 1954.

Williams, William George. *The Ingham Lectures. A Course of Lectures on the Evidences of Natural and Revealed Religion, delivered before the Ohio Wesleyan University, Delaware, OH*. Cleveland: Ingham, Clarke and Company; New York: Nelson & Phillips, 1873.

Wilson, Leonard G., ed. *Benjamin Silliman and His Circle: Studies on the Influence of Benjamin Silliman on Science in America*. New York: Science History Publications, 1979.

Winchell, Alexander. *Pamphlets: Education; Philosophy*. 12 pamphlets in 1 vol. New York: J. Soule and T. Mason, 1858–1889.

Winckelmann, Johannes. *Die Protestantische Ethik II*. 1978. Rpt., Gütersloh, Ger.: Gütersloher Verlagshaus Mohn, 1995.

Winship, Michael P. *Seers of God: Puritan Providentialism in the Restoration and Early Enlightenment*. Baltimore: Johns Hopkins University Press, 1996.

Woods, Leonard. *History of Andover Theological Seminary*. Ed. George S. Baker. Boston: James R. Osgood and Company, 1885.

Young, Robert M. *Darwin's Metaphor: Nature's Place in Victorian Culture*. Cambridge: Cambridge University Press, 1985.

Dissertations Cited

Gundlach, Bradley John. "The Evolution Question at Princeton, 1845–1929." Ph.D. diss., University of Rochester, New York, 1995.

MacPherson, Ryan C. "The *Vestiges of Creation* and America's Pre-Darwinian Evolution Debates: Interpreting Theology and the Natural Sciences in Three Academic Communities." Ph.D. diss., University of Notre Dame, 2003.

McElligott, John F. "Before Darwin: Religion and Science as Presented in American Magazines, 1830–1860." Ph.D. diss., New York University, 1973.

Ogilvie, Marilyn Bailey. "Robert Chambers and the Successive Revisions of the *Vestiges of the Natural History of Creation*." Ph.D. diss., University of Oklahoma, Norman, OK, 1973.

Topham, Jonathan. "The Bridgewater Treatises and British Natural Theology in the 1830s." Ph.D. diss., Lancaster University, 1993.

Conference Papers Cited (with Permission)

Bademan, Bryan. "'Let Us Rise Through Nature Up to Nature's God': Nature and Design in Mid-nineteenth-century Protestant Thought." Paper presented at the Intellectual History Seminar, Dept. of History, University of Notre Dame, Apr. 2001, and, in abbreviated form, to Ecology, Theology, and Judeo-Christian Environmental Ethics, a Lilly Fellows Program National Research Conference, University of Notre Dame, Feb. 2002.

MacPherson, Ryan. "Transformations of the *Vestiges of Creation* in American Periodicals, 1845–1860." Paper presented for Mephistos 2000, University of Oklahoma, 4 Feb. 2000.

———. "The *Vestiges of Creation* Meets the Scientific Sovereignty of God: Natural and Theological Science at Princeton, 1845–1859." Paper presented for the History of Science Society Annual Meeting, Milwaukee, WI, 8 Nov. 2002.

———. "The *Vestiges of Creation* Meets the Sovereignty of God: Pre-Darwinian Evolution at Princeton." Paper presented for the Intellectual History Seminary, Dept. of History, University of Notre Dame, 5 Apr. 2002.

Nartonis, David. "Idealist Philosophy of Science in Nineteenth-Century Harvard." Rev. ms., based on a paper delivered to the Congress of the History of the Philosophy of Science Working Group, Montreal, 21–23 June 2002.

Acknowledgments

This book has been adapted from my doctoral dissertation, defended at the University of Notre Dame in 2003. Prof. Phillip Sloan of the Program in History and Philosophy of Science suggested *Vestiges of the Natural History of Creation* as a topic for my research paper in his Spring 1998 course "The Darwinian Revolution." That paper became the seed from which my dissertation developed, and that dissertation transformed me from being Prof. Sloan's student to being Phil's colleague. I thank Phil for everything that he has done for me between then and now: the classes he taught, the guidance he gave, the research tips he offered, and the encouragement that has never ended.

Prof. Jane Maienschein of Arizona State University introduced me to Phil at the November 1996 History of Science Society meeting in Atlanta. Prof. Marga Vicedo, in turn, had introduced me to Jane. It was in Marga's fall 1995 history of biology class at ASU West that I first encountered *Vestiges*. Both Marga and Jane kept in contact with me during my days at Notre Dame. I am grateful for their support.

The faculty and graduate students in Notre Dame's HPS program and in the Department of History shaped the development of this project in important ways. I am thankful especially for the participants in the weekly Colloquium on Religion and History (CORAH) and the monthly Intellectual History Seminar who provided feedback to earlier presentations of some arguments that appear in this dissertation. I am indebted to graduate students Neil Dhingra, Bryan Bademan, Margaret Abruzzo, and Jonathan Den Hartog, in the History Department; Elizabeth Hayes in HPS; Jerry Park in Sociology; and David Maxwell in Theology—for both their scholarly criticism and their constant friendship.

Among the faculty in Notre Dame's History Deptartment, I thank Profs. Fr. Robert Sullivan, John McGreevy, and Christopher Hamlin (who also teaches in the HPS program) for serving on my dissertation proposal committee. Notre

Dame history professors George Marsden and James Turner served on the committee for the dissertation itself. It was an honor to have my work reviewed by those two scholars, who have contributed so much to our understandings of American religion and American intellectual life. George is to be commended for taking the time to eat lunch with his students each week at CORAH, a venue where questions of faith and history can, simultaneously, be scrutinized and cherished. I thank Jim for the witty emails that kept my spirits high and for his endless insights into the nature of nineteenth-century higher education.

Beyond Notre Dame, Prof. Ronald Numbers, of the University of Wisconsin–Madison, also served on my committee. I thank Ron for coaching me not to write vaguely of "American intellectuals" or "American Protestant thinkers" while he and fellow historian of science and religion Jon Roberts were watching a Badgers game one afternoon. My research began by retracing trails that Ron has blazed through the landscape of American science and religion, and when reviewing my drafts Ron provided both encouragement and challenges. Jon Roberts, I should like to acknowledge, also provided helpful criticism when responding to a paper I delivered at the 2002 History of Science Society meeting.

The long list of archival collections presented in the bibliography reminds me of how little of this dissertation I researched and drafted at Notre Dame. One hundred forty-one days of the year 2002 alone were spent elsewhere, as were several months during the preceding year. I am grateful for the assistance of library personnel who coordinated my visits and guided me in locating some of the more obscure documents. I wish to mention especially the following: Earle Spamer at the Academy of Natural Sciences at Philadelphia; Barbara Trippel Simmons at Amherst College; Robert Cox, Roy Goodman, and Valerie Lutz at the American Philosophical Society; Lisa DeCesare at the Gray Herbarium, Harvard University; Robert Young and Dana Fischer, at the Museum of Comparative Zoology, Harvard; Frances O'Donnell, at the Andover-Harvard Divinity School Library; Jack Eckert at the Countway Library of Medicine, Harvard; Susan Fraser at the New York Botanical Garden Library; Gretchen Oberfranc at Firestone Library, Princeton University; Dan Linke at Mudd Library, Princeton; Robert Benedetto at Henry Luce III Library, Princeton Theological Seminary; Marc Rothenberg and Robert Millikan at the Smithsonian Institution; Michael Frost at the Yale University Archives; and, Toby Appel at the Cushing Medical Historical Library, Yale.

During my travels I also received guidance from medical historian Philip Wilson of Pennsylvania State University, who was a "fellow research fellow" with me at the American Philosophical Society in the fall of 2001. One day when we took an excursion northward, Philip introduced me to the lecture manuscripts collection at Princeton. From that discovery emerged my decision to frame this project in terms of a comparison of how *Vestiges* was received in different educational settings.

While researching at Harvard I met with David Nartonis, a retired physics professor who at the time was pursuing—and publishing—research in the philosophy of science. To him I owe a fair deal of my understanding of the role of idealism at Harvard, as set forth in chapter 3.

Through correspondence and personal contact at the History of Science Society annual meetings, I received helpful suggestions from Dr. James Secord of Cambridge University. Jim's *Victorian Sensation*, a phenomenal study of British reactions to the *Vestiges*, which is discussed in the introduction to this book, provided me with an ambitious goal, as I sought to contribute to the history of American scientific culture something similar to his contribution to the British context. I am grateful for his encouragement, and for several research leads. The plausibility of my dissertation, when proposed in 2001, rested in large part upon the success of Jim's research in demonstrating that if one spends countless hours sifting through archives, a large number of references to *Vestiges* will turn up. It was an honor to follow his example.

Several sources of funding made this dissertation possible. I am indebted to Michael Crowe, a professor emeritus in Notre Dame's HPS program, and Peter Diffley, dean of the Notre Dame Graduate School, plus several of the professors mentioned above, for assisting me with one or more of my grant applications. I thank the donors for their generosity and the application review committees for placing confidence in me to do the work that I had proposed:

- *Bethany Lutheran College*, Faculty Development Grant, 2004.
- *National Science Foundation*, Dissertation Improvement Award (SES-0217398), 2002.
- *Princeton University Libraries*, Visiting Researcher Fellowship, 2002.
- *University of Notre Dame*, Zahm Research Travel Grant, 2002.
- *American Philosophical Society Library*, Resident Research Fellowship, 2001.

    ○ *University of Notre Dame*, Presidential Fellowship, 1997–2002.

Prof. Don Howard, the department chair for Notre Dame's HPS program, also arranged on several occasions for departmental research monies to supplement the grants mentioned above. Furthermore, Don kindly appointed me as a research assistant to himself and Fr. Ernan McMullin, the founder of our HPS program, so that I would have financial support during my final year of writing, yet without the demands of a teaching schedule that would prevent me from completing this project.

Beyond my academic needs, I also was sustained during the course of this project by the pastors and members of Peace Lutheran Church in Granger, Indiana, and several other churches in the areas where I visited for research and recuperation. Several individuals welcomed me into their homes when I needed a weekend of vacation from the long hours of archival research. I thank especially Pastor Bill Kessel of Cottonwood, Arizona, for assuring me that one can complete a Ph.D. without loosing sanity (his doctorate is in anthropology). The members of Peace Lutheran Church, King of Prussia, Pennsylvania, in the greater Philadelphia area, endured the aftermath of 9/11 with me. Marty Sponholz, a physical science professor at Martin Luther College, New Ulm, Minnesota, emailed me with encouragement during those times when I thought I would never complete my dissertation. Those at my local congregation who have provided constant support are too numerous for each to be mentioned, but I would like to thank by name Steve Huffman, who after months of asking "how many pages this week?," volunteered to read the entire manuscript. How satisfying it was to learn that I had made my thesis understandable to an information technology specialist who has never before studied the history of science and religion.

My parents, Donald and Barbara, also have provided much encouragement during my graduate school years, as well as before and after. My grandmother, Emma Mary Hubsch, or "Mamie" as her seven grandchildren knew her, passed away in December 2002, the very evening that I had completed my last day of archival research. Mamie had been a high school and grade school teacher in Lower Merion and Narberth Townships, just outside of Philadelphia. She loved history. Her students sewed American flags that were flown at the United States Capitol Building and in a hospital in South Vietnam. They took annual field trips to the Smithsonian Institution Museums in Washington, D.C.

One morning, when I arrived early at the Smithsonian Institution Archives, I telephoned Mamie while waiting for the door to open. She shared in this project for as long as she lived, and it was to her that I dedicated my dissertation.

Finally, I wish to acknowledge my wife Marie both for smoothing out my writing and for sharing with me the wonderful endeavor of raising our children. Although I had intended to publish this work much sooner, I much prefer to have a home full of children who crave—and receive—my personal attention. Just because a career goal is worthwhile, does not mean it cannot be properly delayed. Dear readers, my family thanks you for waiting patiently.

Index

About the Author

Ryan C. MacPherson holds a Ph.D. in History and Philosophy of Science from the University of Notre Dame. He currently serves as Chair of the History Department at Bethany Lutheran College in Mankato, Minnesota, where he teaches courses in U.S. history, the history of science, and bioethics.

Dr. MacPherson's publications span the fields of history, theology, law, and public policy. His books include *Telling the Next Generation: The Evangelical Lutheran Synod's Vision for Christian Education, 1918–2011 and Beyond* (managing editor, 2011); *The Culture of Life: Ten Essential Principles for Christian Bioethics* (2012); *Studying Luther's Large Catechism: A Workbook for Christian Discipleship* (2012); and, *Rediscovering the American Republic: Biographies, Primary Texts, Charts, and Study Questions—Exploring a People's Quest for Ordered Liberty* (2 vols., 2012–2013). Since 2012, he has served as senior editor of *The Family in America: A Journal of Public Policy*.

Dr. MacPherson is a nationally featured speaker for academic associations, religious organizations, and public policy forums. He has been interviewed on numerous radio programs. A homeschool parent who has taught both children's Sunday school and adult Bible classes, Dr. MacPherson also is the founding president of the Hausvater Project (www.hausvater.org), a nonprofit organization promoting a biblical vision for family, church, and society in the spirit of the Lutheran confessions.

For more information, visit: **www.ryancmacpherson.com**